# Environmental Carcinogens: Polycyclic Aromatic Hydrocarbons

## Chemistry, Occurrence, Biochemistry, Carcinogenicity

Editor

### Gernot Grimmer, Dr. Chem.
Biochemical Institute for
Environmental Carcinogens
Ahrensburg, West Germany

CRC Press, Inc.
Boca Raton, Florida

Library of Congress Cataloging in Publication Data

Luftqualitätskriterien für ausgewählte polyzyklische aromatische Kohlenwasserstoffe.
  English.
  Environmental carcinogens, polycyclic
aromatic hydrocarbons.

  Updated translation of: Luftqualitätskriterien für ausgewählte polyzyklische
aromatische Kohlenwasserstoffe.
  Bibliography: p.
  Includes index.
  1. Hydrocarbons—Toxicology. 2. Aromatic
compounds—Environmental aspects. 3. Carcinogens—Environmental aspects. 4. Environmentally induced diseases—Germany (West)

I. Grimmer, G. (Gernot), 1924-    . [DNLM:
1. Polycyclic hydrocarbons. 2. Air pollutants.
3. Carcinogens, Environmental. WA 754 E313]
RC268.7.H9E58        65.9'511        82-4210
ISBN 0-8493-6561-9          AACR2

This book represents information obtained from authentic and highly regarded sources. Reprinted material is quoted with permission, and sources are indicated. A wide variety of references are listed. Every reasonable effort has been made to give reliable data and information, but the author and the publisher cannot assume responsibility for the validity of all materials or for the consequences of their use.

All rights reserved. This book, or any parts thereof, may not be reproduced in any form without written consent from the publisher.

Direct all inquiries to CRC Press, Inc., 2000 Corporate Blvd., N.W., Boca Raton, Florida, 33431.

© 1983 by CRC Press, Inc.

International Standard Book Number 0-8493-6561-9

Library of Congress Card Number 82-4210
Printed in the United States

# PREFACE

In the Federal Republic of Germany, as in other countries, the increasing industrialization and urbanization involves a growing hazard to the environment. Therfore, there is a definite need for measures to protect man and animal and plant life against harmful environmental influences. The legal bases for such measures are the Federal Emission Control Law and the Technical Instructions for Maintaining Air Purity, an administrative regulation that includes, among others, air quality standards for ten noxious substances. The Federal Environmental Protection Agency (Unweltbundesamt) supports the Federal Ministry of the Interior in solving problems of aar quality control, e.g., by compiling scientific data on the effect of relevant air pollutants. On the basis of these data, proposals for emission standards are then developed.

Limitation of emissions is often necessary when the emission of a toxic substance jeopardizes human health. In the case of carcinogenic substances, preventive guidelines should be laid down even if it is not scientifically justifiable to regard them as authoritative. Their use may considerably reduce the carcinogenic risk to man.

The group of compounds called polycyclic aromatic hydrocarbons (PAH) is under strong suspicion of contributing, as an air pollutant, to the frequency of the incidence of lung cancer. This statement is based on observations in occupational medicine, animal bioassays, and epidemiologic studies. Thus a reduction of the PAH content in the atmosphere would most probably decrease the frequency of lung cancer.

The present survey comprises today's knowledge of environmental pollution—in particular of the atmosphere—by PAH and of the biological effects of this class of substance, putting special emphasis on their carcinogenic activity. The research data and conclusions derived therefrom are meant to assist the goverment of the Federal Republic of Germany in determining an air quality standard.

We wish to thank in particular all those scientists who helped to compile this documentation as well as the members of the working group "Investigations on the Carcinogenic Burden by Air Pollution in Man" and "Inventory and Biological Impact of Polycyclic Carcinogens in the Environment" who prepared a large amount of the cited research data.

Dr. Heinrich von Lersner<br>
President of the Federal<br>
Environmental Protection<br>
Agency

# INTRODUCTION

## D. Schmähl

Environmental protection is an urgent demand of our time. The present industrial society has confronted physicians, ecologists, and biologists with new problems arising from this very society. Nevertheless, it should be mentioned that in the Western Hemisphere the earliest law on environmental protection was drafted more than 700 years ago by the Hohenstaufen Emperor Friedrich II in the "Constitution of Melfi". It is relevant to the subject matter of the present volume that this historical constitution which Friedrich II intended to be a clean-air act read: "We endeavour to keep the health of the air given to us by God clean by providing preventive measures as far as possible. We therefore decree that nobody is allowed to rinse flax and hemp in waters less than one mile away from dwellings because this would have a detrimental effect on the air quality."

Whereas in former times the main concern was to keep the air free from odors and pathogenic agents of infectious diseases (malaria = malair = bad air), the working group "Investigations on the carcinogenic burden by air pollution in man" was initiated by the German Federal Ministry of the Interior in 1969 to investigate the problems arising from carcinogenic agents contained in the air. These agents were to be identified chemically, tested biologically, and their hazard to man was to be assessed. The initiative of the Federal Ministry of the Interior has to be seen against the background that for many years the incidence of cancer of the respiratory tract has been growing. Although we know that this increase can be explained predominantly by inhalation of tobacco smoke, other sources such as heating installations, incinerators, industrial plants, vehicle exhausts, coal power plants, etc. emit potentially carcinogenic compounds into the atmosphere which may etiologically play a role in carcinogenesis. This is mainly supported by epidemiological studies in man which are dealt with in a later chapter.

According to present knowledge, cancer of the respiratory tract must be considered in most cases as the product of inhaled carcinogens. Since air is our most vital food — in the truest sense of this word — the responsible legislative authorities must strive to find ways and means of determining the potential carcinogens in the air we breathe and either to eliminate them or to reduce them to a level which, according to our present state of knowledge, causes as little risk as possible. However, those versed in this field know that it is futile to propose a "zero value" for the concentration of carcinogenic substances in the air since this is not feasible in our highly industrialized and technological world. Furthermore, we have to bear in mind that the air contains natural substances that are considered to be carcinogenic, e.g., the natural level of radioactivity and some metals and dusts. Due to the given limitations of analytical detection, a "zero value" per se is not appropriate.

The working groups "Investigations on the carcinogenic burden by air pollution in man" and "Inventory and biological impact of polycyclic carcinogens in the environment" were initiated with the aim of advising the legislative authorities in their legislative action. At the start of our investigations, we had only very little reliable scientific information which might have served as a basis for legislative initiatives. The scientists of all participating institutions and their colleagues in the ministries agreed that it was impossible to detect, identify, and assess in the air all the potential carcinogens that lay within the scope of these working groups. We therefore decided to first direct our attention to the polycyclic aromatic hydrocarbons (PAH) because this class of compounds comprises carcinogens which occur in the atmosphere — sometimes in considerable amounts — and also ubiquitously, even though in varying concentrations. Other inhalation carcinogens, such as asbestos fibers or N-nitroso compounds could be left out, all the more so since we knew that other German and foreign groups have dealt and are dealing with just these substances.

The results obtained by our working groups are included in the present documentation, which has been accomplished on the initiative of the German Federal Environmental Protection Agency and concerns the complex problems inherent in the carcinogenic activity of polycyclic aromatic hydrocarbons. These problems are dealt with not only from the restricted vantage point of the two working groups, but also an attempt is made to assess the risk of PAH as inhalation carcinogens in general, putting adequate emphasis on the results obtained in model investigations carried out with automobile exhaust condensates by one of the working groups.

Polycyclic aromatic hydrocarbons such as benzo[a]pyrene have been proved to be carcinogenic in a large number of animal species when applied to the skin or injected subcutaneously. Man is not an exception to this rule. Thus, in 1775, the English physician Percivall Pott made the first corresponding observation in chimney sweeps whose skin had come into intensive contact with soot, and PAH were identified as the carcinogenic substances contained therein. The occupational bronchial carcinoma of gasworkers and coke plant workers also seems to be induced by PAH generated in the furnaces. The biological effect of benzo[a]pyrene was demonstrated experimentally in human skin in 1939. From our present-day knowledge, these PAH exhibit predominantly a local carcinogenic activity, i.e., they induce malignant tumors at the site of application and not (or only exceptionally) in distant organs.

Thus it is to be expected that after inhalation they display their carcinogenic activity in the respiratory tract. That this may be the case is shown in investigations on tobacco smoke carcinogenesis. Here the carcinogenic substances can be localized predominantly in the PAH-containing fraction of the tobacco smoke. It seems to be permissible to reason by analogy that what is true for the tobacco smoke condensate is also applicable to noxious substances derived from other sources of combustion.

Today it is relatively easy to obtain an exact determination of hydrocarbon concentrations in various media by a variety of analytical techniques. Furthermore, it is not difficult to investigate certain hydrocarbons in simply designed experiments, e.g., with rats and mice, with regard to their carcinogenic activity on skin, when applied topically or subcutaneously. It is, however, considerably more difficult to test the suspected compounds with an adequate inhalation technique, because the safety of individuals working with these substances has to be guaranteed, and animals used in inhalation experiments should have a structure of the respiratory tract and reaction to harmful inhalatory substances similar to man. This already hints at the essential difficulty of all experimental investigations, i.e., the extrapolation of experimental findings to the human situation.

Despite the problems and difficulties mentioned, there is no alternative to animal experiments, since for ethical reasons experiments with humans are out of the question, and prospective epidemiologic studies involving humans would take many years or even decades. This brings us to a second great difficulty inherent in our subject, i.e., the long latency period required for the induction of malignant tumors by chemical substances. We have to accept this as a fact. Dealing with the assessment of long-term effects is naturally far more difficult than evaluating short-term effects. Even today satisfactorily proven results are rare in the field of the toxicological effects of long-term inhalation because this field of science is comparatively new.

A third difficulty has to be mentioned still: as a rule the etiology of tumors in man concerns a combination of effects, i.e., the tumor in question was induced not by *one* specific noxious substance but by a great number of them. In addition, the constitutional predisposition of the individual and the individual modes of reaction within the respiratory tract (e.g., destruction of the ciliary function due to nonspecific influences) may play a role. Beside the substance-related toxicological point of view, we therefore have to take into account individual-specific considerations which may vary from individual to individual.

Despite these difficulties and shortcomings, the present documentation attempts to give a survey of the significance of carcinogenic hydrocarbons as potential hazards to man. We have tried to present — as comprehensively as possible — the results of chemical, biological, and epidemiological studies, as well as interpretations given by other groups of scientists, in considering the above problems. Furthermore, the present volume is suited to introduce the interested reader to the problems of experimental and practical oncology, to explain the methods and their necessary fundamentals, and to show him the manifold difficulties which quite often inhibit the assessment of the results obtained. In this respect the present documentation exceeds the narrow field of hydrocarbons and can provide mental stimulations for critical initiatives in other fields.

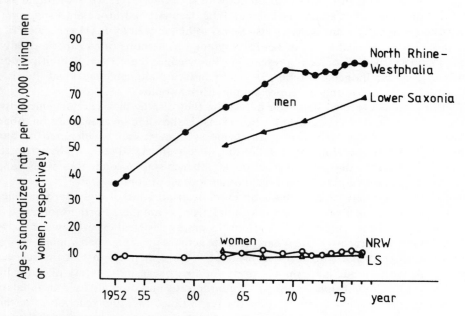

FIGURE 1. Mortality due to malignant tumors of the respiratory organs in North Rhine-Westphalia and Lower Saxonia from 1952—1976[1].

Figure 1 shows the increase of bronchial cancer deaths from 1952 to 1976 using the German federal states North Rhine-Westphalia and Lower Saxony as examples. The observed death rates demonstrate that despite intensive efforts in the therapeutic field considerably more people died of lung cancer annually (1974: about 24,000) than, e.g., in road accidents (1974: about 14,600 deaths) in West Germany. This fact justifies the demand for giving primary prevention of bronchial cancer a higher ranking.

Finally, a word of explanation: due to the subject matter of this study we have had to give qualitative evaluations in many places, resulting perhaps in discrepancies which, however, are due to the method of presentation only.

## REFERENCE

1. Pott, F. and Dolgner, R., Zur Problematik einer Grenzwertfindung für PAH, *Staub-Reinhalt. Luft*, 39, 12, 1979.

## THE EDITOR

**Gernot Grimmer, Dr. Chem.**, earned his Doctoral degree in chemistry from the University of Hamburg, West Germany in 1952. In 1959 he was appointed university lecturer in biochemistry at the University and nominated for apl. Professor in 1969. Since then he has served as chief of the Biochemical Institute of Environmental Carcinogens in Hamburg-Ahrensburg and as a teacher at the University of Hamburg.

Professor Dr. Grimmer has published over 100 articles, mostly related to polycyclic aromatic hydrocarbons, in journals of chemistry and biochemistry. He is also the editor of *Biochemistry*, a textbook for students of medicine and biochemistry, published by Bibliographisches Institut, Mannheim, West Germany in 1969.

Professor Dr. Grimmer is a member of the German Society of Biological Chemistry, the German Chemical Society, the German Society of Research on Mineral Oil and Coal Chemistry, the German Pharmaceutical Society, the German Society of Fat Science, and the Commission on Food Chemistry.

# CONTRIBUTORS

**H. Brune, M.D.**
Consulting Institute for Preventive
 Medicine and Environmental
 Toxicology
Hamburg, West Germany

**R. P. Deutsch-Wenzel, Ph.D.**
Consulting Institute for Preventative
 Medicine and Environmental
 Toxicology
Hamburg, West Germany

**S. Dobbertin, D.V.M.**
Federal Environmental Agency
West Berlin, West Germany

**G. Grimmer, Dr. Chem.**
Professor
Biochemical Institute for
 Environmental Carcinogens
Ahrensburg, West Germany

**M. Habs, M.D.**
Institute for Toxicology and
 Chemotherapy
German Cancer Research Center
Heidelberg, West Germany

**J. Jacob, Ph.D.**
Biochemical Institute for
 Environmental Carcinogens
Ahrensburg, West Germany

**J. Misfeld, Ph.D.**
Professor
Institute for Mathematics
University of Hannover
Hannover, West Germany

**U. Mohr, M.D.**
Professor
Department of Experimental Pathology
Medical University
Hannover, West Germany

**G. Oberdörster, D.V.M.**
Associate Professor
Department of Radiation Biology and
 Biophysics
University of Rochester Medical Center
Rochester, New York

**F. Pott, M.D.**
Professor
Medical Institute for Environmental
 Hygiene
University of Dusseldorf
Dusseldorf, West Germany

**D. Schmähl, M.D.**
Professor
Institute for Toxicology and
 Chemotherapy
German Cancer Research Center
Heidelberg, West Germany

**P. Schneider, M.Sc.**
Department of Experimental Pathology
Medical University
Hannover, West Germany

**D. Steinhoff, M.D.**
Institute for Experimental Toxicology
Bayer AG
Wuppertal-Elberfeld, West Germany

# TABLE OF CONTENTS

Chapter 1
Environmental Carcinogens: A Risk for Man? Concept and Strategy of the
Identification of Carcinogens in the Environment ........................... 1
G. Grimmer and J. Misfeld

Chapter 2
Chemistry ................................................................ 27
G. Grimmer

Chapter 3
Occurrence of PAH ....................................................... 61
G. Grimmer and F. Pott

Chapter 4
Behavior of PAH in the Organism ........................................ 129
F. Pott, B. Oberdörster, J. Jacob, and G. Grimmer

Chapter 5
Biological Activity ...................................................... 157
D. Schmähl, R. P. Deutsch-Wenzel, H. Brune, P. Schneider, U. Mohr, M. Habs,
 F. Pott, and D. Steinhoff

Chapter 6
Epidemiology ........................................................... 221
J. Misfeld

Chapter 7
Extrapolation of Experimental Results to Man ........................... 237
D. Schmähl and M. Habs

Chapter 8
Conclusions ............................................................ 247
S. Dobbertin

Index .................................................................. 251

Chapter 1

# ENVIRONMENTAL CARCINOGENS: A RISK FOR MAN? CONCEPT AND STRATEGY OF THE IDENTIFICATION OF CARCINOGENS IN THE ENVIRONMENT

## TABLE OF CONTENTS

1.1 Methods for Proof of the Existence of Carcinogens in Ambient Air — G. Grimmer ................................................................. 2

1.2 Biological Detection Systems for Carcinogens: Methods of Analysis of Carcinogenic Substances in Environmental Samples — G. Grimmer ......... 3
    1.2.1 Analysis of the Percentage Amounts that Cause Biological Effects ... 5
    1.2.2 Example of an Analysis of the Proportions of Biological Impact: The Carcinogenic Components of Automobile Exhaust ............... 7
        1.2.2.1 Fractionation of the Condensate of Automobile Exhaust .... 7
        1.2.2.2 Other Animal Experiment Models Used as a Detection System ............................................................. 10
        1.2.2.3 Analysis of the Proportions of Activities in Mutagenicity Assays ............................................................ 11
    1.2.3 Carcinogenic Components of Automobile Exhaust Condensate from Gasoline Engines ............................................... 11
        1.2.3.1 Intratracheal Installation ................................. 11
        1.2.3.2 Topical Application ....................................... 11
        1.2.3.3 Identification of Carcinogenic Components ............. 11
    1.2.4 Carcinogenic Components of Exhaust Condensate from Diesel Engines ................................................................. 12
    1.2.5 Carcinogenic Components of Cigarette Smoke .................... 12
    1.2.6 Rates of the Biological Effect of BaP in Various Matrices ......... 15
    1.2.7 Rates of the Biological Effect of All Carcinogenic PAH in Various Matrices ............................................................. 15

1.3 The "Representative PAH Concentration" of a City — G. Grimmer ........ 15
    1.3.1 PAH Produced by Various Emittants ........................... 15
    1.3.2 Local and Temporary Variations of the PAH Concentration in a City ................................................................. 16
        1.3.2.1 The Temporary Variation on a Selected Collecting Station in the City During the Day and the Year ................... 16
        1.3.2.2 The Local Variation in Different Areas of the City ........ 18

1.4 Fundamental Mathematical and Statistical Principles for Analyzing the Proportional Activity — J. Misfeld ..................................... 21

References ................................................................. 24

## 1.1 METHODS FOR PROOF OF THE EXISTENCE OF CARCINOGENS IN AMBIENT AIR

### G. Grimmer

Based on several epidemiological investigations and research projects of occupational medicine (Chapter 6), we have to assume that some malignant diseases, for instance, in the bronchi, buccal cavity, larynx, bladder, and other organs, are induced by exogenous risk factors.[23a] Exogenous factors are, e.g., smoking habits or air pollution at the place of residence or at work. Epidemiological investigations permit the detection of correlation between the frequency of cancer and the number of cigarettes smoked per day or the level of air pollution. "Air pollution" which is considered a cause of cancer cannot be attributed to a single emitter, but mostly constitutes a mixture of emissions from different sources.

A comparison of the lung cancer mortality in the eleven countries of West Germany leads to significant regional differences in the mortality rates as shown in Table 1.[18a] Of specific interest is the fact that the regional mortality rates of 35- to 64-year-old males differ significantly in 1975 (Bavaria 61.5/100,000; Baden-Württemberg 60.6/100,000 and Saarland 105.9/100,000). Besides smoking habits, air pollution must be taken into consideration as an explanation for this difference. Thus, lung cancer seems to be provoked by environmental carcinogens in West Germany. According to the regional differences it is to be supposed that carcinogens are present in cigarette smoke as well as in emissions from various sources.

An epidemiological study does not permit us to state the individual compounds or chemical classes by which the effect has been induced. To be able to take specific steps to reduce the incidence of these diseases, the type of emitter or carcinogen involved has to be known. So far we can only surmise which carcinogens might have been responsible in individual cases. But we are still far from an exact assessment of the role of the individual known classes of carcinogenic substances, such as polycyclic aromatic hydrocarbons, nitrogen-containing aromatic compounds, aromatic amines, nitrosamines, or inorganic compounds.

Epidemiological methods do not produce results which permit us to decide whether the observed increase in the frequency of human lung cancer, for instance, might be ascribed to the benzo[a]pyrene content of the atmosphere. It is also not possible to decide whether the effect of benzo[a]pyrene is increased by certain accompanying substances or whether other carcinogenic substances participate in this effect. We therefore cannot specify the proportional balance of the carcinogenic activity of such complex mixtures as cigarette smoke or airborne particulate matter sampled in a highly polluted area.

Epidemiological studies are methods of detecting the existence of risk factors which induce bronchogenic carcinomas. These studies have demonstrated that (1) smoking habits and (2) air pollution in polluted areas are the primary risk factors for the development of malignancies of the respiratory tract. These two factors have an influence on the morbidity rate.

Studies of the working area also are suitable for revealing correlations between the morbidity rate and the concentration of a product and the period of exposure. Unlike air pollution mixtures which are produced by several emitters, it is easier in this case to establish a correlation. As in the "monocausal" case of cigarette smoke, inhalation of the product or production-related exhaust gases, is to be suspected as a cause, or the effect can be ascribed to dermal contact with the product. However, in studies on working places too, a single substance can be detected as causative in only a few cases,

Table 1
## MORTALITY RATES PER 100,000 INHABITANTS FOR ICD 162, LUNG — MALE[a]

| Age group | Year | S-H | Ham | LS | Brem | NRW | Hess | R-P | B-W | Bav | Saar | B | WG |
|---|---|---|---|---|---|---|---|---|---|---|---|---|---|
| 35—64 years | 1955 | 58.5 | 96.2 | 47.3 | 76.2 | 73.4 | 54.1 | 58.5 | 52.5 | 60.2 | 77.2 | 77.8 | 63.4 |
| | 1965 | 84.3 | 95.5 | 70.7 | 81.3 | 82.4 | 69.3 | 83.7 | 62.1 | 67.1 | 98.6 | 88.2 | 75.8 |
| | 1975 | 72.9 | 88.6 | 66.2 | 85.4 | 88.3 | 72.8 | 80.5 | 60.6 | 61.5 | 105.9 | 85.9 | 74.9 |
| Older than 64 years | 1955 | 127 | 264 | 121 | 247 | 166 | 122 | 112 | 120 | 122 | 111 | 245 | 146 |
| | 1965 | 289 | 464 | 275 | 438 | 317 | 278 | 266 | 241 | 256 | 337 | 461 | 299 |
| | 1975 | 403 | 529 | 382 | 464 | 472 | 370 | 346 | 326 | 339 | 529 | 538 | 403 |
| Total | 1955 | 33.0 | 59.6 | 28.4 | 50.8 | 41.9 | 31.1 | 31.5 | 30.2 | 33.0 | 38.2 | 51.4 | 36.5 |
| | 1965 | 62.2 | 96.2 | 53.4 | 75.6 | 57.0 | 52.8 | 54.2 | 41.1 | 47.8 | 61.9 | 109.5 | 55.7 |
| | 1975 | 65.1 | 83.4 | 60.8 | 75.0 | 76.9 | 61.2 | 60.1 | 52.9 | 54.7 | 88.2 | 84.4 | 65.4 |

*Note:*  
S-H Schleswig-Holstein  
Ham Hamburg  
LS Lower Saxony  
Brem Bremen  
NRW North Rhine Westphalia  
Hess Hesse  
R-P Rhineland-Palatinate  
B-W Baden-Württemberg  
Bav Bavaria  
Saar Saarland  
B Berlin (West)  
WG West Germany  

[a] Standardized, using the age structure of the resident population of West Germany in 1965.

e.g., asbestos, etc. It is of course not possible to determine the percentage rate at which risk factors such as "coke production", "chimney sweeping", or "raw paraffin processing" participate in the total effect.

Generally, the carcinogenicity of a sample is caused by a mixture of single carcinogenic compounds. Each of these individual substances contributes a part to the total carcinogenicity. The question arises: how large is the percentage rate of this compound of the total effect of the sample?

## 1.2 BIOLOGICAL DETECTION SYSTEMS FOR CARCINOGENS — METHODS OF ANALYSIS OF CARCINOGENIC SUBSTANCES IN ENVIRONMENTAL SAMPLES

During incomplete combustion of fossil fuels, such as hard coal, brown coal, heating oil, diesel and gasoline fuel, etc., not only combustion gases ($CO_2$, CO, $NO_x$, $SO_2$) and water are formed, but also other components which are condensable at ambient temperature. All these substances are emitted during combustion processes. The condensable components are thus a part of the emissions; they form a complex mixture of uncombusted fuel, partly oxidized compounds, pyrolysis products, soot, and flue ash.

**Which Substances Contained in this Mixture are Carcinogenic?**

In animal experiments, condensates from cigarette smoke, coal heating and exhaust from vehicles can be proved to have a carcinogenic effect.

The isolation of the individual substances which are responsible for the carcinogenic activity of the total mixture is possible only if the activity of the total mixture (emission condensate) can be compared with that of individual fractions of the mixture by means of suitable biological tests. The biological test is used as a detection system which can identify individual carcinogenic substances or their mixtures. If several substances participate in the total carcinogenic activity, the problem will be solved when a mixture of the pure substances — used at the same concentrations as they occur in the condensate — has the same activity as the total condensate mixture.

In this manner it would be possible to investigate the condensable components of flue gases of hard coal, brown coal, heating oils, vehicle emissions from gasoline and Diesel engines, and emissions from coking plants, refuse incinerating plants, and other industrial installations. Cigarette smoke condensates may also be separated into their carcinogenic components. In few cases only has it been investigated in detail whether the alleged carcinogenic effect of all these products of incomplete combustion can be attributed to the same substances or at least to the same chemical classes of substances.

Not all biological tests are suited for analyzing the percentage rates of activity. Animal test models which result in malignant tumors at the site of application after long-term treatment are primarily suited for the detection of carcinogenic effects. These animal test models have to meet a number of requirements in order to be suitable as detection systems for carcinogens contained in a multicomponent mixture.

**Criteria to Be Met by a Biological Detection System for the Isolation of Carcinogenic Components of a Multicomponent Mixture (Condensate)**

The detection systems (e.g., animal test models) used for the detection of carcinogenic components and assessment of their percentage rate in the total activity ought to have the following properties as essential prerequisites:

1. A clear dose-response relationship for the individual carcinogenic substance as well as for the total mixture (e.g., a condensate of combustion gases).
2. An additive effect of the individual activities of two or more carcinogens of a mixture. This would be the simplest possibility because the test result could easily be interpreted. The larger the mutual interaction of the active substances is, the more difficult will be the interpretation of the test results.
3. No, or only a minor, modification of the carcinogenic effect by inactive accompanying substances (e.g., no inhibition and no promotion of the carcinogenic effect by the large excess of noncarcinogenic substances in the mixture).

So far we do not know whether such a test model exists nor whether the requirements to be met by such a model are really present in the human lung.

**Animal Test Models for the Investigation of Carcinogenic Activities**

Various animal test models are being used for the investigation of carcinogenic activities. They are described in detail in Chapter 5. The carcinogenic effect of an emission condensate can be detected only if the carcinogenic components are present at a tumor-inducing concentration. In Table 2 this theoretical detection limit is compared, using benzo[a]pyrene in several animal test models.

In Table 2, the following models are compared with one another:

1. Subcutaneous application in mice
2. Topical application to the dorsal skin of mice
3. Implantation into the lung of rats
4. Instillation into the trachea of hamsters

The first two lines of Table 2 indicate those amounts of BaP which induce tumors in 20% of the animals after a single or several applications. Lines 3 and 4 give the maximal applicable total doses of an emission condensate. Thus, for instance, not more than 50 mg of condensate can be dropped onto the skin if the diameter of the treated skin area is to be as small as possible in order to achieve a high condensate concentration per unit surface area. Since, however, the topical application has to be repeated twice weekly, 104 × 50 mg of condensate can be applied in the course of 1 year. Sub-

Table 2
COMPARISON OF THE DETECTION LIMIT OF BaP IN FOUR
ANIMAL TEST MODELS

|  | Subcutaneous connective tissue (mouse) | Topical skin (mouse) | Implantation lung (rat) | Instillation trachea (hamster) |
|---|---|---|---|---|
| BaP individual dose ($\mu$g)[a] | 5 | 1.5 | 10 | 10—200 |
| BaP annual dose ($\mu$g)[a] | 5 | 150 | 10 | 500—10,000 |
| Matrix max. ID (mg) | 500 | 50 | 20 | 10 |
| Matrix max. AD (mg) | 500 | 5000 | 20 | 500 |
| Detection limit (ppm) | 10 | 30 | 500 | 1000 |

*Note:* ID = Individual dose, AD = annual dose.

[a] Individual dose needed to induce about 20% carcinomas.

cutaneous administration or implantation into the lung, on the other hand, cannot be repeated as often as desired. The detection limit of a highly active carcinogen can be derived from the well-detectable effective dose and the maximum amount of condensate which can be applied. This value is stated in the last line of Table 2. Thus, the most sensitive test method for PAH is the subcutaneous administration of the carcinogen. In this model it can still be detected when the total activity of all carcinogenic components of the condensate corresponds to the activity of 5 $\mu$g BaP (in 500 mg of tricaprylin).

The theoretical considerations described earlier are applicable only if the stated prerequisites have been met: dose-response relationship, additivity, and no inhibition of the carcinogenic activity by secondary substances.

If a noncarcinogenic substance potentiates the effect of a carcinogen which is contained in a mixture of many individual substances, such as any condensate, the simple method outlined earlier of classifying the components as carcinogenic or noncarcinogenic is complicated to such an extent that a satisfactory classification becomes practically impossible. A suitable animal test model is the key to the detection of all carcinogenic components of a complex mixture, and to an assessment of the percentage amount which individual carcinogens contribute to the total activity. Since, however, all the available models are subject to certain defects, it is preferable that several animal models be used for the solution of scientific problems.

### 1.2.1 Analysis of the Percentage Amounts that Cause Biological Effects

The purpose of the analysis of the proportions of biological activities of an emission condensate is to identify all the carcinogenic components and to determine their concentrations (quantitative proportions) and their contribution to the total carcinogenic activity of the condensate. Thus, the activity of the condensate can be defined as the biological activity of all the carcinogenic components of the condensate. This aim is achieved when the carcinogenic activity of the condensate observed in an animal test model corresponds to the activity of the mixture of the contained pure substances, i.e., it can be simulated by pure substances. In this mixture the pure substances are contained in the same quantitative proportions as in the condensate, i.e., they are dosed in the same concentrations.

This balance of the activity rate of all contained carcinogens is, of course, applicable

Table 3
CORRELATION OF THE
AMOUNT OF AUTOMOBILE
EXHAUST CONDENSATE
(285 μg BaP PER GRAM)
AND THE INCIDENCE OF
CARCINOMA

| Amount (mg/year) | Animals with carcinomas (%) |
|---|---|
| 53 | 1.3 |
| 106 | 15.0 |
| 158 | 29.7 |
| 316 | 60.0 |
| 438 | 71.8 |

*Note:* Topical application to mice.

only to the respective animal model. The profile of activities (percentage amount of the individual substances in the total activity) of the carcinogenic components of a condensate may be different in different animal test models.

The test models are selective detectors for substances with carcinogenic effects. An essential criterion for a suitable carcinogen-specific detector, however, is, for instance, a clear-cut correlation between the amount of substance and the response frequency (Table 3).

From Table 3 we can see that

1. The incidence of carcinoma depends on the amount of the carcinogen administered.
2. To provoke skin carcinoma in 60% of the experimental animals with pure benzo[a]pyrene, about 1 mg of this substance is required. The same tumor frequency can be achieved with 315 mg of vehicle exhaust condensate of a gasoline engine. These 315 mg, however, contain only 90 μg BaP. Thus, more than ten times the amount of BaP present in the exhaust would be necessary to get the same carcinogenic effect as that of 1 mg pure BaP. The presence of BaP can therefore explain only 9% of the carcinogenic effect of the automobile exhaust condensate.

## Fractionation of a Condensate

An inventory of all carcinogenic components of a condensate is one of the prerequisites necessary to explain the overall biological effect. For this purpose, each carcinogenic substance contained in the condensate has to be isolated, identified, and assessed quantitatively.

There is a clear concept for the isolation of carcinogenic substances from a condensate, i.e., a complex mixture of carcinogenic and noncarcinogenic substances. The condensate has to be separated into carcinogenic and carcinogen-free fractions by means of suitable methods, such as liquid-liquid extraction or chromatographic methods, etc. The fractions are tested for carcinogenicity in animal test models. In the simplest case the activity of a single carcinogenic fraction corresponds to that of the unchanged initial material. Then the noncarcinogenic components have to be eliminated from the biologically active fraction by a suitable process. This procedure has to be repeated until a pure carcinogenic substance has been isolated. Fractionation of a condensate involves, however, the risk of losing the biologic activity during the necessary separation steps. It is therefore recommended to control a possible loss of activity by testing a reconstituted condensate obtained by recombining all fractions, in addition to testing the initial material and the individual fractions. The reconstituted condensate should have the same activity as the unseparated initial material.

The separation of a condensate into carcinogen-containing and carcinogen-free fractions results in the detection of a carcinogenic class of substances and finally in the isolation of individual carcinogenic substances in this class.

### 1.2.2 Example of an Analysis of the Proportions of Biological Impact: The Carcinogenic Components of Automobile Exhaust

The carcinogenic activity of automobile exhaust condensate (AEC) observed after repeated application to the dorsal skin of mice has been described repeatedly.[5,6,25,43] This animal test model is therefore well suited for proving carcinogenic behavior. Tumorigenesis has also been induced by AEC in other biological test systems, such as subcutaneous application[33] or instillation into the trachea of hamsters.[31] Analogous to carcinogenic hydrocarbons, automobile exhaust condensate applied topically to the skin of mice induced suppression of the sebaceous glands within a few days.[6] Although this sebaceous gland suppression shows a high correlation (approximately 90%) with the carcinogenic activity of polycyclic aromatic hydrocarbons or corresponding nitrogen-containing aromatic compounds, it is not a criterion of carcinogenic activity.

Taking, for instance, the application to mouse skin as a detection system for carcinogenic activity, increasing doses of automobile exhaust condensate induced an increasing tumor incidence. Thus, approximately 53 mg, 158 mg, and 438 mg of condensate (annual dose) induced malignant tumors in 1.3, 29.7, and 71.8% of the animals (80 animals), respectively[6] (see Table 3).

It is necessary to determine the dose-response relationship by testing two or more doses in order to be able to assess the percentage rate of impact which the subfractions have in the total activity.

#### 1.2.2.1 Fractionation of the Condensate of Automobile Exhaust

The experiments were started on the working hypothesis that the carcinogenic activity of automobile exhaust condensate (AEC) is based — in whole or in part — on the contained polycyclic aromatic hydrocarbons (PAH). Consequently, the AEC was resolved into a PAH-containing and a PAH-free fraction. If the PAH-free fraction had a carcinogenic effect in the biological tests, the conclusion would have to be drawn that further carcinogenic classes, e.g., nitrogen-containing aromatic compounds (carbazole or acridine derivatives), participated in the carcinogenic action of AEC.

The AEC was separated into a PAH-containing and a PAH-free fraction in two steps. In a third step, the concentrated PAH fraction was separated according to molecular size. The fraction of PAH with more than three rings could be separated gas chromatographically into individual compounds. The suspected carcinogenic PAH of this fraction were finally combined to a mixture of 15 pure substances at the corresponding percentage amounts, and their carcinogenic activity was compared with that of AEC.

The initial material was separated according to the following criteria:

1. Hydrophobic-hydrophilic substances
2. Aromatic-aliphatic substances
3. PAH with two and three rings — PAH with four to seven rings

### Table 4
### CARCINOGENIC ACTIVITY OF AEC (TOPICAL APPLICATION). SEPARATION INTO HYDROPHOBIC AND HYDROPHILIC COMPONENTS

| Weight (%) | Designation | PAH content | Tumor yield (%) Dose 1 | Tumor yield (%) Dose 2 |
|---|---|---|---|---|
| 100 | Exhaust condensate | + | 15 | 60 |
| 67 | Methanol-water phase | − | 1 | 3 |
| 33 | Cyclohexane phase | + | 14 | 61 |

*Note:* Dose 1 and 2 correspond to 1.05 and 3.15 mg AEC, respectively, administered twice weekly. The stated tumor yield is a median value (probit).

Finally the activity of the initial AEC was compared with that of a mixture of the benzo[a]pyrene and the other carcinogenic components contained in the AEC.

**Hydrophobic-hydrophilic substances** — The first step in the process was distribution between methanol, water, and cyclohexane (9:1:10). The hydrophilic phase was extracted several times with cyclohexane. Thus all PAH contained in the methanol-water phase were extracted except for a small residue of about 1 to 2% (see Table 16, Chapter 2). The results of the subsequent biological testing by topical application (80 CFLP mice per dose group) are presented in Table 4. Both fractions were dosed proportionally. Table 4 demonstrates that AEC and its hydrophobic components induced comparable rates of carcinomas. Only a minor proportion of the carcinogenic effect of AEC cannot be attributed to PAH.[6,20,30]

**Aromatic-aliphatic substances** — In the subsequent step of the process the hydrophobic phase is again dissolved in cyclohexane and extracted with nitromethane five times to eliminate all PAH from the cyclohexane solution. The result obtained in this series of experiments is described in Table 5. Both fractions were applied at doses which correspond to those in the condensate. Table 5 demonstrates that the AEC and the aromatic fraction induced comparable rates of carcinomas, whereas the aliphatic fraction is inactive.[6,20,30]

**PAH with two and three rings — PAH with four to seven rings** — In the third step of the process the aromatic components were resolved according to molecular size by means of chromatography on Sephadex® LH 20 with isopropanol. Aromatic compounds with two and three rings were separated from those with four to seven rings. Table 6 presents the results of the animal experiments after 83 weeks. The fractions were dosed proportionally. Table 6 demonstrates that the carcinogenic activity of the total AEC can be attributed to the contained PAH with four to seven rings. The result of both test series is illustrated in Figure 2. Whereas the methanol phase (67% by weight), the cyclohexane phase (15.3% by weight), and the fraction which contains PAH with two and three rings (13% by weight) have practically no carcinogenic effects, all the carcinogenic compounds of automobile exhaust condensate are comprised in the PAH fraction which contains PAH with four to seven rings (4% by weight).

**Comparison of AEC with benzo[a]pyrene and a mixture of the carcinogenic PAH extracted from AEC** — Finally a mixture of 15 pure substances was produced in which

Table 5
CARCINOGENIC ACTIVITY OF AEC.
SEPARATION OF THE HYDROPHOBIC
COMPONENTS INTO AROMATIC AND
ALIPHATIC SUBSTANCES

| Weight (%) | Designation | PAH content | Tumor yield (%) Dose 1 | Dose 2 |
|---|---|---|---|---|
| 100 | AEC | + | 15 | 60 |
| 15.3 | Cyclohexane phase | − | 1 | 0 |
| 17.7 | Nitromethane phase | + | 15 | 60 |

*Note:* Dose 1 and 2 correspond to 1.05 and 3.15 mg AEC, respectively, administered twice weekly. AEC = mean value.

Table 6
CARCINOGENIC ACTIVITY OF AEC.
SEPARATION OF THE PAH FRACTION
ACCORDING TO THE NUMBER OF RINGS
(MOLECULAR SIZE)

| Weight (%) | Designation | Tumor yield (%) Dose 1 | Dose 2 |
|---|---|---|---|
| 100 | AEC | 9 | 61 |
| 13 | PAH with 2 and 3 rings | 0 | 0 |
| 4 | PAH with 4 to 7 rings | 6 | 61 |

*Note:* Dose 1 and 2 correspond to 1.05 and 3.15 mg AEC, respectively, administered twice weekly.

the PAH were contained at the same quantitative ratio as in the AEC. The mixture consisted of the following PAH: benzo[c]phenanthrene (0.26), cyclopenta[cd]pyrene (6.20), benz[a]anthracene (0.30), chrysene (0.70), benzo[b]fluoranthene (0.56), benzo[k]fluoranthene (0.20), benzo[j]fluoranthene (0.30), benzo[a]pyrene (1.00), 11H-cyclopenta[qrs]benzo[e]pyrene (0.47), 10H-cyclopenta[mno]benzo[e]pyrene (0.16), dibenz[a,j]anthracene (0.35), indeno[1,2,3-cd]pyrene (0.70), dibenz[a,h]anthracene (0.08), PAH-M300 A (0.23), PAH-M300 B (0.21). It was attempted to simulate the carcinogenic activity of AEC by applying the above mixture of 15 PAH. Some of the compounds had been extracted from AEC directly. Their weight percentage amounted to about 0.7% of the AEC, while benzo[a]pyrene accounted for 0.06%. The evaluation of the animal experiments revealed that this mixture of 15 pure substances produced the same tumor rate as the AEC only if the administered doses were twice as high as the amounts contained in AEC. Consequently, the mixture of pure substances can account for only half the carcinogenic activity of AEC. Presumably, some carcinogenic PAH contained in the four- to seven-ring fraction of AEC were not identified. One of these may be, for instance, anthanthrene, the carcinogenic activity of which has only recently been described in tests of topical application.[9] To produce the same tumor incidence, the doses of pure benzo[a]pyrene, however, have to be ten times higher

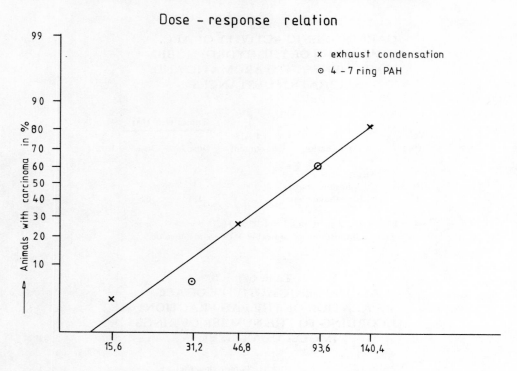

FIGURE 2. Tumorgenic activity of AEC and its fractions after topical application (CFLP mice). The PAH fraction (4—7 rings) was dosed according to its percentage rate in AEC (approximately 4% of the AEC). AEC = automobile exhaust condensate.

than those contained in the AEC since benzo[a]pyrene accounts for only about 10% of the carcinogenic activity of the total condensate.

In summarizing, it can be stated that

1. The total carcinogenic activity of AEC can be attributed to the contained PAH with four to seven rings. This fraction accounts for 4% by weight of the AEC.
2. The amount of benzo[a]pyrene contained in AEC accounts for about 10% of the carcinogenic effect of AEC.
3. A mixture of 15 individual PAH which are contained in AEC accounts for about 50% of the carcinogenic effect of AEC observed in experiments of topical application to mouse skin. Probably not all PAH which are responsible for this effect were included in this mixture of 15 PAH.

*1.2.2.2 Other Animal Experiment Models Used as a Detection System*

The cited fractions of AEC were also tested for biological activity in other animal experiment models. Thus, they were, for instance, injected subcutaneously in mice.[33] In this model, the carcinogenic activity of AEC from gasoline engines cannot easily be ascribed to the contained PAH, unlike the topical application, because unknown components of AEC reduced considerably the PAH-induced formation of sarcomas. After elimination of the hydrophilic components, the remaining AEC exerted a highly carcinogenic effect.

In this experiment the sebaceous gland suppression test can also be used as a biological detection system in a manner analogous to the topical application method. In this model, the destruction of sebaceous glands of the dorsal skin of mice is measured.

This destruction is effected by carcinogenic PAH or nitrogen-containing polycyclic heterocycles, whereas noncarcinogenic PAH do not affect the sebaceous glands. As in the corresponding long-term experiment in which skin carcinomas are formed after repeated topical application of an acetone solution of AEC, three topical applications result in a clear correlation between the applied dose and the degree of sebaceous gland suppression within 6 days. Therefore, this test may be used as a biological detection system for the isolation of individual active substances contained in AEC mixtures. PAH-containing fractions which produced carcinomas after several months of topical application to mouse skin yielded analogous results and effected a dose-related suppression of sebaceous glands.[6]

*1.2.2.3 Analysis of the Proportions of Activities in Mutagenicity Assays*

Mutagenicity tests, such as, for instance, the reverse mutation of histidine-deficient mutants of *Salmonella typhimurium* (Ames test) have been used for the identification of mutagens in complex mixtures. Several authors reported on the separation of air pollutants into PAH-containing and PAH-free fractions.[10,14,32,38-40] Other investigations concerned mixtures of PAH which were formed during the conversion of solid fossil fuels such as coal, tarry sands, etc. into liquid or gaseous products. In these conversion products, too, the fraction containing benzene derivatives and aromatic compounds with two rings did not exert a mutagenic effect. The effect is induced by those fractions which contained PAH with four and five rings. These results were obtained partly without activation and partly after addition of microsome fractions.[22] These results could even be obtained, although to a lesser extent, without microsomal activation.

**1.2.3 Carcinogenic Components of Automobile Exhaust Condensate from Gasoline Engines**

The carcinogenic activity of automobile exhaust can be proven in various animal experiments.[5,6,25,31,33,43] The best method of simulating the carcinogenic hazard to man would doubtless be the direct inhalation of automobile exhausts. This is, however, impossible because of the high toxicity of these gases (about 2 to 5% CO in exhaust of gasoline engines).

*1.2.3.1 Intratracheal Instillation*

Instillation of AEC (2.5 mg or 5 mg, every 2 weeks) into the trachea of Syrian golden hamsters produced multiple lung adenomas in all experimental animals after 30 to 60 weeks. This animal experiment clearly demonstrated the carcinogenic effect of automobile exhausts.

*1.2.3.2 Topical Application*

A preferred animal experimental model for the detection of carcinogenic effects is the topical application of the dissolved carcinogenic substance, or mixture of substances, to mouse skin. A solution of exhaust condensate from gasoline engines applied topically twice weekly (dose range: 0.5 to 4.7 mg) produced dose-related skin carcinomas in 1 to 72% of the experimental animals.[5,6] The BaP content of the AEC amounted to about 285 µg BaP per gram (for details see Chapter 1.2.2).

*1.2.3.3 Identification of Carcinogenic Components*

In the animal experiment of topical application to mouse skin, the investigation of the separated PAH-containing and PAH-free fractions[19,20] for carcinogenic activity revealed that more than 90% of the carcinogenic activity of AEC could be attributed to the PAH-containing fraction, whereas the corresponding doses of the PAH-free fraction (about 96% by weight) accounted for only a very low residual activity.[5,30] A synthetic mixture of 15 carcinogenic PAH detected in the PAH fraction of AEC simulated more than 50% of the carcinogenic potency of AEC. The ranking of BaP in the carcinogenic activity was about 8 to 10%.[7]

Thus, automobile exhaust condensate was the first complex mixture of noxious substances in which the majority of all contained carcinogenic substances could be detected and the carcinogenic activity be ascribed to individual components (for details see Chapter 1.2.2).

### 1.2.4 Carcinogenic Components of Exhaust Condensate from Diesel Engines

Exhaust condensate was obtained from a passenger car driven by a diesel engine in the so-called "Europa test" (ECE* regulation 15). The acetone-dissolved condensate was tested for carcinogenic activity only by topical application to the dorsal skin of mice. In this animal model the condensate proved to have a dose-related effect (3 doses, 80 mice per group). The carcinogens contained in this exhaust condensate have not been identified yet. It is, however, to be expected that PAH also account for a considerable proportion of the carcinogenic activity of diesel exhaust.

Diesel exhaust condensates have a lower PAH content than exhaust condensates of gasoline engines of comparable capacity. The exhaust condensates used in the animal experiment mentioned earlier had a concentration of 20 to 30 $\mu$g BaP per gram. If exhaust condensates of gasoline and diesel engines are applied topically at doses which contain identical amounts of BaP, their carcinogenic effects are comparable.[8]

### 1.2.5 Carcinogenic Components of Cigarette Smoke

Daily smoking of nontipped cigarettes over 2 years induced invasive lung tumors in 10 out of 20 beagles.[2,23] Instillation of cigarette smoke condensate into the trachea of rats, rabbits, or hamsters did not induce carcinomas in the trachea of experimental animals.[12,15,44] Shabad[36] alone reported epidermal carcinomas in the trachea of rats (2 out of 43 experimental animals) after instillation of a total of 500 to 600 mg of condensate. This finding has not yet been confirmed by other authors. Instillation of a PAH-containing fraction, which was highly carcinogenic in topical application to mouse skin, also did not produce any clearly malignant neoplastic changes despite the very high doses administered.[37]

The carcinogenic activity of tobacco smoke condensate was observed for the first time during topical application of the condensate to mouse skin.[41,45] In the meantime this result has been confirmed repeatedly. A dose-related effect on the skin could be determined for this type of application.[11] This result has permitted the determination of the rate of activity of PAH and other carcinogenic compounds contained in cigarette smoke condensate (CSC).

Although it is possible to state which components of the particle phase of cigarette smoke are carcinogenic, this statement has to be restricted to findings obtained by dropping the condensate and its fractions onto mouse skin. A possible activity of the gaseous phase of cigarette smoke is not taken into account.

Several authors investigated cigarette smoke condensate for its carcinogenic components, using the topical application as a detection system.[13,16,17,34,42] When using a very cautious separation method[18] it was possible to localize about 50% of the total carcinogenic activity of cigarette smoke condensate in the fraction of PAH with three to seven rings. This fraction, which also contained other substances besides PAH,

---

\* Economic Commission for Europe.

Table 7
COMPARISON OF THE
CARCINOGENIC ACTIVITY OF
CIGARETTE SMOKE
CONDENSATE (CSC),
RECONSTITUTED CSC (r-CSC),
AND PROPORTIONALLY
DOSED PAH FRACTIONS
(ACTIVITY RATE)[a]

| Activity rate | | Weight | |
|---|---|---|---|
| r-CSC | PAH fraction | (%) | Ref. |
| 0.80 | 1.0 | 6.8 | 16 |
|  | 0.5 | 1.1 |  |
| — | 0.66 (0.73) | 6.5 | 17 |
|  | 0.41 (0.47) | 0.4 |  |
| — | 0.67 (0.66) | 24.5 | 27 |
|  | 0.67 (0.63) | 3.5 |  |

*Note:* Weight percent (%) = weight percent of the PAH fraction which also contains compounds other than PAH. The rates of activity of the reconstituted CSC and the PAH fraction are referred to the original CSC (1.00). The given figures represent the rate of carcinogenicity (malignant tumors); figures in brackets give the total rate of malignant and benign tumors. Weekly doses: Dontenwill et al.:[16,17] 150 mg and 225 mg or proportional (fractions); Lee et al.;[27] 300 mg and 600 mg or proportional (fractions).

[a] Original condensate = 1.00.

accounted for 0.4% by weight of the total condensate.[17] Table 7 shows that the highest activity was localized in the lowest weight fraction.

Many working groups have dealt with the problem of identifying the carcinogenic components of CSC and reported their results in recent years.[1,17,24,26,27] So far, a separation (fractionation) of cigarette smoke condensate without partial loss of biologic activity has not been described. A measurable loss of activity, for instance, occurs predominantly during the chromatographic fractionation with inorganic column material. When comparing the CSC (neutral components) with the condensate which was reconstituted from the individual fractions (r-CSC), Bock et al.[4] recorded a reduction of the carcinogenic activity of the latter to about one fourth of that of the initial material. The biological activity is only slightly reduced, however, by organic packing materials such as Sephadex® LH 20 or Bio-Beads SX-12®.

Good evidence of the careful separation (fractionation) of CSC is given by the comparison of the biological activity of the initial and the reconstituted condensate. A fractionating process during which no, or only a low, reduction of activity occurs, permits reaching a balance of the activities of all the fractions. In Table 7 the activities of CSC, r-CSC, and the PAH-containing fractions (which also contain compounds other than PAH) are compared.

## Table 8
### COMPARISON OF THE CARCINOGENIC ACTIVITY OF CIGARETTE SMOKE CONDENSATE (CSC), RECONSTITUTED CSC (r-CSC), AND PROPORTIONALLY DOSED PAH FRACTIONS

| Carcinoma and tumor incidence (%) | | | Weight | |
|---|---|---|---|---|
| CSC | r-CSC | PAH fraction | (%) | Ref. |
| 22 (40) | 10 (60) | not tested | 2.6 | 24 |
| 22 (60)[a] | 22 (56)[a] | 12 (52)[a] | 5.1 | 3 |
| (32) | (7.5) | (7.5) | 5.6 | 26 |
| (57)[a] | (52)[a] | (20)[a] | 5.6 | |

[a] After pretreatment with 7.12-dimethylbenz[a]anthracene. Figures in brackets refer to total tumor incidence (carcinomas and papillomas). Since only one dose was tested, the rate of activity could not be calculated.

As shown in Table 7, only a rate X of the original cigarette smoke condensate is needed to obtain the same rate (%) of tumor-bearing animals in this experimental group when applying proportional doses of this fraction of the condensate. Only 80 mg of the original, unfractionated CSC, for instance, are needed to induce the same effect as 100 mg r-CSC. If only one class of carcinogen, such as the PAH, were responsible for the cutaneous carcinogenic activity of CSC and no loss of activity occurred during fractionation, CSC and the PAH-containing fraction would have identical activity when administered at proportional doses. Table 7 shows that PAH alone usually do not achieve this rate of activity.[16,17,27] The observed rates of activity rank between 0.41 and 1.0 (related to the carcinoma yield).

Other investigators tested CSC, r-CSC, and the PAH-containing fractions only at one dose level. Consequently, it is not possible to calculate the rate of activity by corresponding mathematical transformation. In this case the loss of biological activity can be assessed only by comparing the tumor rates of the individual experimental groups. As can be seen in Table 8, there is a considerable loss of activity during the separation processes used.[3,4,24,26]

Identical biological effects of CSC and r-CSC could be recorded only if the experimental mice had been pretreated with subthreshold doses of 7,12-dimethylbenz[a]anthracene. Under these experimental conditions, the PAH-containing fraction never reached the same rate of carcinomas as the CSC. In one case[3] at least a comparable rate of tumors was reported.

Analogous to the findings summarized in Table 8, further results obtained with and without pretreatment with 7,12-dimethylbenz[a]anthracene indicate that besides PAH other classes of carcinogenic chemicals, for instance, nitrogen-containing polycyclic aromatic compounds, contribute to the activity observed in this test model. The tumor rates observed in pretreated animals might possibly be explained by assuming that some of the complete carcinogens (initiating carcinogens) are eliminated or destroyed during fractionation. On the other hand, the cited experimental results permit the conclusion that promoting substances (promotors) are also contained in CSC.[35]

The experimental results presently available suggest that a fraction which contains, among others, PAH with more than three rings, accounts for 50% of the carcinogenic activity of CSC observed after topical application to the dorsal skin of mice.

## Table 9
## PERCENTAGE OF CARCINOGENIC ACTIVITY OF BaP RELATED TO THE TOTAL CARCINOGENICITY

|  | Carcinogenicity of the BaP portion (%) |
|---|---|
| Automobile exhaust condensate (gasoline engine) | 9.6 |
| Automobile exhaust condensate (diesel engine) | 16.7 |
| Domestic hard coal | ca. 6[a] |
| Domestic brown coal (briquets) | ca. 9[a] |
| Lubricating oil from cars (used) | 18 |
| Sewage sludge (extract) | 22.9 |
| Cigarette smoke condensate | ca. 1 |

[a] Preliminary, not yet finished.

**Rate of BaP on the carcinogenicity of CSC** — In several publications the activity of the total CSC and the contained BaP was compared. When using long-term topical application to the dorsal skin of mice as a model, the BaP contained in CSC accounted for only 1 to 2% of the carcinogenic activity of CSC.

### 1.2.6 Rates of the Biological Effect of BaP in Various Matrices

Comparison of other pyrolysis condensates in this test model results in very different rates for the biological effects of BaP (Table 9). Thus, BaP counts for a minor part of the total carcinogenic effect, if the effect of the condensate is compared with that of the BaP content of the sample. In extracts of sewage sludge, BaP accounts for almost 23% of the carcinogenic effect; in cigarette smoke condensate, however, it accounts for only 1%.

### 1.2.7 Rates of the Biological Effect of All Carcinogenic PAH in Various Matrices

What part of the total effect of a condensate or a matrix do other PAH have? This is a general question for all matrices. If the topical application onto the mouse skin is used as a test system, it can be stated:

If the topical application model is used as an indicator for the carcinogenicity of substances, by far the greater part of the carcinogens has to be looked for among the PAH.

## 1.3 THE "REPRESENTATIVE PAH CONCENTRATION" OF A CITY

### 1.3.1 PAH Produced by Various Emittants

To estimate the average burden of PAH produced by various environmentally relevant sources we have to ask: What amounts of PAH are formed during the combustion of 1 kg fuel under standard conditions? Table 11 shows, using BaP as an example, that there are considerable differences. During combustion, 1 kg of fuel yields between 0.0001 and 100 mg BaP.

Table 11 also shows the annual fuel consumption of West Germany. From the two columns it can be estimated that PAH are predominantly formed by domestic coal heating.

**Table 10**
**THE PERCENTAGE OF TUMOR-PRODUCING ACTIVITY OF THE PAH FRACTION (MORE THAN 3 RINGS) OF THE TOTAL CARCINOGENICITY**

|  | Carcinogenicity explained by PAH portion (%) |
|---|---|
| Automobile exhaust condensate | 91 |
| Domestic hard coal flue gas | ca. 90[a] |
| Used lubricating oil | 70 |
| Cigarette smoke condensate | ca. 50 |

[a] Preliminary, not finished.

**Table 11**
**EMISSION OF BENZO[a]PYRENE PRODUCED BY 1 kg FUEL**

|  | BaP emission (mg/kg) | Consumption[a] per year (tons) |
|---|---|---|
| Hard coal briquets (domestic heating) | 5.0—380.0 | 1,128,000 |
| Oil heating (light-oil) (domestic heating) | 0.0001 | 42,600,000 |
| Passenger cars | 0.10 | 20,400,000 |
| Cars with diesel engines | 0.03 | 7,650,000 |
| Power plants | 0.005 | 26,800,000 |

[a] In West Germany.

### 1.3.2 Local and Temporary Variations of the PAH Concentration in a City

To estimate the environmental burden of an area, a city, or a country, it is not sufficient to know only the total amount of PAH emitted, but to also know the local PAH concentration and the variation in concentration during specific time periods. That is, to measure the real PAH burden of the inhabitants of an area, it is necessary to compare local concentrations of several selected PAH and to record the temporary variation in PAH profiles of different areas in a city during the year.

#### 1.3.2.1 The Temporary Variation on a Selected Collecting Station in the City During the Day and the Year

The variation of the concentrations during a day for several PAH in a city is shown in Figure 3.[21] This is a residential area, heated by domestic coal. The concentration of the PAH is plotted along the ordinate and the groups of lines represent the various PAH. The seven lines represent the collecting hours, as explained in the corner of the figure. For example, in the case of BaP the difference in the concentration between the first 2 hr in the morning and the following hours is very great; in the early morning the concentration is more than five times higher. This is also confirmed by long-term measurements. Figure 4 shows the variation of the BaP concentration, measured each week in the same area, during a period of 12 months. The height of the lines represent the concentration of BaP. The number of weeks in 1978 and 1979 are plotted along the abscissa. The highest BaP concentration — 300 ng BaP per cubic meter — was

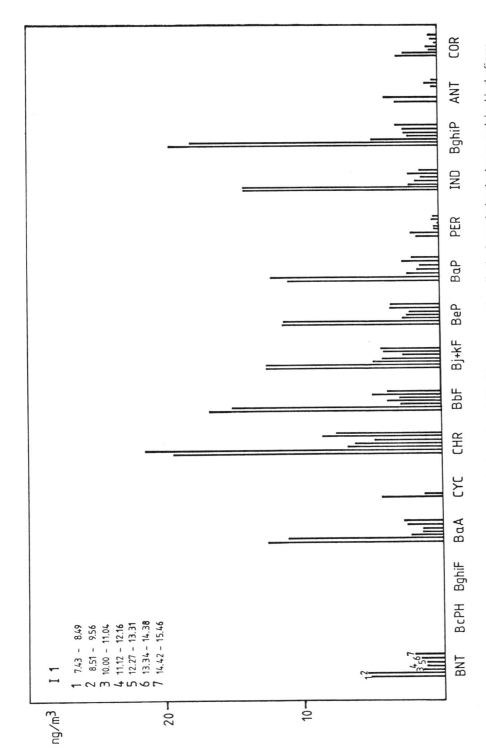

FIGURE 3. Residential area, chiefly heated by domestic coal. The 7 lines represent the collecting hours during the day, as explained in the figure.

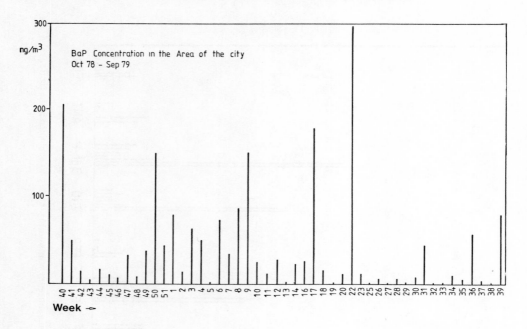

FIGURE 4. Benzo[a]pyrene concentration in the same area of the city from October 1978 to September 1979. (Collecting period: 1 hr per week.)

recorded in the 22nd week of 1979, the last week of May. The lowest concentration was detected 2 weeks later, in June.

The concentration of PAH at a selected place in the city varies greatly even in a residential area. However, it is possible to sum the total amount and thus to calculate the average concentration for this place.

*1.3.2.2 The Local Variation in Different Areas of the City*

The second problem is the local differences in different areas in the city. Figure 5 shows the average concentration of BaP in the four areas, burdened by typical sources.

In each area about 50 samples of 15 PAH were collected during the period of October 1978 to September 1979; Figure 5 demonstrates the results of BaP determinations. In area I, burdened principally by domestic coal pollution, the average value over 50 weeks was 15.4 ng BaP per cubic meter, and the range was 0.3 to 72.5 ng BaP per cubic meter. The other areas, II, III, and IV, show averages of 6, 31, and 40.4 ng BaP per cubic meter, respectively.

The most polluted area — 40 ng BaP per cubic meter — surrounds the coke plant. About 3 km away from this area, the annual concentration drops off to one third.

Completely different from this pattern is the distribution of cyclopenta[cd]pyrene (CYC), a PAH chiefly produced by automobiles with gasoline engines (Figure 6). As was to be expected, the tunnel carrying automobile traffic shows the highest annual concentration. The average concentration is 88 ng CYC per cubic meter, and the range is 0.1 to 440 ng/m³. Surprisingly, at a distance of about 4 km, the concentration of CYC drops off to 1.6 ng/m³. This is about 1/50 of the concentration in the tunnel. The same is also true for the other areas.

1. The distribution of different PAH in the city may not correlate with each other.
2. The different emittants produce different PAH profiles.

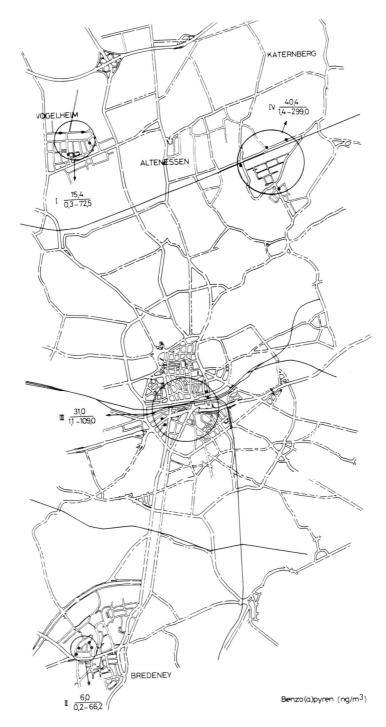

FIGURE 5. Map of an industrial city in which four areas polluted by typical emittants are labeled: (I) residential area chiefly heated by domestic coal, (II) a residential area heated chiefly by oil fuel, (III) a tunnel with automobile traffic, and (IV) an area surrounding a coke plant. The distribution pattern of benzo[a]pyrene (ng/m³) in the city. Average value over 50 weeks, and below the line the range (ng BaP per cubic meter).

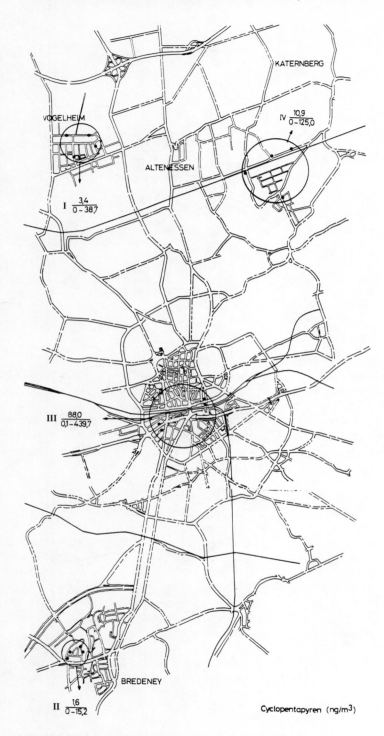

FIGURE 6. The distribution pattern of cyclopenta[cd]pyrene (ng/m³) in the city (for explanation see Figure 5).

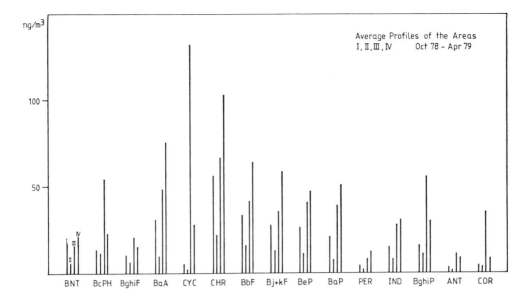

FIGURE 7. The average profile of the areas I, II, III, and IV (see Figure 5). Averages of about 30 weeks from October 1978 to April 1979.

Figure 7 shows a comparison between the concentrations of the four areas. The ratios of the concentrations within the four areas are similar in the case of CYC, BghiP, or COR (coronene). These PAH predominate in area III. In the case of CHR, BbF, Bj + kF, BeP, as well as in the case of BaP, the ratios of the concentrations in the four areas are very similar. In this figure, area IV shows the highest concentrations.

From this representation it seems that there are two chief sources of emission in the city:

1. Pyrolysis and combustion of coal and briquets
2. The emission of automobile exhaust especially in the tunnel

In the last case, CYC is the PAH with the highest concentration. This is in contrast to the emissions from coal combustion, which produces only small amounts of CYC.

The composition of the PAH mixture, the PAH profile, in some cases differs from area to area of the city. In the traffic area (tunnel) CYC predominates. This is not the case in the other areas. BNT, which originates from pyrolysis or combustion of hard and brown coal, is typical for these emission sources. Perhaps the different PAH profiles are helpful in recognizing the main sources of air pollution.

## 1.4 FUNDAMENTAL MATHEMATICAL AND STATISTICAL PRINCIPLES FOR ANALYZING THE PROPORTIONAL ACTIVITY

### J. Misfeld

As explained above, it is an essential objective of animal experiments, carried out using carcinogens, to reach the following conclusions:

(a) Substance B is $\varrho$-times as effective as substance A. If substance A is a mixture of several components including substance B, the following conclusion is desired:
(b) X % of the activity of A can be attributed to B.

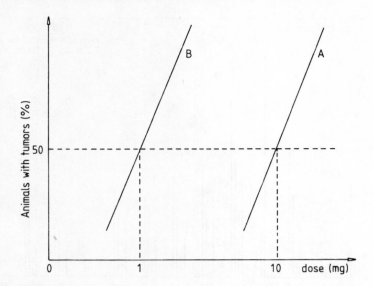

FIGURE 8. Definition of the relative potency: B is ten times as potent as A.

Therefore a basic prerequisite for deriving such conclusions is an experimental setup which permits an exact assessment of the dose-response relationships of the test substances. This is not explained here in detail because it has been described in the literature.[29] In many cases it has proved useful to transform the dose axis logarithmically, according to the Weber-Fechner law, and the response axis by means of the probit function.

As an example, in the investigations discussed later, the dose-response relationship of the two substances being compared, A and B, can often be represented graphically as parallel straight lines. In this case we define: let a dose $d_A$ of substance A and a dose $d_B$ of substance B cause the same response. Then $\varrho := d_B/d_A$ is called the relative potency of A to B. We say A is $\varrho$-times as potent as B.

*Note*

According to definition, $\varrho$ is the quotient of equally effective doses of both substances; therefore $\varrho$ is called the relative potency (see Figure 8). This describes the procedure necessary to arrive at a statement of the type (a), as mentioned earlier.

**The Concept of "Proportional Activity"**

Starting from the assumption that substance A is a mixture which includes B and its complement C, it follows that

$$A = B + C$$
$$1 = b + (b - 1) \quad \text{(in units of weight)} \qquad (1)$$

It is postulated that in units of weight the dose-response relationships are parallel lines. $d_A$ units of weight of A and $d_B$ units of weight of B are needed to induce identical effects. It follows, therefore,

$$\rho(\text{B vs. A}) = \frac{d_A}{d_B} \tag{2}$$

The effect of A is of course also due to the existence of B in A. Assuming that the effect of A is based only on B, it follows:

$$b \cdot d_A = d_B \tag{3}$$

or

$$b \cdot \frac{d_A}{d_B} = b \cdot \rho = 1$$

Instead of Equation 3, however, Equation 4 is generally applicable:

$$b \cdot d_A < d_B \tag{4}$$

or

$$b \cdot \rho < 1$$

The value $b \cdot \varrho$ defines the extent to which the effect of A can be attributed to B.

*Definition:* Let A be a compound in the mixture M, the percentage weight of A in M being a and the relative potency A to M being $\varrho_{A/M}$. Then $R = a \cdot \varrho_{A/M}$ is called the percentage potency of M that can be calculated (by means of similar action) from A.

*Note*

It is possible that $b \cdot \varrho > 1$. In the case of a significant deviation, this indicates a "nonadditive action" of the components of A. These complicated reaction mechanisms will not be discussed here.

*Example*

If substance B is contained in A at a rate of 0.05 this means that

$$A = B + C$$
$$1 = 0.05 + 0.95$$

and $d_A = 10$ mg, $d_B = 1$ mg, consequently

$$\rho = \frac{d_A}{d_B} = 10$$

and

$$b \cdot \rho = 0.05 \cdot 10 = 0.5$$

Thus 50% of the potency of A can be calculated (by means of similar action) from B. For further examples see Misfeld and Weber[28] and Misfeld and Timm.[30]

## REFERENCES

1. Akin, F. J., Snook, M. E., Severson, R. E., Chamberlain, W. J., and Walters, D. B., Identification of polynuclear aromatic hydrocarbons in cigarette smoke and their importance as tumorigens, *J. Natl. Cancer Inst.*, 57, 191, 1976.
2. Auerbach, O., Hammond, E. C., Kirman, D., and Garfinkel, L., II. Pulmonary neoplasms, *Arch. Environ. Health*, 21, 754, 1970.
3. Bock, F. G., Swain, A. P., and Stedman, R. L., Bioassay of major fractions of cigarette smoke condensate by an accelerated technic, *Cancer Res.*, 29, 584, 1969.
4. Bock, F. G., Swain, A. P., and Stedman, R. L., Carcinogenesis assay of subfractions of cigarette smoke condensate prepared by solvent-solvent separation of the neutral fraction, *J. Natl. Cancer Inst.*, 49, 477, 1972.
5. Brune, H., Experimental results with percutaneous applications of automobile exhaust condensates in mice, *IARC Sci. Publ.*, 16, 41, 1977.
6. Brune, H., Habs, M., and Schmähl, D., The tumor-producing effect of automobile exhaust condensate and fractions thereof, *J. Environ. Pathol. Toxicol.*, 1, 737, 1978.
7. Brune, H., unpublished data.
8. Brune, H. et al., unpublished data.
9. Cavalieri, E., Mailander, P., and Pelfrene, A., Carcinogenic activity of anthracene on mouse skin, *Z. Krebsforsch.*, 89, 113, 1977.
10. Commoner, B., Carcinogens in the environment, *Chem. Technol.*, 76, 1977.
11. Davies, R. F., Lee, P. N., and Rothwell, K., A study of the dose response of mouse skin to cigarette smoke condensate, *Br. J. Cancer*, 30, 146, 1974.
12. Davies, B. R., Whitehead, J. K., Gill, M. E., and Lee, P. N., Response of rat lung to tobacco smoke condensate and fractions derived from it administered repeatedly by intratracheal instillation, *Br. J. Cancer*, 31, 453, 1975.
13. Day, T. D., Carcinogenic action of cigarette smoke condensate on mouse skin, *Br. J. Cancer*, 21, 56, 1967.
14. Dehnen, W., Pitz, N., and Tomingas, R., The mutagenicity of air-borne particulate pollutants, *Cancer Lett.*, 4, 5, 1977.
15. Della Porta, G., Kolb, L., and Shubik, P., Induction of tracheobronchial carcinomas in the Syrian golden hamster, *Cancer Res.*, 18, 592, 1958.
16. Dontenwill, W., Elmenhorst, H., Harke, H. P., Reckzeh, G., Weber, K. H., Misfeld, J., and Timm, J., Experimentelle Untersuchungen über die tumorerzeugende Wirkung von Zigarettenrauch-Kondensaten an der Mäusehaut. III. Mitteilung: Untersuchungen zur Identifizierung und Anreicherung tumorauslösender Fraktionen, *Z. Krebsforsch.*, 73, 305, 1970.
17. Dontenwill, W., Chevalier, H. J., Harke, H. P., Klimisch, H. J., Brune, H., Fleischmann, B., and Keller, W., Experimentelle Untersuchungen über die tumorauslösende Wirkung von Zigarettenrauch-Kondensaten an der Mäusehaut. VI. Mitteilung: Untersuchungen zur Franktionierung von Zigarettenrauch-Kondensat, *Z. Krebsforsch.*, 85, 155, 1976.
18. Elmenhorst, H. and Grimmer, G., Polycyclische Kohlenwasserstoffe aus Zigarettenrauch-Kondensat. Eine Methode zur Fraktionierung grosser Mengen für Tierversuche, *Z. Krebsforsch.*, 71, 66, 1968.
18a. Frentzel-Beyme, R., Leutner, R., Wagner, G., and Wiebelt, H., *Cancer Atlas of the Federal Republic of Germany*, Springer-Verlag, Berlin, 1979.
19. Grimmer, G., Analysis of automobile exhaust condensates, in Air Pollution and Cancer in Man, *IARC Sci. Publ.*, No. 16, 29, 1977.
20. Grimmer, G. and Böhnke, H., The tumor-producing effect of automobile exhaust condensate and fractions thereof. I. Chemical studies, *J. Environ. Pathol. Toxicol.*, 1, 661, 1978.
21. Grimmer, G., Naujack, K.-W., and Schneider, D., Comparison of the profiles of polycyclic aromatic hydrocarbons in different areas of a city by glass-capillary-gas-chromatography in the nano-gram range, *Int. J. Environ. Anal. Chem.*, 10, 265, 1981.
22. Guerin, M., Epler, J. L., Griest, W. H., Clark, B. R., and Rao, T. K., Polycyclic aromatic hydrocarbons from fossil fuel conversion processes, in *Carcinogenesis*, Vol. 3, Jones, P. W. and Freudenthal, R. I., Eds., Raven Press, New York, 1978, 21.
23. Hammond, E. C., Auerbach, O., and Kirman, D., Effects of cigarette smoking on dogs. I. Design of experiment, mortality, and findings in lung parenchyma, *Arch. Environ. Health*, 21, 740, 1970.
23a. Hammond, E. C., Garfinkel, L., Seidman, H., and Lew, E. A., "Tar" and nicotine content of cigarette smoke in ratio to death rates, *Environ. Res.*, 12, 263, 1976.
24. Hoffmann, D. and Wynder, E. L., A study of tobacco carcinogenesis. XI. Tumor initiators, tumor accelerators, and tumor promoting activity of condensate fractions, *Cancer (Philadelphia)*, 27, 848, 1971.

25. Kotin, P., Falk, H. L., and Thomas, M., Aromatic hydrocarbons. II. Presence in particulate phase of gasoline-engine exhaust and carcinogenicity of exhaust extracts, *Arch. Indust. Hyg. Occup. Med.*, 9, 164, 1954.
26. Lazar, P., Chouroulinkov, I., Izard, C., Moree-Testa, P., and Hemon, D., Bioassay of carcinogenicity after fractionation of cigarette smoke condensate, *Biomedicine,* 20, 214, 1974.
27. Lee, P. N., Rothwell, K., and Whitehead, J. K., Fractionation of mouse skin carcinogens in cigarette smoke condensate, *Br. J. Cancer,* 35, 730, 1977.
28. Misfeld, J. and Weber, K. H., Tierexperimente mit Tabakrauchkondensaten und ihre statistische Beurteilung, *Planta Med.*, 22, 281, 1972.
29. Misfeld, J. and Timm, J., Mathematical planning and evaluation of experiments, in Air Pollution and Cancer in Man, *IARC Sci. Publ.,* No. 16, 11, 1977.
30. Misfeld, J. and Timm, J., The tumor-producing effect of automobile exhaust condensate and fractions thereof, *J. Environ. Pathol. Toxicol.,* 1, 747, 1978.
31. Mohr, U., Reznik-Schüler, H., Reznik, G., Grimmer, G., and Misfeld, J., Investigations on the carcinogenic burden by air pollution in man. XIV. Effect of automobile exhaust condensate on the Syrian golden hamster lung, *Zentralbl. Bakteriol. Parisitenkd. Infektionskr. Hyg. Abt. 1 Orig. Reihe B,* 163, 425, 1976.
32. Pitts, J. N., Grosjean, D., Mischke, T., Simmon, V. F., and Poole, D., Mutagenic activity of airborne particulate organic pollutants, *Toxicol. Lett.,* 1, 65, 1977.
33. Pott, F., Tomingas, R., and Misfeld, J., Tumors in mice after subcutaneous injection of automobile exhaust condensates, *IARC Sci. Publ.,* No. 16, 79, 1977.
34. Roe, F. J. C., Salaman, M. H., and Cohen, J., Incomplete carcinogens in cigarette smoke condensate: tumor promotion by a phenolic fraction, *Br. J. Cancer,* 13, 623, 1959.
35. Schmeltz, I., Tosk, J., Hilfrich, J., Hirota, N., Hoffmann, D., and Wynder, E. L., Bioassay of naphthalene and alkylnaphthalene for co-carcinogenic activity. Relation to tobacco carcinogenesis, in *Carcinogenesis,* Vol. 3, Jones, P. W. and Freudenthal, R. I., Eds., Raven Press, New York, 1978, 47.
36. Shabad, L. M., Review of attempts to induce lung cancer in experimental animals by tobacco smoke, *Cancer (Philadelphia),* 27, 51, 1970.
37. Simons, P. J., Lee, P. N., and Roe, F. J. C., Squamous lesions in lungs of rats exposed to tobacco-smoke-condensate fractions by repeated intratracheal instillation, *Br. J. Cancer,* 37, 965, 1978.
38. Talcott, R. and Wei, E., Brief communication: airborne mutagens bioassayed in Salmonella typhimurium, *J. Natl. Cancer Inst.,* 58, 449, 1977.
39. Teranishi, K., Hamada, K., and Watanabe, W., Mutagenicity in Salmonella typhimurium mutants of the benzene-soluble organic matter derived from airborne particulate matter and its five fractions, *Mutat. Res.,* 56, 273, 1978.
40. Tokiwa, H., Morita, K., Takeyoshi, H., and Ohnishi, Y., Detection of mutagenic activity in particulate air pollutants, *Mutat. Res.,* 48, 237, 1977.
41. Wynder, E. L., Graham, E. A., and Croninger, A. B., Experimental production of carcinoma with cigarette tar, *Cancer Res.,* 13, 855, 1953.
42. Wynder, E. L. and Wright, G., A study of tobacco carcinogenesis, *Cancer (Philadelphia),* 10, 255, 1957.
43. Wynder, E. L. and Hoffmann, D., Study of air pollution carcinogenesis. III. Carcinogenic activity of gasoline engine exhaust condensate, *Cancer (Philadelphia),* 15, 103, 1962.
44. Wynder, E. L. and Hoffmann, D., Experimental tobacco carcinogenesis, *Adv. Cancer Res.,* 8, 249, 1964.
45. Wynder, E. L. and Hoffmann, D., *Tobacco and Tobacco Smoke,* Academic Press, New York, 1967.

## Chapter 2

# CHEMISTRY

### G. Grimmer

## TABLE OF CONTENTS

2.1 Choice of Relevant Polycyclic Aromatic Hydrocarbons ..................28
    2.1.1 Mechanisms of Formation ........................28
    2.1.2 Number of PAH Detected in the Environment ................30
    2.1.3 Criterion of Choice: Carcinogenicity........................30
    2.1.4 Criterion of Choice: Frequency of Occurrence..................31

2.2 Nomenclature of PAH......................................31
    2.2.1 Rule 1: "Base Components" ................................31
    2.2.2 Rule 2: "Orientation"..................................31
    2.2.3 Rule 3: "Numbering".................................35
    2.2.4 Rule 4: "Lettering"...................................35
    2.2.5 Rule 5: "Condensed Systems".............................35

2.3 Methods of PAH Detection..................................39
    2.3.1 Introduction........................................39
    2.3.2 Methods of Chemical Analysis.............................40
    2.3.3 Extraction.........................................42
    2.3.4 Extraction Methods...................................42
        2.3.4.1 Extraction Group: Fats, Oils .....................43
        2.3.4.2 Extraction Group: Water Samples..................43
        2.3.4.3 Extraction Group: Meat Products ..................44
        2.3.4.4 Extraction Group: Leaves, Seeds, Fruits.............44
        2.3.4.5 Extraction Group: Samples Containing Inorganic Components ........................................44
        2.3.4.6 Extraction Group: Soot- and Graphite-Like Materials 44
    2.3.5 Enrichment........................................45
        2.3.5.1 Liquid-Liquid Partition of Biphase Solvent Mixtures ..46
            2.3.5.1.1 Hydrophilic/Hydrophobic (Polar/Nonpolar)................................46
            2.3.5.1.2 Aliphatic/Aromatic...................46
        2.3.5.2 Chromatographic Methods.......................46
            2.3.5.2.1 Silica Gel, Alumina, etc. as Column-Packing Material........................46
            2.3.5.2.2 Sephadex® LH 20, 60, Acetylcellulose as Column-Packing Material.............46
        2.3.5.3 Summary and Appraisal of Enrichment Methods .....49
    2.3.6 Separation Methods for the Chromatographic Separation and Quantitative Assessment of PAH Mixtures ....................49
    2.3.7 Gas Chromatography with High-Performance Columns ..........50
        2.3.7.1 Glass Capillary Columns........................50
        2.3.7.2 Packed High-Performance Columns (about 10 m) ....52
    2.3.8 Gas Chromatography with Packed Columns (<5 m)..............54
    2.3.9 Thin-Layer Chromatography...............................54

| | | | |
|---|---|---|---|
| 2.3.10 | High-Performance Liquid Chromatography (HPLC) | | 55 |
| 2.3.11 | Final Evaluation of Methods Used for Separating Complex PAH Mixtures | | 55 |
| | 2.3.11.1 | Gas Chromatography with Glass Capillaries | 55 |
| | 2.3.11.2 | Gas Chromatography with Packed High-Performance Columns | 56 |
| | 2.3.11.3 | Gas Chromatography with Packed Columns (<5 m) | 56 |
| | 2.3.11.4 | Thin-Layer Chromatography | 56 |
| | 2.3.11.5 | High-Performance Liquid Chromatography (HPLC) | 56 |

References ... 57

## 2.1 CHOICE OF RELEVANT POLYCYCLIC AROMATIC HYDROCARBONS

In order to protect men, animals, plants, and our habitat against noxious environmental pollution, it is necessary to identify the substances causing this noxious effect and to measure the frequency of their occurrence in the environment. According to our present knowledge, damage is caused by means of irradiation as well as by certain organic and inorganic substances which are classified under the general term "noxious matter". However, without knowing the concentrations at which damage is first detectable, measurement of substances occurring in the environment do not have any practical use.

Noxious substances in air and food are hazards to human health. It is particularly difficult to detect them if their properties are of a long-term toxic nature. A number of substances belonging to the chemical class of polycyclic aromatic hydrocarbons (PAH) complies with the criteria of noxious substances: they occur ubiquitously in the environment and exhibit a cancer-inducing effect in various animal species. They are also suspected of causing certain cancers in man if they are present in the environment in sufficient quantity. On the basis of findings in various animal species and their action on cultures of animal and human tissue, it is possible to state which PAH have carcinogenic properties. It is the task of chemistry to detect known hazardous PAH in the environment and to determine their concentration.

PAH are formed during incomplete combustion or pyrolysis of organic materials containing carbon and hydrogen. Materials that do not contain any hydrogen atoms, e.g., blast furnace coke or polytetrafluoroethylene, do not produce any PAH during thermal decomposition under an inert protective gas or an oxygen-containing atmosphere. PAH are also formed during the formation of fossil fuel from preformed structures in the process of carbonization.

### 2.1.1 Mechanisms of Formation

The generation of PAH during thermal decomposition of organic materials is based on at least two different mechanisms: (1) incomplete combustion or, to be more precise, pyrolysis and (2) a process of carbonization, e.g., during formation of mineral oil or coal. The more completely an organic material, e.g., a fossil fuel, can be combusted with the appropriate amount of oxygen to carbon dioxide and water, the smaller is the amount of PAH formed as a by-product. During an incomplete combustion, an oxygen deficiency occurs in microranges and thus pyrolysis conditions build up, i.e., in these microranges the amount of oxygen necessary for a complete combustion is lacking. The more homogeneously the fuel and the necessary amount of air are intermixed, the more complete the combustion. In general only gases can be mixed so homogeneously that no microranges with oxygen deficiency occur. Liquid drops and particulate fuel can — strictly speaking — never comply with this prerequisite. It should be mentioned that even mixing of gases requires a finite time (e.g., the gasoline engine).

Thus during "incomplete" combustion a pyrolysis of fuel takes place in microranges beside the "complete" combustion to $CO_2$ and $H_2O$. When considering pyrolysis, thermal decomposition under a protective gas, as a separate process, the interrelations of temperature, type of material, amount of PAH, and profile of PAH can best be seen. This mechanism of formation which takes place predominantly at higher temperatures, is mostly effected via $C_2$-fractions and yields unsubstituted base components in the majority of cases. The base components which at this temperature are thermodynamically most stable are formed in corresponding quantitative ratios. Irrespective of the type of material to be burned, surprisingly similar profiles are formed at a defined temperature. For example, thermal decomposition of pit coal, cellulose, tobacco, and also of polyethylene, polyvinyl chloride, and other plastic materials which is carried out at 1000°C (in air or nitrogen), yields the same PAH in very similar ratios (see Chapter 3). In a model installation, under comparable conditions, similar PAH profiles are formed from different organic materials. Consequently, PAH profiles seem to depend predominantly on temperature.

The amount of PAH formed under inert gas depends on the material as well as the temperature. Thus under comparable conditions, e.g., at 700°C in a nitrogen stream, 1 g of glucose, dry tobacco, and paraffin wax yields 886 ng, 752 ng, and 66,600 ng BaP, respectively, in the distilled pyrolysate.[16a] That the amount of PAH formed is dependent on temperature is shown in Table 12, which shows the variation in BaP formed with temperature of pyrolysis. Under comparable conditions, 1 g of tobacco yields 44 ng of BaP at 400°C and 183,500 ng of BaP at 1000°C, i.e., more than the thousandfold amount. At 1000°C, the temperature of maximum formation is not yet reached in the case of this heterogenous organic material.[16a] Similarly to BaP, the formation of other base components (unsubstituted) is also temperature-dependent, but the formation of alkyl-substituted PAH detectable in tobacco smoke is not so dependent on temperature.

The various processes of carbonization that occur, e.g., during the generation of mineral oil and coal, also produce such alkyl-substituted base components; these particular PAH mixtures formed in mineral oils at low temperatures (about 200°C), over a period of millions of years, differ from those PAH mixtures formed during incomplete combustion. Here, too, the PAH fractions of mineral oils of different origin, e.g., paraffinic or naphthenic oils, have a quite similar composition. The correlation between the increase of this PAH fraction and the degree of maturity of the oil suggests an abiologic temperature-dependent formation of substances of this class. The profile of PAH contained in mineral oil is clearly different from the profile of PAH obtained by incomplete combustion or pyrolysis of organic material, i.e., PAH mixtures contained in the exhaust gases from automobiles or oil heating plants clearly differ from the PAH profiles of mineral oil. For instance, mineral oil frequently contains only one of several possible isomeric PAH, and one form is the predominant isomeric form. This applies, for example, to the two isomeric benzopyrenes. In mineral oil benzo[e]pyrene is the predominant form, whereas the amount of the other isomer, benzo[a]pyrene, is not very great. In exhaust gas condensate from gasoline engines, diesel engines, or oil heating installations the ratio of the two isomeric benzopyrenes is about 1:1. Phenanthrene occurs 50 times more frequently in mineral oil than anthra-

Table 12
FORMATION OF BENZO[a]PYRENE FROM BLEND
TOBACCO (100 g) AS A FUNCTION OF
TEMPERATURE (1000 m$\ell$ N$_2$/MIN)

| Temperature (°C) | Condensate (g) | Nitromethane extract (g) | BaP (µg) | BeP (µg) | e/a |
|---|---|---|---|---|---|
| 400 | 18.6850 | 1.8820 | 4.4 | 3.2 | 0.73 |
| 500 | 18.8966 | 1.9260 | 12.9 | 8.4 | 0.65 |
| 600 | 15.2240 | 1.7201 | 32.0 | 19.0 | 0.60 |
| 700 | 11.2810 | 1.5622 | 56.0 | 29.4 | 0.53 |
| 700 | 10.3490 | 1.4931 | 88.6 | 45.2 | 0.51 |
| 800 | 7.8900 | 1.2725 | 270.0 | 155.0 | 0.57 |
| 900 | 4.9921 | 1.2441 | 1820.0 | 824.0 | 0.45 |
| 900$^a$ | 5.0120 | 1.3330 | 4725.0 | 2015.0 | 0.43 |
| 1000$^a$ | 4.6941 | 1.2320 | 18350.0 | 6710.0 | 0.36 |

$^a$ At 1500 m$\ell$/min to avoid a back reaction of the condensate.

cene does. This ratio differs considerably from that in automobile exhaust gas, which, on the basis of investigations with 100 passenger cars, is 4:1.[31]

In this connection it is noticeable that the main components of mineral oil can be freely arranged as a series of substances which differ from each other by ring closure via a C$_2$ or C$_4$ bridge. The most frequent three-, four-, five-, and six-ring compounds of mineral oil can be arranged in the following series:[31] phenanthrene→triphenylene→benzo[e]pyrene→dibenzo[fg,op]tetracene ( = 1,2,6,7-dibenzopyrene).

A series deriving from anthracene by linear condensation of C$_4$ fragments (e.g., tetracene, pentacene) is almost completely lacking. However, quite often a further benzene ring is added and a five-membered ring is formed. Thus, naphthalene (in positions 1 and 8) combines with benzene to form fluoranthene, and, analogously, benzo[b]fluoranthene and indeno[1,2,3-cd]pyrene are formed of phenanthrene and pyrene, respectively.

These examples lead to the assumption that the PAH contained in mineral oil are derived from preformed structures in the course of a process of carbonization at temperatures between 200° and 300°C during a period of several million years. According to investigations on more recent marine deposits,[39] PAH are produced from carotinoids which were kept at temperatures between 65° and 200°C for up to 2 months.

Today it is still an open question whether in addition to these two mechanisms of formation, biosynthesis in microorganisms or plants is possible. In any case, the amounts of PAH possibly formed in biosynthesis will hardly be of importance in comparison with the quantities formed abiotically.

### 2.1.2 Number of PAH Detected in the Environment

During the incomplete combustion or pyrolysis of organic material, several hundred different PAH are formed. Until 1967, e.g., 106 PAH were described as contained in pit coal tar.[46] In other matrices, such as cigarette smoke, about 280 PAH,[64] automobile exhaust gas, 146,[29] exhaust from fuel oil, 108 (without bicyclic aromatic hydrocarbons),[36] and mineral oil samples, about 150,[31] polycyclic aromatic hydrocarbons were identified on the basis of their characteristic mass and UV spectra.

### 2.1.3 Criterion of Choice: Carcinogenicity

It is impossible to select out of the previously mentioned several hundred PAH those which have a possible harmful effect on human health. At the moment such a causal relation cannot be proven in man. Also in the methods used on animal test models such as topical application, subcutaneous injection, oral administration, and intratracheal instillation, the information needed for assessing the carcinogenicity of many PAH is often fragmentary. Furthermore, it is to be anticipated that a classification of PAH with regard to their carcinogenic activity (i.e., according to the dose which induces the same tumor rate) will be different for different test systems.

### 2.1.4 Criterion of Choice: Frequency of Occurrence

From the above it follows that during incomplete combustion of fossil fuels, which is the main source of PAH emission, a number of specific, thermodynamically stable compounds is formed, the quantitative relationship of which shows a certain temperature dependency. Therefore, the formation of the same main components is to be expected from the pyrolytic process. Leaving mineral oil and coal aside, i.e., PAH mixtures formed by carbonization, then the PAH occurring most frequently in the environment are unsubstituted base components (without side chains) consisting of several rings. Table 13 comprises data of PAH (with more than 3 rings) which occur most frequently in the environment. Table 14 lists those PAH which occur only seldom, or not at all, in the environment. Some of them, such as methylcholanthrene or 7,12-dimethylbenz[a]anthracene are often used as model substances in animal experiments.

## 2.2 NOMENCLATURE OF PAH

The designations given correspond to the proposals of the International Union of Pure and Applied Chemistry (IUPAC) and are generally used nowadays. The formerly applied designations which are characterized by the use of numbers for describing the ring fusion are added in parentheses in Tables 13 and 14.[7a] Table 13 gives a selection of PAH occurring most frequently, and in the highest concentrations, in the environment. Some other PAH which also have to be considered as base components occur only in infrequent emissions and are not included here.

To facilitate understanding, the IUPAC rules are summarized below. This seems necessary all the more so, since the nomenclature used in the literature often leads to serious misunderstandings. Details of the rules for nomenclature have been published several times.[40,56]

### 2.2.1 Rule 1: "Base Components"

Only a selected number of "base components" are used in the nomenclature. Higher condensed rings are then derived from these base components. The base components used should (1) have as many rings as possible and (2) occur as far as possible from the beginning of the list of names below: pentalene, indene, naphthalene, azulene, heptalene, biphenylene, as-indacene, s-indacene, acenaphthylene, fluorene, phenalene, phenanthrene, anthracene, fluoranthene, acephenanthrylene, aceanthrylene, triphenylene, pyrene, chrysene, naphthacene, pleiadene, picene, perylene, pentaphene, pentacene, tetraphenylene, hexaphene, hexacene, rubicene, coronene, and trinaphthylene, heptaphene, heptacene, pyranthrene, and ovalene.

### 2.2.2 Rule 2: "Orientation"

The PAH formula is oriented so that (1) the greatest number of rings are in a horizontal row and (2) a maximum number of rings are above and to the right of the horizontal row.

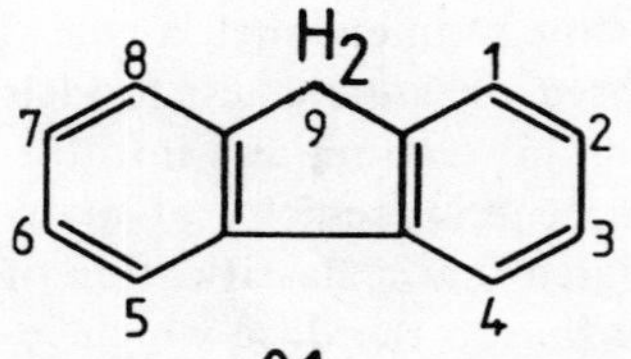


01

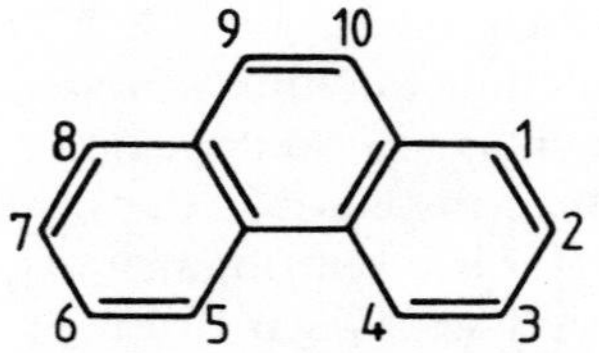


02

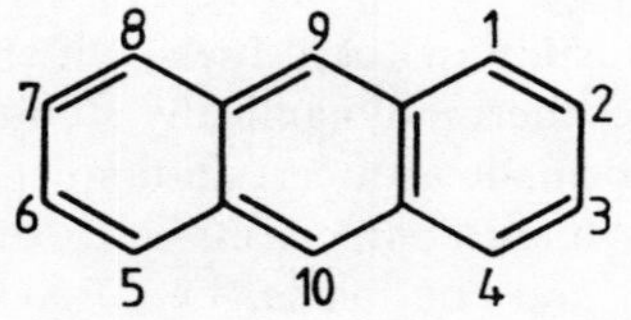


03

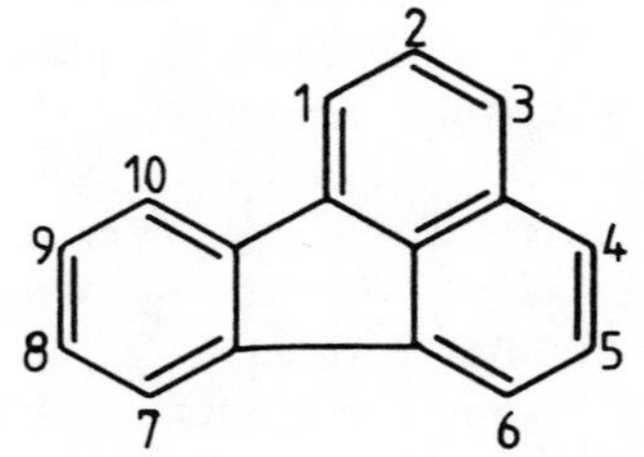


04

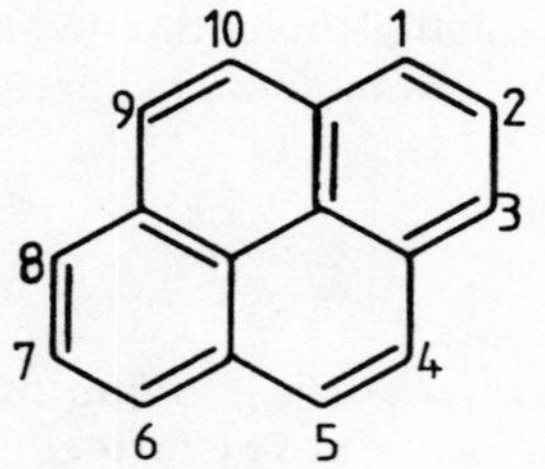


05

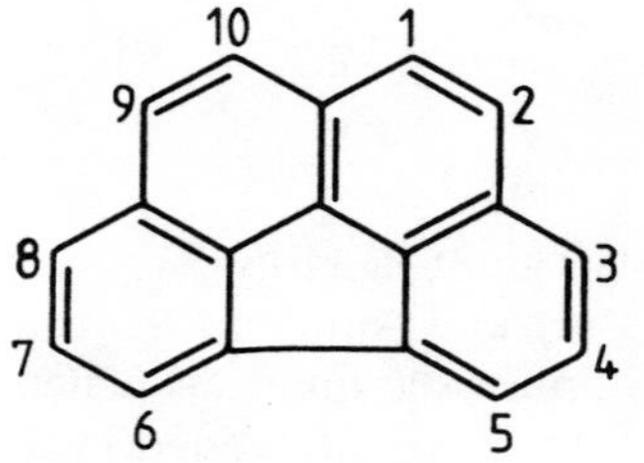


06

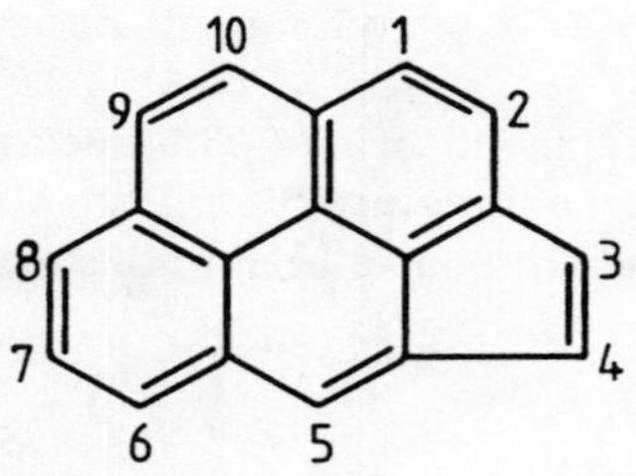


07

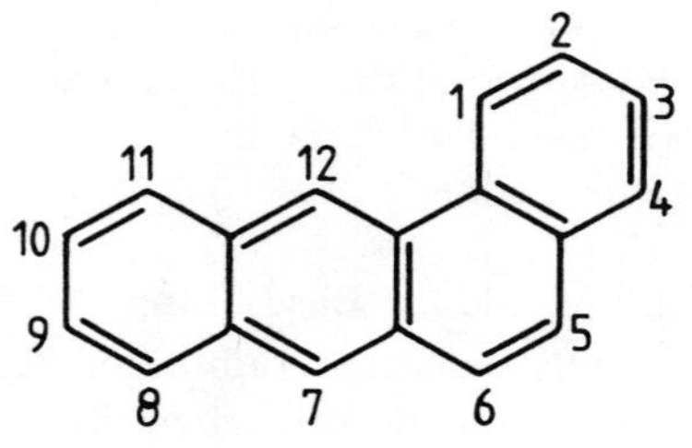


08

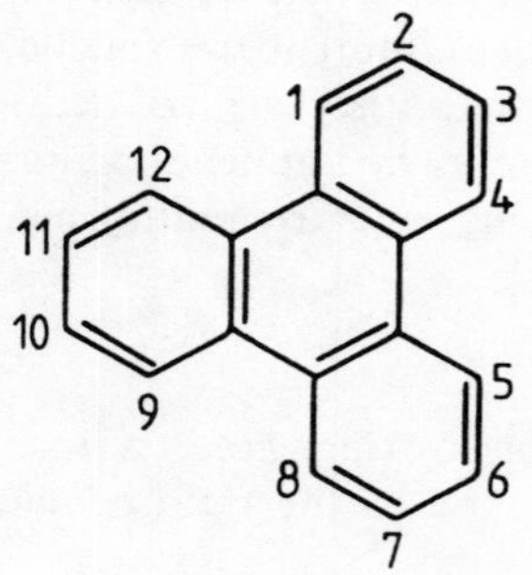


09

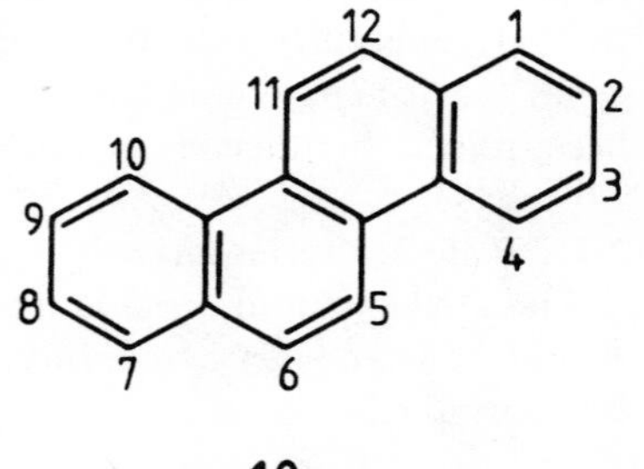


10

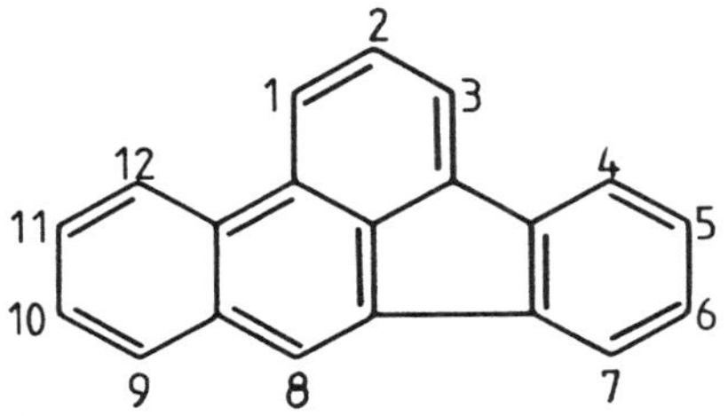

11

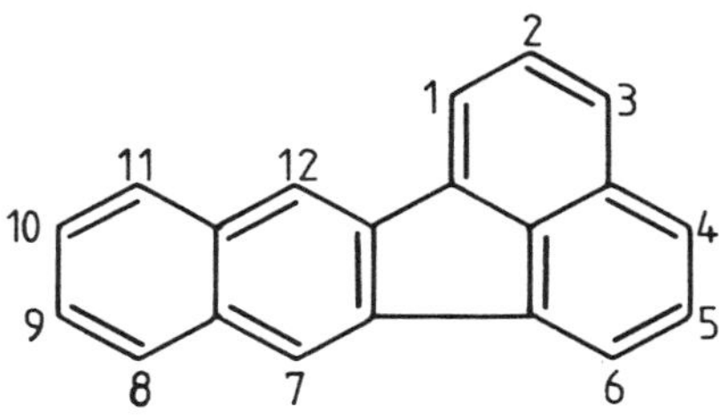

12

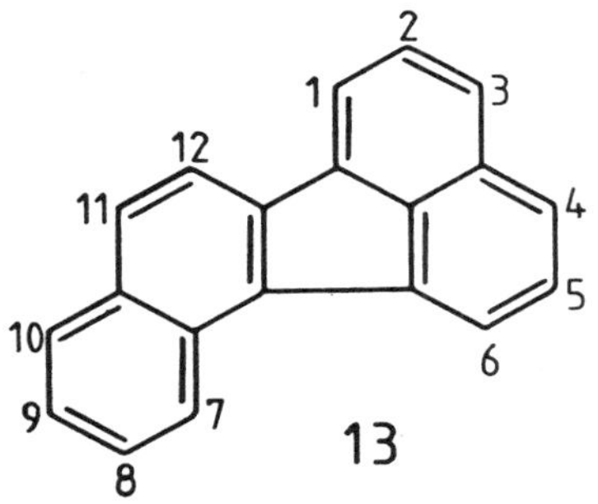

13

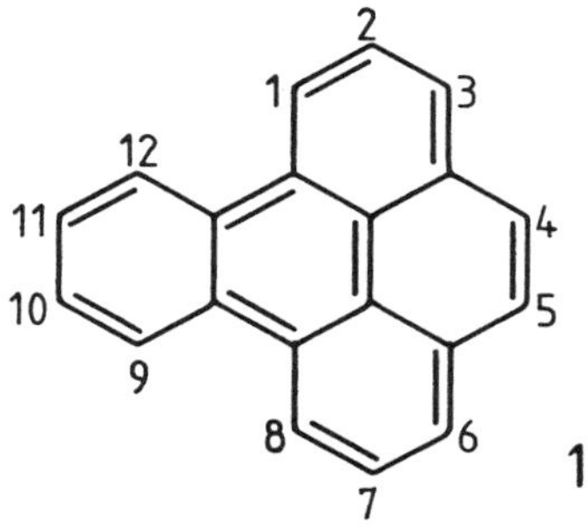

14

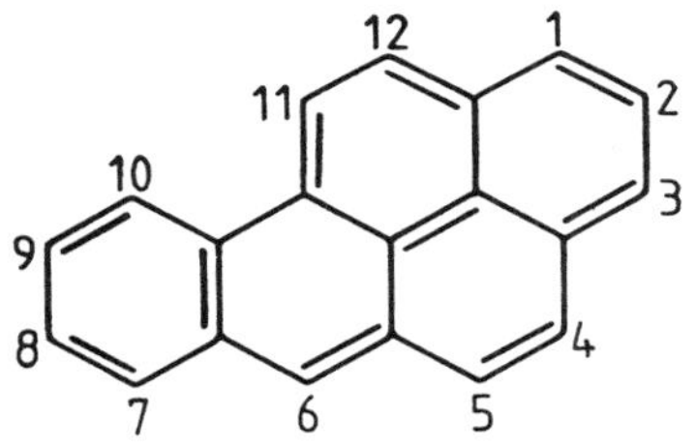

15

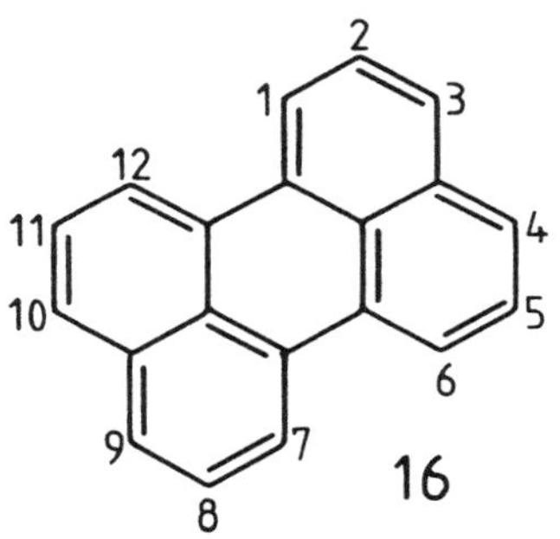

16

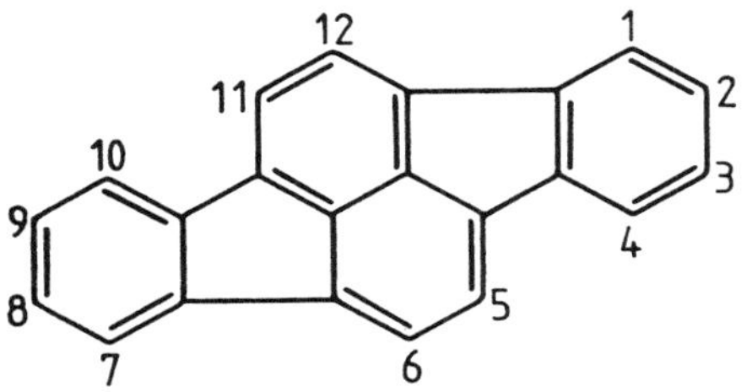

17

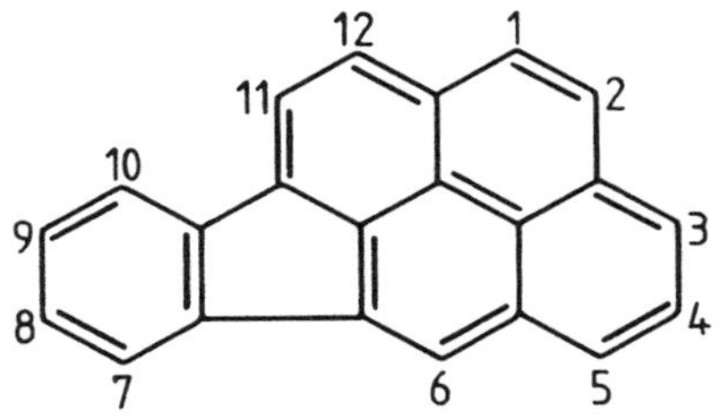

18

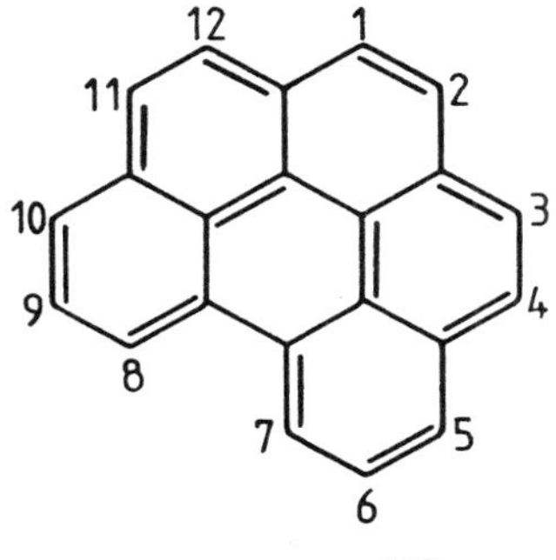

19

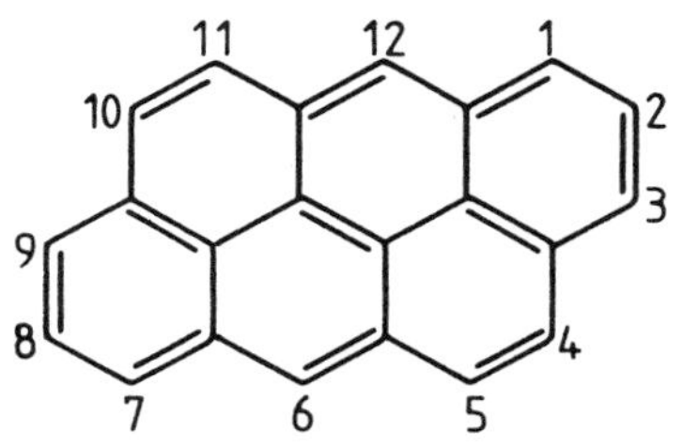

20

21

22

23

24

25

26

27

28

29

30

### 2.2.3 Rule 3: "Numbering"

The first carbon atom that belongs to the uppermost ring and is not engaged in ring fusion with another ring is given the number C-1. Numbering continues in a clockwise direction omitting those carbon atoms which do not carry a hydrogen atom. These are designated by adding Roman letters (e.g., 1a, 2a, etc.).

### 2.2.4 Rule 4: "Lettering"

The bond between C-1 and C-2 is designated as side "a". The designation of the other peripheral sides continues in clockwise direction in alphabetical order.

### 2.2.5 Rule 5: "Condensed Systems"

Components not mentioned in Rule 1 are described as products of condensation (e.g., by benzo-). In this case the smaller component according to Rule 1 is placed as prefix before the name of the larger base component (e.g., naphthoperylene). The numbering of carbon atoms of the smaller component participating in the fusion are placed between the two components in brackets. These numbers are placed within the brackets before the side designation of the larger component and separated by a hyphen. If several smaller components are fused with a large one (e.g., dibenz[a,h]anthracene), the sides are separated by a comma. If a smaller component is fused with a larger one by several common faces, then the letters are not separated by a comma (e.g., benzo[ghi]perylene).

The following examples are meant to illustrate the IUPAC rules:

## Table 13
## POLYCYCLIC AROMATIC HYDROCARBONS OCCURRING MOST FREQUENTLY IN THE ENVIRONMENT

| IUPAC designation (and the old nomenclature) | Molecular formula | Mol wt | Bp (°C) |
|---|---|---|---|
| 01 Fluorene | $C_{13}H_{10}$ | 166 | 293 |
| 02 Phenanthrene | $C_{14}H_{10}$ | 178 | 338.4 |
| 03 Anthracene | $C_{14}H_{10}$ | 178 | 340 |
| 04 Fluoranthene (FLU) | $C_{16}H_{10}$ | 202 | 383.5 |
| 05 Pyrene (PYR) | $C_{16}H_{10}$ | 202 | 393.5 |
| 06 Benzo[$ghi$]fluoranthene(BghiF) (= 2,13-benzofluoranthene) | $C_{18}H_{10}$ | 226 | 431.8 |
| 07 Cyclopentadieno[$cd$]pyrene (CP) = cyclopenta[$cd$]pyrene | $C_{18}H_{10}$ | 226 | 439 |
| 08 Benz[$a$]anthracene (BaA) = tetraphene (= 1,2-benzanthracene) | $C_{18}H_{12}$ | 228 | 437.5 |
| 09 Triphenylene (TRI) | $C_{18}H_{12}$ | 228 | 438.5 |
| 10 Chrysene (CHR) | $C_{18}H_{12}$ | 228 | 441 |
| 11 Benzo[$b$]fluoranthene (BbF) = benz[$e$]acephenanthrylene (= 3,4-benzofluoranthene) | $C_{20}H_{12}$ | 252 | 481.2 |
| 12 Benzo[$k$]fluoranthene (BkF) (= 11,12-benzofluoranthene) | $C_{20}H_{12}$ | 252 | 481 |
| 13 Benzo[$j$]fluoranthene (BjF) (= 10,11-benzofluoranthene) | $C_{20}H_{12}$ | 252 | 480 |
| 14 Benzo[$e$]pyrene (BeP) (= 1,2-benzopyrene) | $C_{20}H_{12}$ | 252 | 492.9 |
| 15 Benzo[$a$]pyrene (BaP) (= 3,4-benzopyrene) | $C_{20}H_{12}$ | 252 | 495.5 |
| 16 Perylene (PER) | $C_{20}H_{12}$ | 252 | 497[a] |
| 17 Indeno[1,2,3-$cd$]fluoranthene (IF) (= peri-phenylenefluoranthene) | $C_{22}H_{12}$ | 276 | 531[a] |
| 18 Indeno[1,2,3-$cd$]pyrene (IP) (= 2,3-o-phenylenepyrene) | $C_{22}H_{12}$ | 276 | 534[a] |
| 19 Benzo[$ghi$]perylene (BghiP) (= 1,12-benzoperylene) | $C_{22}H_{12}$ | 276 | 542[a] |
| 20 Anthanthrene (ANT) | $C_{22}H_{12}$ | 276 | 547[a] |
| 21 Coronene | $C_{24}H_{12}$ | 300 | 590[a] |

[a] Estimated from the gas-chromatographic retention time.

*Example 1:* Benzo[$a$]pyrene, benzo[$e$]pyrene

According to Rule 2, pyrene is oriented in the following way:

According to Rules 3 and 4, the numbering and designation of sides is carried out as follows:

## Table 14
## POLYCYCLIC AROMATIC HYDROCARBONS WHICH OCCUR NOT AT ALL OR ONLY IN LOW CONCENTRATIONS IN THE ENVIRONMENT

| | IUPAC designation (and the old nomenclature) | Molecular formula | Mol wt | Bp (°C) |
|---|---|---|---|---|
| 22 | Benzo[a]fluorene (= 1,2-benzofluorene) | $C_{17}H_{12}$ | 216 | 407 |
| 23 | Benzo[b]fluorene (= 2,3-benzofluorene) | $C_{17}H_{12}$ | 216 | 402 |
| 24 | Benzo[c]fluorene (= 3,4-benzofluorene) | $C_{17}H_{12}$ | 216 | 406 |
| 25 | 7,12-Dimethylbenz[a]anthracene (= 9,10-dimethyl-1,2-benzanthracene) | $C_{20}H_{16}$ | 256 | — |
| 26 | 3-Methyl-1,2-dihydro-benz[j]aceanthrylene (= 3-methylcholanthrene) | $C_{21}H_{16}$ | 268 | 506[a] |
| 27 | Dibenz[a,c]anthracene = benzo[b]triphenylene (= 1,2,3,4-dibenzanthracene) | $C_{22}H_{14}$ | 278 | 535[a] |
| 28 | Dibenz[a,h]anthracene (= 1,2,5,6-dibenzanthracene) | $C_{22}H_{14}$ | 278 | 535[a] |
| 29 | Dibenzo[a,j]anthracene (= 1,2,7,8-dibenzanthracene) | $C_{22}H_{14}$ | 278 | 531[a] |
| 30 | Dibenzo[a,l]pyrene = dibenzo[def,p]chrysene (= 1,2,3,4-dibenzopyrene) | $C_{24}H_{14}$ | 302 | 595[a] |
| 31 | Dibenzo[a,i]pyrene = benzo[rst]pentaphene (= 3,4,9,10-dibenzopyrene) | $C_{24}H_{14}$ | 302 | 594[a] |
| 32 | Dibenzo[a,h]pyrene = dibenzo[b,def]chrysene (= 3,4,8,9-dibenzopyrene) | $C_{24}H_{14}$ | 302 | 596[a] |
| 33 | Dibenzo[a,e]pyrene = naphtho[1,2,3,4-def]chrysene (= 1,2,4,5-dibenzopyrene) | $C_{24}H_{14}$ | 302 | 592[a] |
| 34 | Dibenzo[e,l]pyrene = dibenzo[fg,op]tetracene (= 1,2,6,7-dibenzopyrene) | $C_{24}H_{14}$ | 302 | 592[a] |

[a] Estimated from the gas-chromatographic retention time.

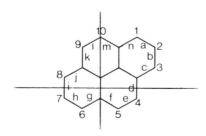

Benzo[a]pyrene and benzo[e]pyrene are formed by fusing the benzene rings with pyrene at face "a" or face "e", respectively.

38    *Environmental Carcinogens: Polycyclic Aromatic Hydrocarbons*

Benzo (a) pyren (BaP)                    Benzo (e) pyren (BeP)

Now both structural formulas have to be reoriented according to Rule 2 with corresponding renumbering and redesignation of the sides:

Benzo (a) pyrene                         Benzo (e) pyrene

*Example 2:* Benzo[a]fluoranthene (BaF), benzo[b]fluoranthene (BbF), benzo-[j]fluoranthene (BjF), benzo[k]fluoranthene (BkF).

According to Rules 2, 3, and 4, fluoranthene is oriented, numbered, and lettered as follows:

Fluoranthene

whereas, according to Rule 2, the orientation of BjF and BkF does not have to be changed after the attachment of the benzene ring, the structural formulas of BaF and BbF have to be reoriented:

BjF                                      BkF

BaF

BbF

*Note:* Benzo[a]fluoranthene is a synonym of benz[a]aceanthrylene. The same applies to benzo[b]fluoranthene and benz[e]acephenanthrylene. According to Rule 1 the second designation is to be preferred in both cases, since the base component "fluoranthene" is given first.

Some of the PAH listed in Tables 13 and 14 are available at a purity of 99% (and more) from the Commission of the European Communities, Community Bureau of Reference BCR, Rue de la Loi 200, B-1049 Brussels, Belgium.

## 2.3 METHODS OF PAH DETECTION

### 2.3.1 Introduction

Today there are two methods for assessing the carcinogenic properties of environmental samples:

1. The chronic toxic testing of these samples, using appropriate experimental animals
2. The chemical analysis of the carcinogenic constituents of samples, so far as these are known

The testing of samples, or a suitable extract, in animal experiments for carcinogenic properties is carried out by feeding, topical application, subcutaneous injection, intratracheal application, etc. and, if suitable doses are used, gives clear evidence of tumor-producing properties. The evidence obtained in animal experiments covers all the classes of carcinogenic substances contained in the suspected sample if they exhibit activity in the chosen experimental animal model (e.g., PAH, nitrogen-containing polycyclic aromatic compounds, nitrosamines, aromatic amines, mycotoxins, or carcinogenic inorganic compounds). The disadvantages of this test system are the long duration of the experiments (several months) and the correspondingly high costs.

For the chemical analysis, it is necessary that the substances inducing malignant tumors in animal experiments are known. Not all PAH cause cancer. Consequently, all compounds belonging to a class of noxious substances have to be tested in several animal models in order to assess their activity. In the case of the several hundred PAH, such comparative investigations are hardly feasible. Economic reasons demand a restriction to those PAH which occur most frequently and in highest concentrations in the environment.

The assessment of tumor-inducing activities of a sample by means of chemical analytical data thus requires the carcinogenic evaluation of the most essential components and is bound to the corresponding animal model. According to our present knowledge, the classification of all PAH — starting with the PAH not inducing any tumors at all up to the PAH inducing cancer at a very low dose — is not identical for all experimental animal models. Therefore the quantitative assessment of carcinogenic activity of a sample by means of chemical analytical data permits, at best, the statement that the respective sample shows a correspondingly strong, moderate, or no effect in the animal model used. For this chemical analytical statement on carcinogenic properties of a sample based on the contained PAH, it is necessary that all the carcinogenic PAH can be identified and that the activity of individual compounds is not suppressed or potentiated by other constituents of the sample. Despite these restrictions, such a total inventory of the carcinogenic effect — as can be established in animal experiments by a defined tumor rate — is possible in some cases by simple addition of the carcinogenic effect of about 20 individual substances. Further experiments are required to show whether these findings will remain limited to the condensates from automobile exhaust gas, using only the repeated topical application to mouse skin as test model (see Chapter 1.2.2).

Advantages of the chemical analytical identification of PAH are the short time needed and the good reproducibility. The utilization of individual PAH, such as benzo[a]pyrene, for the assessment of carcinogenic activity of a sample is permissible only if all PAH contained in the type of sample under investigation are present in certain definite quantitative ratios, i.e., have the same profile. However, only a PAH profile analysis can prove whether the quantitative proportion of benzo[a]pyrene is identical in all samples of the same sample type.

In general the PAH profiles of different types of environmental samples show only little similarity. If, for example, the carcinogenic activity of automobile exhaust gas condensate is compared with the activity of benzo[a]pyrene contained in this condensate in topical application to the dorsal skin of mice, it is found that benzo[a]pyrene accounts only for about 10% of the total activity, i.e., in order to obtain the same effect with pure benzo[a]pyrene one has to use ten times as much benzo[a]pyrene as is contained in the condensate.[5] The same applies to the particulate matter of cigarette smoke: the benzo[a]pyrene content of this condensate accounts only for approximately 1 to 2% of its effect after topical application of the condensate.[10]

These results obtained in animal experiments clearly prove that it is not sufficient to define merely the benzo[a]pyrene content of a sample or product. It is necessary rather to record the whole profile of all carcinogens. In the case of automobile exhaust gas these are exclusively PAH, whereas in the case of cigarette smoke condensate the PAH fraction accounts only for about 50% of the carcinogenic activity. Thus in the latter case the carcinogenic effect is caused also by substances of other chemical classes, such as, e.g., nitrogen-containing aromatic compounds.

In order to evaluate the carcinogenic activity of a sample on the basis of experimental results, as many as possible of the carcinogens contained in the mixture have to be identified. Only when investigating several samples of the same type, the benzo[a]pyrene content may be used as standard of comparison for the carcinogenicity — provided that the percentages of activity of benzo[a]pyrene are identical.

## 2.3.2 Methods of Chemical Analysis

If all carcinogenic PAH contained in a sample are to be identified, it is an additional prerequisite that all PAH contained can be completely extracted. This can only be done without difficulty if the sample is completely soluble in an appropriate solvent. An insoluble residue involves the risk that PAH are partly or completely adsorbed to the residue or included therein. In all presently known chemical processes of determination it is first necessary for a quantitative assessment that the PAH mixture be extracted from the matrix without losses in order to obtain a final pure PAH fraction. The PAH fraction usually consists of more than 100 individual compounds.

In order to be able to identify individual PAH in such a complex mixture, separation methods of high resolution are needed, since in most cases the identification and quantitative assessment of individual compounds is possible only after complete separation. The required extensive separation can be achieved only by separation methods which yield separation values of several tens of thousands. Such figures, i.e., the total number of height equivalents of theoretical plates (HETP) of a separating column, are achieved by only a few separation methods. At present the best separation results are obtained with glass capillary columns in gas chromatography: in routine separation these give a value of more than 70,000 HETP. Packed high-capacity columns produce in most cases only about 25,000 HETP, even at a length of 10 m, whereas the separation parameters obtained in routine operation with high-pressure liquid chromatography do not exceed 20,000. For the most part, separation by column chromatography on Sephadex® LH 20, aluminum oxide, silica gel, or cellulose powder does not satisfactorily separate the main and secondary components of a PAH mixture extracted from a sample. Separation parameters of several thousands, achievable with the usual methods of thin-layer chromatography, do not produce satisfactory separations if they are employed unidimensionally.

The quantitative assessment of individual PAH obtained is then carried out by means of appropriate detection systems which either detect substance-specific properties (e.g., UV or fluorescence, aromaticity in the electron capture detector [ECD]) or directly respond to the quantity of substance (e.g., carbon content in the flame ionization detector [FID], halogen content in the ECD, etc.). The advantage of the FID is that nonidentified PAH can also be detected quantitatively by their signal areas, because the signal areas of all PAH peaks correspond quantitatively to the introduced amount of substance. Thus, e.g., the recording strip of a FID gives a direct reproduction of the composition of the mixture, not needing a correction by substance-specific factors, such as extinction coefficients.

Low-voltage mass spectrometry (LVMS) is also a highly resolving separation method which records the masses of substances. Since by this method the compounds are separated according to their molecular mass, all isomers of the same mass are recorded jointly, and compounds with different carcinogenicity but identical mass cannot be distinguished. Thus the strongly carcinogenic benzo[a]pyrene has the same molecular formula as the inactive benzo[e]pyrene or perylene and the four isomeric benzofluoranthenes which have different carcinogenic activities. The total amount of these seven isomers determined by LVMS does not give any evidence of the carcinogenic activity of the mixture. A distinction by this method will be possible only in combination with capillary gas chromatography. As additional information, the LVMS records the heteroatoms (N, S, O), thus permitting a simple differentiation against pure PAH. Here, too, the individual signals (masses) in the fingerprint of a class of heterocycles do not correspond to individual substances, but to the sum of compounds with the same molecular formula.

It is the objective of the chemical analytical identification of PAH to assess the carcinogenic activity of an environmental sample by means of chemical analytical data. This is based on the assumption that the carcinogenic properties of most frequently occurring PAH are known from certain animal experiments, that the individual PAH have additive activities with little mutual influence, and that their activities are not potentiated or suppressed by impurities. With these reservations, the chemical analytical inventory of all carcinogens contained in a sample permits the statement that in a specific animal model the carcinogenic activity of a sample corresponds to the x-fold activity of its benzo[a]pyrene content. The decisive advantage of the chemical analytical method is the short period of time needed. Contrary to animal experiments which give results only after 1 to 2 years, it permits an assessment after only several hours.

If, for example, the PAH composition of cigarette smoke is compared with that of mineral oil, not only are different PAH found in the two materials, but the few common PAH are also contained at completely different quantitative ratios.

It is not possible to describe the multitude of known determination methods for all types of environmental samples. In general all these chemical analytical determination methods have the following four process steps in common: extraction, isolation of the PAH mixture, separation, and quantitative detection, including identification and quantitative determination of the PAH.

### 2.3.3 Extraction

The choice of the extraction process depends largely on the type of environmental sample to be analyzed. As stated earlier, problems will not be encountered if the sample (matrix) is completely soluble in an organic solvent, such as acetone, cyclohexane, etc. This applies, for example, to fats or vegetable oils as well as mineral oil products. If in the case of heterogenously composed materials this precondition is not complied with, two problems arise. The extraction may be incomplete because (1) the PAH cannot be separated from the cells in the case of biological samples, or they are included, for example, in soot samples, and (2) the insoluble matrix constituents contained in dust, sediment, soil, soot samples, etc. adsorptively retain the PAH. This also applies to protein-containing foodstuffs, such as samples of smoked fish. Here the completeness of PAH extraction depends on the adsorption-desorption equilibrium between the insoluble residue and the solvent. In most cases acetone is the best extracting agent compared, for instance, with methanol or nonaromatic hydrocarbons, because it seems to be better suited to cleave the charge-transfer complexes. The extraction of soot and active carbon samples is particularly difficult. Aromatic hydrocarbons such as toluene or xylene are the most effective extracting agents for this type of sample which has a structure similar to that of PAH. As shown in Table 15, the amount of PAH extractable from diesel engine soot varies, depending on the solvent, from 3% up to the quantitative recovery of benzo[b]chrysene, which is not originally present in diesel engine soot but is added as a reference standard, and thus characterizes the extraction of PAH from the soot surface as equilibrium between adsorbing material and toluene solution. A further characteristic finding of this experiment is the fact that the adsorption equilibrium depends largely on the molecular size. Thus benzo[c]phenanthrene is separated in the first xylene extraction to a considerably greater degree than is benzo-[ghi]perylene.

The problem of incomplete extraction can often be solved by complete saponification of the matrix with alcoholic potassium hydroxide solution, because the homogeneous solution formed can directly be processed in the usual manner (e.g., protein-containing foodstuffs such as meat, fish, poultry, yeast, single-cell protein, etc.). Thus only about 25% of benzo[a]pyrene or other PAH contained in a smoked-fish sample can be extracted by multiple extraction using boiling methanol, although the fish protein is cooked to a homogeneous pulp. If this protein pulp which is not soluble in methanol is saponified with methanolic alkali, the amount of benzo[a]pyrene obtained after usual clean-up of the clear saponified solution is three times that obtained after the preliminary extraction with boiling methanol.[25] If the residue which is not soluble in the extracting agent can be rendered completely soluble by an alkaline saponification, this control of the completeness of extraction can be carried out without difficulty and is to be recommended.

### 2.3.4 Extraction Methods

Due to differences in solubility and adsorption of different sample types (matrices), it is not possible to propose a generally usable extraction method. The safest way of

Table 15
AMOUNT OF THE FIRST EXTRACTION[a] OF AIR POLLUTION[b] BY
DIFFERENT SOLVENTS[c] FOLLOWED BY A SECOND EXTRACTION
WITH TOLUENE[c]

|  | Methanol | Acetone | Cyclohexane | Xylene | Toluene |
|---|---|---|---|---|---|
| Benzo[b]chrysene[d] (300 ng) | 3.1 | 24.6 | 81.2 | 92.8 | 96.4 |
| Benzo[b]naphtho (2,1-d) thiophene | 65.1 | 70.3 | 53.7 | 78.2 | 91.2 |
| Benzo[c]phenanthrene | 65.2 | 74.2 | 53.9 | 89.3 | 93.3 |
| Benz[a]anthracene + chrysene | 57.6 | 70.3 | 70.2 | 82.5 | 89.5 |
| Benzofluoranthenes [b + k + j] | 20.3 | 44.2 | 46.0 | 81.3 | 92.2 |
| Benzo[e]pyrene | 20.6 | 43.7 | 46.9 | 81.1 | 88.2 |
| Benzo[a]pyrene | 19.5 | 42.1 | 44.2 | 80.5 | 90.1 |
| Indeno[1,2,3-cd]pyrene | 0.5 | 8.0 | 21.3 | 72.1 | 91.0 |
| Benzo[ghi]perylene | 0.5 | 8.7 | 17.7 | 61.6 | 89.2 |

*Note:* Both extracts = 100%.

[a] % of both extractions.
[b] Preferentially diesel engine emission, samples of 20 mg.
[c] 100 ml, 1 hr refluxed.
[d] % recovery of the added internal standard benzo[b]chrysene (300 ng).

avoiding problems of adsorption consists of completely dissolving the sample, without leaving a residue, in appropriate solvents or solvent mixtures, with subsequent extraction of the sample solution by means of a second immiscible solvent. Since many sample types which contain inorganic constituents or graphite-like structures cannot be dissolved homogeneously, there are several extraction methods suited for the individual sample types. Below they are summarized in six groups:

1. Fats, oils
2. Water samples
3. Meat products, etc.
4. Leaves, seeds, fruits
5. Samples containing inorganic components
6. Soot- and graphite-like materials

*2.3.4.1 Extraction Group: Fats, Oils*

Sample types are vegetable and animal fats such as margarine, butter, clarified butter, olive, sunflower-seed, palm-kernel, linseed, cottonseed, rapeseed oils, etc. and mineral oil products such as fuels for gasoline and diesel engines, heating oils, lubricating oil, used lubricating oil, metal-cutting oil, liquid and solid paraffin, microwax, ointment bases, etc.[19,24,25,38,70] These sample types are soluble in cyclohexane or other aliphatic solvents without residue. The solvent selected should be immiscible with PAH-selective solvents such as dimethylformamide, dimethylsulfoxide, nitromethane, etc.

*2.3.4.2 Extraction Group: Water Samples*

Sample types are drinking water, ground water, river water, lake water, waste water, etc. For extracting the PAH, the water samples are added to organic solvents such as cyclohexane,[35,45,61] heptane,[42] benzene,[3,69] 1,1,2-trifluorotrichloroethylene,[30] ether,[69] chloroform,[53] mixtures of cyclohexane and alcohol,[53] etc.

Due to possible difficulties involved in the complete desorption of PAH, the adsorptive isolation of PAH with active carbon[2] is not recommended. The adsorption of PAH to Tenax® GC yielded rates of desorption exceeding 90%.[49]

If the sample is clouded by colloids, particularly soot-containing colloids, an adsorptive binding of PAH to the colloidal phase may take place. In these cases the water samples cannot be extracted completely with the solvents listed earlier.

### 2.3.4.3 Extraction Group: Meat Products

Sample types are high-protein foodstuff such as meat, poultry, fish, yeast, single-cell proteins, and animal tissues. These samples can mostly be dissolved without residue after 2 to 4 hr of saponification with 2 N methanol-potassium hydroxide solution (methanol + water at a 9:1 ratio).[17,25] With a high content of methanol in the saponified solution, the PAH can be extracted with cyclohexane without strong emulsification.

### 2.3.4.4 Extraction Group: Leaves, Seeds, Fruits

Sample types are green vegetables such as spinach, cabbage, kale, borecole, various kinds of lettuce, tea; seeds of grain, flour, baked products, beans, peas, lentils, coffee beans, spices and herbs, potatoes, turnip cabbage, root vegetables, all kinds of fruits, etc. The high-carbohydrate samples cannot generally be homogenized by alkaline saponification. After careful mechanical cutting of the sample with rotating knives (to a homogenized mash) the PAH can be extracted in the presence of suitable solvents such as acetone. The filtrate can then be processed in the usual way.

### 2.3.4.5 Extraction Group: Samples Containing Inorganic Components

Sample types are airborne particulate matter, automobile exhaust gas, heating furnace exhaust gas, soil and sediment samples, and sewage sludge (freeze-dried). Acetone and dimethylformamide are particularly suited for extracting sediment samples,[24,28] freeze-dried sewage sludge,[30] or filters loaded with automobile emissions.[23] In case of necessity, the completeness of extraction with these solvents can be checked by an additional extraction with boiling xylene, in particular if the sample might have contained soot particles. For extracting filters loaded with airborne particulate matter by means of cyclohexane — a weak eluting solvent — the use of ultrasound during extraction[62] and sublimate under reduced pressure[52,54,71] has been proposed. In the case of some soot samples, the use of benzene is advantageous.[43,63] A critical study on the extraction of samples of airborne particulate matter was presented by Stanley et al.[66]

### 2.3.4.6 Extraction Group: Soot- and Graphite-like Materials

Sample types are soot from diesel engines, active carbon, technical soots, lignite, peat (freeze-dried), etc. Whereas fuliginous emissions of gasoline engines or oil heating installations can easily be extracted exhaustively with the usual solvents such as acetone, dimethylformamide, etc., soot emitted by a diesel engine at full load cannot be eluted with these solvents. The same applies to a number of technical carbon blacks. PAH added to the extraction solution are adsorbed completely to the soot surface. Only by using boiling xylene or toluene, can an equilibrium between PAH adsorbed to the soot surface and those dissolved in the solvent be established. This equilibrium may be checked by adding PAH which are not originally contained in the soot sample, such as benzo[b]chrysene, or dibenz[a,j]anthracene as internal reference standards.

Another problem involved in the complete extraction of different types of carbon black is concerned with the availability of PAH included in the matrix. By extensive grinding, e.g., in colloid mills, it is possible to establish a correlation between particle size of ground material and amount of PAH which are extractable with xylene. These data may permit a conclusion on the PAH fractions included in the soot sample.

## 2.3.5 Enrichment

Usually several separation processes have to be carried out in succession to isolate the PAH included in the extraction solution. The number of separation processes needed, such as liquid-liquid partition, adsorption and partition chromatography, molecular distillation, etc., depends on the type of matrix and the PAH concentration in this matrix. Thus the separation of PAH contained in samples of airborne particulate matter or water can be achieved by simple purification in chromatographic columns, whereas the isolation of the PAH fraction from cigarette smoke condensate requires several effective purification steps.

The enrichment process aims at the complete separation of impurities from PAH. The effectiveness of a separation process used for isolating PAH can be assessed by determining (1) the weight of the PAH-containing fraction in comparison with the PAH-free fraction and (2) the rate at which PAH can be transferred into the PAH-containing fraction. In the case of biphase liquid-liquid partition for instance, this rate can be estimated on the basis of the distribution coefficients. The total isolation process for a specific matrix is optimal when a very pure PAH fraction is obtained in only a few particularly selective separation steps (enrichment steps). This fraction ought to contain all the PAH of the sample at the highest possible percentage. Under these conditions yield corrections are not necessary if a reference standard was added prior to the sample extraction. The reference standard, which has to be a PAH, should of course not differ from other compounds of this class of noxious substances with respect to its solution properties. If the addition of labeled standard compounds is to be avoided, the chosen standard should not be contained in the sample. This applies, for instance, to benzo[*b*]chrysene in the case of samples investigated so far.

The selective isolation of PAH from the soluble total sample or from the extraction solution is based on two characteristic properties of PAH: (1) their lipoid solubility, i.e., their hydrophobic character, and (2) their aromaticity. Since the majority of accompanying substances do not have these two properties, a separation of the sample into a PAH-containing and a PAH-free fraction can be accomplished for all types of matrices investigated so far.

The detection limit of the individual method determines the sample size needed. In most cases the detection method is adapted to the separation process used for separating the PAH mixture (PAH fraction) into individual PAH. Several methods are used at present for the quantitative assessment of the PAH fraction: separation by gas chromatography or capillary gas chromatography in combination with the FID, high-pressure liquid chromatography (HPLC) in combination with a UV ray or fluorescence detector, and thin-layer chromatography which is mostly carried out two-dimensionally in combination with reflection measuring and spot extraction.

If samples exceeding 1 g are needed to identify PAH, the first step should preferably be a liquid-liquid partition. The sample, e.g., a vegetable oil, is dissolved in cyclohexane and the solution is extracted with a mixture of dimethyl formamide and water (9:1). This is possible because the dimethylformamide-water mixture is hardly miscible with cyclohexane. When repeating the extraction of the cyclohexane phase with a fresh dimethylformamide-water mixture (9:1), a total of 94 to 99% of each PAH contained in the oil sample may be concentrated in this phase. Since — starting with 1 kg of oil (triglyceride) — only about 2 g (0.2%) of the oil is extracted from the cyclohexane phase together with the PAH, the separation method of liquid-liquid partition is a very effective isolation step in this case. The concentration factor achieved with this simple method is 1:500.

The most frequently used methods of isolation are described below. This survey is not meant to be complete, but includes the most common separation methods as examples.

### 2.3.5.1 Liquid-Liquid Partition of Biphase Solvent Mixtures
#### 2.3.5.1.1 Hydrophilic/Hydrophobic (Polar/Nonpolar)

One example is the already mentioned extraction of water samples with immiscible organic solvents (second extraction group). PAH can also be completely extracted from acetone or methanol solutions after dilution with water (fourth and fifth extraction groups). Extraction of the sample by using cyclohexane after saponification with methanol-potassium hydroxide solution (methanol + water at a 9:1 ratio) is a similar process step (third extraction group).

#### 2.3.5.1.2 Aliphatic/Aromatic

Exemplary for the separation of aromatic compounds from nonaromatic, nonpolar impurities are the ternary mixture of cyclohexane + dimethylformamide + water (first extraction group) and the binary mixtures of isooctane + dimethylsulfoxide[38] and cyclohexane + nitromethane.[15]

The distribution coefficients and the amounts of PAH obtainable by single or multiple extraction are given in Table 16 for the biphase solvent mixtures mentioned earlier.[18,25]

### 2.3.5.2 Chromatographic Methods
#### 2.3.5.2.1 Silica Gel, Alumina, etc. as Column-Packing Material

The low polarity of PAH permits the elution of compounds of this chemical class from highly polar compounds in the matrix by using nonpolar solvents such as cyclohexane, isooctane, benzene, or mixtures of these solvents. The activity of inorganic adsorbents such as silica gel, alumina, florisil,[38] etc., depends on the water content. Thus a reproducible retention volume of alumina with a certain particle size can be achieved only at an exactly defined water content. A water content reduced by 0.2% (total content: about 4.0%) yields a considerably increased elution volume. Table 17 gives elution volumes of some PAH, using alumina and silica gel as packing material (water content: 10%) and cyclohexane as eluant.[16]

When using highly active silica gel at a low water content, the contained heavy metal may initiate decomposition reactions of the PAH. This reaction can be avoided by initially washing the silica gel with 5 $N$ HCl.

The strong influence of the water content on PAH adsorption and the low adsorptive capacity (overloading) of inorganic packing material are serious disadvantages if an exactly reproducible separation of PAH mixtures is to be achieved. These column packings can be utilized without difficulty for a rough separation of PAH mixture and accompanying substances.

Thin-layer plates of pure inorganic material should not be used for preliminary separation since decomposition of PAH on silica gel plates has been occasionally reported.[62]

#### 2.3.5.2.2 Sephadex® LH 20, 60, Acetylcellulose as Column-Packing Material

In the last few years hydrophobic Sephadex® has predominantly been used as an organic adsorbent in column chromatography because its retention volumes are not dependent on the water content, and even at high PAH concentrations the elution volumes do not change. It is particularly suitable for isolating PAH because nonaromatic, nonpolar compounds such as paraffins, triglycerides, etc. are eluted with alcohol directly behind the solvent front, whereas aromatic compounds are considerably retarded according to the number of rings. This effect can be seen in Table 18.[19] Sephadex® LH 20 or 60 are, furthermore, used as a carrier of the stationary phase in partition chromatography. Thus a mixture of dimethylformamide and water (85:15) can easily be adsorbed to Sephadex® LH 20 with subsequent addition of hexane (sat-

Table 16
DISTRIBUTION COEFFICIENTS OF POLYCYCLIC AROMATIC
HYDROCARBONS IN CYCLOHEXANE + METHANOL + WATER,
NITROMETHANE + CYCLOHEXANE, AND DIMETHYLFORMAMIDE +
WATER + CYCLOHEXANE

|  | CH/Me[a] | CH %[b] | NM/CH[c] | NM %[d] | DMF/CH[e] | DMF %[f] |
|---|---|---|---|---|---|---|
| Phenanthrene | 3.44 | 95 | 1.53 | 94 | 2.70 | 85 |
| Anthracene | 3.09 | 94 | 1.27 | 92 | 2.36 | 83 |
| Pyrene | 3.24 | 94 | 1.64 | 95 | 3.32 | 86 |
| Fluoranthene | 2.83 | 93 | 1.66 | 95 | 3.66 | 88 |
| Benz[a]anthracene | 3.15 | 94 | 1.74 | 95 | 4.30 | 90 |
| Chrysene | 2.70 | 94 | 1.85 | 96 | 3.73 | 88 |
| Benzo[e]pyrene | 5.50 | 98 | 2.10 | 97 | 7.16 | 94 |
| Benzo[a]pyrene | 6.00 | 98 | 2.05 | 96 | 6.88 | 93 |
| Perylene | 3.90 | 96 | 1.99 | 96 | 5.39 | 92 |
| Anthanthrene | 4.40 | 96 | 1.80 | 95 | 7.64 | 94 |
| Benzo[ghi]perylene | 4.60 | 97 | 1.54 | 94 | 7.41 | 94 |
| Dibenz[a,h]anthracene | 6.40 | 98 | 2.00 | 96 | 8.70 | 95 |
| Benzo[b]chrysene | 5.00 | 97 | 2.03 | 96 | 7.02 | 93 |
| Coronene | 5.30 | 97 | 1.69 | 95 | 9.30 | 95 |

[a] CH/Me = distribution coefficient in the mixture of cyclohexane + methanol + water (10:9:1).
[b] CH % = percentage of PAH in cyclohexane phase after two extractions.
[c] NM/CH = distribution coefficient in the mixture of nitromethane + cyclohexane (1:1).
[d] NM % = percentage of PAH in nitromethane phase after three extractions.
[e] DMF/CH = distribution coefficient in the mixture of dimethylformamide + water + cyclohexane (9:1:5).
[f] DMF % = percentage of PAH in DMF-water phase after single extraction.

urated with dimethyl formamide and water) as a mobile elution phase.[26] The advantage of this type of chromatography is the very high capacity of the impregnated Sephadex® column which permits separation of up to 1 g of a mixture of substances on 10 to 20 g of Sephadex® LH 20. The partition chromatography is preferably used for some problematic matrices such as fresh or used lubricating oil, metal-cutting oil, crude mineral oil, or cigarette smoke condensate. In these cases large amounts of methyl and dimethyl derivatives of three- and four-ring compounds are present as well as unsubstituted base components. These derivatives can be separated in the process mentioned.

The described separation methods are adequate enrichment steps for most sample types and yield a gas-chromatographically pure PAH fraction which can be analyzed by various methods of determination.

For all methods of analysis, checking the completeness of extraction and the yield of the isolation procedure by adding one or more reference standards is recommended. Table 19 gives an overview of established processes for isolating PAH from different sample types (matrices).

Some of these methods were tested with respect to their repeatability (identical personnel, equipment, laboratory, and time) and their reproducibility (different personnel, laboratories, and equipment) in several collaborative studies. Such collaborative studies were carried out by the IUPAC and DGMK.* They are concerned with sample types of the extraction groups 1, 3, and 5:

* See Table 19.

Table 17
ELUTION VOLUMES OF PAH, USING CYCLOHEXANE (ml) SILICA GEL[a] AND ALUMINA[b]

| PAH | SiO$_2$ | Al$_2$O$_3$ |
|---|---|---|
| Anthracene | 40—90 | 20—35 |
| Phenanthrene | 40—95 | 19—34 |
| Pyrene | 45—105 | 34—55 |
| Fluoranthene | 57—105 | 40—70 |
| Benz[a]anthracene | 67—130 | 108—150 |
| Chrysene | 75—140 | 106—154 |
| Benzo[a]pyrene | 90—140 | 205—320 |
| Benzo[e]pyrene | 95—160 | 235—350 |
| Perylene | 84—170 | 240—360 |
| Anthanthrene | 84—160 | 430—600 |
| Benzo[ghi]perylene | 105—180 | 490—690 |
| Dibenz[a,h]anthracene | 125—220 | 680—890 |
| Coronene | 130—260 | 1150—1620 |

[a] 20 g, water content: 10%
[b] 15 g, water content: 4%.

Table 18
ELUTION VOLUMES OF PAH ADSORBED TO SEPHADEX® LH 20[a]

| PAH | Volume (ml) |
|---|---|
| Aliphatic hydrocarbons | 20 — 35 |
| Phenanthrene/anthracene | 38 — 50 |
| Pyrene/fluoranthene | 48 — 65 |
| Benz[a]anthracene/chrysene | 60 — 78 |
| Benzofluoranthenes [b + j + k] | 70 — 90 |
| Benzo[a]pyrene | 73 — 93 |
| Benzo[e]pyrene | 78 — 98 |
| Indeno[1,2,3-cd]pyrene/perylene | 81 — 105 |
| Anthanthrene/benzo[ghi]perylene | 89 — 118 |
| Benzo[b]chrysene | 84 — 115 |
| Coronene | 105 — 140 |

[a] 10 g, isopropyl alcohol.

Group 1:  IUPAC. Sample type: sunflower seed oil, dotted with 7 PAH (range μg/kg); method: Grimmer et al.[23a,25] Type of PAH identification: gas-chromatographic profile analysis. Evaluated: 13 analyses.

Group 1:  DGMK. Sample types: (1) lubricating oil, (2) used lubricating oil, (3) heating fuel oil EL; dotted or evaluated: (1) 5 PAH, (2) 11 PAH, (3) 8 PAH; method: Grimmer et al.[26] Type of PAH identification: gas-chromatographic profile analysis. Evaluated: 30 analyses each.

Group 3:  IUPAC. Sample type: minced beef, dotted with 4 PAH (range μg/kg); method: Howard et al.[38b] Type of PAH identification: UV spectrometry and/or fluorescence. Evaluated: 20 analyses.

Group 3:  IUPAC. Sample type: minced beef, dotted with 8 PAH (range μg/kg); method: Grimmer et al.[23a] Type of PAH identification: gas-chromatographic profile analysis. Evaluated: 12 analyses.

Table 19
ISOLATION SCHEMES FOR DIFFERENT SAMPLE TYPES (MATRICES)

| Extraction group | 1 | 2 | 3 | 4 | 5 | 6 |
|---|---|---|---|---|---|---|
| Sample type (matrix) | Fats, oils, plants | Mineral oil products | Water | Meat products | Leaves, seeds, etc. | Samples with inorganic components | Soots |
| Methods tested in collaborative studies, proposed by | IUPAC[a] | DGMK[b] | | IUPAC | | DGMK/VDI[c] | |
| Extraction or dissolution in | CH[d] | CH | CH | KOH[d] | Ac[d] | Ac,(Xy)[d] | Xy |
| Filtration and concentration | − | − | − | − | + | + | + |
| Distribution: methanol + water + cyclohexane (9:1:10) | − | − | − | + | + | − | − |
| Distribution: cyclohexane + dimethylformamide + water (10:9:1; 1:1:1) | + | + | − | + | + | + | + |
| Chromatography with silica gel | + | + | + | + | + | + | + |
| Chromatography with SEPHADEX® LH 20 | + | − | + | + | + | + | + |
| Partition chromatography SEPHADEX® | − | + | − | − | − | − | − |

[a] IUPAC = International Union of Pure and Applied Chemistry.
[b] DGMK = Deutsche Gesellschaft für Mineralölwissenschaft und Kohlechemie (German Society for Mineral Oil Science and Coal Chemistry).
[c] VDI = Verein Deutscher Ingenieure (Society of German Engineers).
[d] CH = cyclohexane, KOH = methanol KOH, Ac = Acetone, Xy = xylene.

Group 5: DGMK. Sample type: automobile exhaust gas condensate, filters loaded with automobile exhaust gas emissions; method: Grimmer et al.[23] Type of PAH identification: gas-chromatographic profile analysis. Evaluated: about 120 analyses in 5 collaborative studies.

*2.3.5.3 Summary and Appraisal of Enrichment Methods*

Isolation of a fraction which contains all or certain PAH of the sample can be achieved by an appropriate combination of several process steps. The more effective the method is, the smaller the number of steps needed to concentrate certain PAH in one fraction, the higher the PAH yield in this fraction, and the lower the amount of impurities contained in the isolated PAH fraction.

All PAH or individual components of this concentrated fraction can be identified in general by any method that permits the separation and quantification of the individual components of the mixture. In most cases the PAH mixture obtained from a sample can be assessed quantitatively by means of gas chromatography and thin-layer chromatography, as well as high-performance liquid chromatography only if the separation power of the corresponding method is adequate.

## 2.3.6 Separation Methods for the Chromatographic Separation and Quantitative Assessment of PAH Mixtures

As mentioned earlier, the PAH fraction isolated from most sample types is a complex mixture of more than 100 individual compounds. The best separation results for such a PAH mixture are obtained with high-performance gas-chromatographic columns whose number of separation stages exceeds 25,000 HETP. In general, a flame ionization detector (FID) is used for the quantitative assessment because it offers the advantage of showing the signal areas of unknown PAH linearly proportional to the mass of carbon combusted and thus gives a direct presentation of the quantitative composition (PAH profile analysis). PAH profile analysis is a prerequisite for listing an inventory of all carcinogenic substances of this class. Furthermore, it permits the identification of the source of emission on the basis of characteristic individual compounds or typical profiles.

If only main components have to be detected, separation processes with a low separating capacity (number of separation stages) will be sufficient, for instance, high-pressure liquid chromatography, thin-layer chromatography, or column chromatography with inorganic or organic carrier material.

Of these methods which can be used for separation and quantitative evaluation of PAH mixtures, only those are considered here which in practice are used for separating "natural" mixtures. Separation processes which were described as methods for separating mixtures of standard substances cannot be taken into account, since quite frequently it is very difficult to adapt these methods to mixtures obtained from samples.

Out of the large number of determination methods, those whose efficiency was proven in collaborative studies of several laboratories should preferably be used. Methods which meet this requirement are presented in the comparative listing of isolation methods (see Table 19).

Sawicki et al.[58] published a comparative study on methods of determination which had been used until 1967. By introducing gas-chromatographic and mass-spectrometric determination methods the methodological efficiency was improved considerably so that nowadays the methods formerly described are hardly used for PAH analyses. At present two methods of determination are predominantly employed for PAH analysis: the gas-chromatographic profile analysis with high-performance columns including the FID as a detector, and if necessary, a mass spectrometer for identification, and the two-dimensional thin-layer chromatography including UV ray or fluorescence detection. Column-chromatographic methods which identify up to 14 individual compounds by UV spectrometry[22] are scarcely used nowadays. Mention must be made of HPLC of which we know but little with respect to its applicability to the separation of PAH mixtures obtained from environmental samples. In comparison with gas chromatography in high-performance columns, the separation power of HPLC is clearly lower so that it seems applicable to special types of samples only. The low-voltage mass spectrometry by which individual compounds are identified on the basis of their masses, without needing separation of the PAH mixture, is only applicable to special investigations. For instance, it is suitable for characterizing heterocyclic compounds (N, S, O).

## 2.3.7 Gas Chromatography with High-Performance Columns
### 2.3.7.1 Glass Capillary Columns

The capillary gas chromatography had already been used more than 10 years ago for analyzing PAH included in airborne particulate matter. In the last few years it has been improved and become a routine method.[1,6,7,19,50] The separating capacity of this method is illustrated in Figure 9.

The PAH mixture obtained from airborne particulate matter can be separated into more than 100 components.[48] Capillary gas chromatography is also used for separating PAH mixtures from automobile exhaust gas,[11,19,20] heavy crude oil fractions,[14] and soil samples.[4]

For economical reasons the samples should be as small as possible. Consequently the total sample is to be fed directly into the column, i.e., splitless. This can be achieved

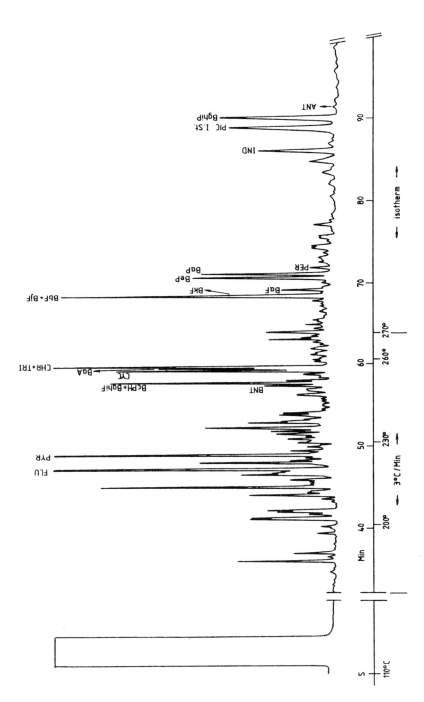

FIGURE 9. Separation of a mixture of polycyclic aromatic hydrocarbons, extracted from airborne particulate matter, which was collected in a German city (October 1980). Glass capillary (0.25 mm × 25 m) coated with polydimethylsiloxane (CP sil 5), splitless injected at 110°C, temperature program 110° to 270°C, 3°C/min, Helium, FID recording.

by special charging techniques[33,34] which prevent the majority of the sample from being blown into the air without being utilized. Thus the total PAH content of the sample is used for determination. If a flame ionization detector is employed to record substances, the normal working range lies between 1 and 20 ng per PAH (e.g., benzo[a]pyrene). The detection limit (defined as triple-noise bandwidth) is about 0.1 ng. The detection sensitivity is improved somewhat by utilizing an electron capture detector (ECD). Since the ECD responds differently to each individual PAH, the quantitative composition of a mixture cannot be recorded directly nor can the amount of an unknown PAH be determined.

The main problem preventing a routine utilization of capillary gas chromatography lies in the difficulty of coating the glass walls permanently with phases such as polysiloxanes Silicone OV 101, OV 17, SE 30, SE 52 etc., Dexil 300 GC etc.). This is largely dependent on the surface structure of the glass used. The phases should release as little material as possible at working temperatures up to 280°C (column bleeding).

PAH are normally identified by comparing their retention time with that of corresponding reference substances. In case of doubt, mass-spectrometric investigations should be carried out additionally (GC-MS combination). Routine assays with the same type of sample usually produce similar PAH profiles so that an identification with the GC-MS combination is required only in special cases.

When using glass columns in all-glass devices, the necessary separation temperature (250 to 280°C) does not cause measurable decomposition of PAH, neither in capillary columns nor in packed high-performance columns. If a PAH profile analysis of high accuracy is required, the number of separation stages, the symmetry of FID signals, and the quotient between signal profile and sample weight should be checked with a test solution prior to starting a series of analyses. The quotient for PAH differs only little in a well-adjusted FID.

Disadvantages of the capillary gas chromatography are the high qualifications required from the operating personnel and the high costs of acquisition, as well as of operation, of a gas chromatograph.

*2.3.7.2 Packed High-Performance Columns (about 10 m)*

The term "high-performance columns" defines columns which have separation stages exceeding 20,000 HETP. In general packed columns have about 2000 to 2900 HETP/m under favorable conditions; this necessitates the use of columns of about 10 m in order to achieve the necessary separation capacity. In practice this number of separation stages means that about 95 to 105 completely separated peaks can be obtained between the retention of fluoranthene and that of coronene.

Packed high-performance columns yield a maximum separation capacity of about 50,000 separation stages. This corresponds to about 160 completely separated signals between fluoranthene and coronene. Figure 10 shows the separation chromatogram of a PAH fraction (four to seven rings) obtained from automobile exhaust gas condensate.[27] Separation stages of 20,000 HETP are sufficient to separate most PAH mixtures isolated from various sample types such as airborne particulates,[72] meat products, fats, oils, and yeasts,[25] automobile exhaust gas condensate,[23] sediment samples,[24] mineral oil products,[27] sewage sludge,[30] and crude oil samples.[31] As shown in Figure 10, a satisfactory separation of the isomeric pair benzo[e]pyrene/benzo[a]pyrene is possible even when using Silicone OV 101. The separating of the isomeric benzofluoranthenes, or of chrysene and triphenylene, however, cannot be achieved. This method, too, offers the advantage of producing a direct picture of the quantitative composition of the PAH mixture when employing the FID.

The routine working range lies between 10 and 200 ng per component. An overloading of up to 50 times this amount does not lead to disturbances. The detection limit

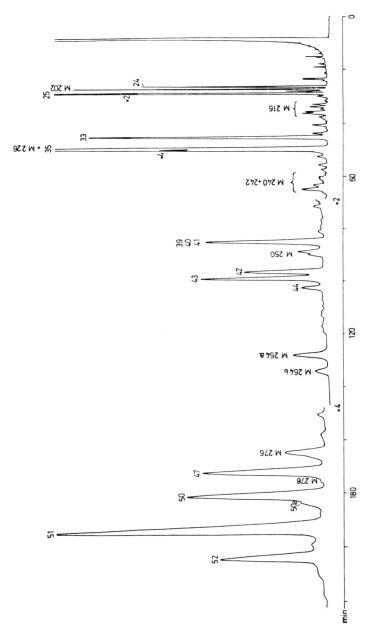

FIGURE 10. Gas chromatogram of a PAH fraction isolated from automobile exhaust gas, Conditions: 2 mm × 20 m, 5% Silicone OV 101 on GasChrom Q 125-150 μm 270°, preliminary pressure: 8.5 bar. (24) fluoranthene, (25) pyrene, (33) benzo[ghi]fluoranthene, (35) chrysene + cyclopenta[cd]pyrene, (39-41) benzofluoranthenes [b + j + k], (42) benzo[e]pyrene, (43) benzo[a]pyrene, (44) perylene, (M264a) 11H-cyclopenta[qrs]benzo[e]pyrene, (M264b) 10H-cyclopenta-[mno]benzo[a]pyrene, (47) indeno[1,2,3-cd]pyrene, (50) reference standard — benzo[b]chrysene, (51) benzo[ghi]perylene, (52) anthanthrene.

(e.g., for benzo[a]pyrene) is about 1 ng. Consequently, the sample size has to be chosen so that about three times this amount is contained in the sample, since the isolation step mostly yields about 70%, and only about 50% of the isolated amount of PAH can be charged into the column (solution volume: about 5 $\mu\ell$ acetone, butan-2-one, or dimethylformamide).

An advantage of this process over the glass capillary gas chromatography is that sample injection causes no problems and does not require high technical skills. Furthermore, the column can easily be regenerated by simple discharging and repacking of carrier material.

### 2.3.8 Gas Chromatography with Packed Columns (<5 m)

The separation of isolated PAH mixtures by means of commonly used packed columns is of course not satisfactory. This applies even if temperature programs are used. Quite often columns with Dexil packing are utilized to investigate PAH mixtures isolated from airborne particulate matter.[47] When using a column of 3.6 m (Dexil 300 GC), it is possible to separate the PAH contained in a sample of airborne particulate matter (from phenanthrene to coronene) into 27 identifiable PAH.[9] The same authors separated this sample into 19 identifiable signals by means of high-pressure liquid chromatography. However, these signals cannot be quantified, due partly to the high background level. A number of investigators restrict the use of normal columns of 2 to 3 m to the identification of benzo[a]pyrene contained in airborne particulate matter.

A selective separation of some critical PAH pairs such as chrysene/benz[a]anthracene/triphenylene/cyclopentadieno[cd]pyrene is possible with nematic phases, as e.g., N,N-bis-(p-methoxybenzylidene)-$\alpha\alpha$-bis-p-toluidine.[41] A separation of the isomeric benzofluoranthenes is also possible in part.[32] However, as was to be expected, further compounds cannot be separated when investigating a PAH mixture isolated from automobile exhaust gas, since a larger number of separation stages would be necessary than those verifiable with these phases. A substantial disadvantage of this method is the strong bleeding of currently available nematic phases, which makes their routine utilization practically impossible.

Another gas-chromatographic process has been proposed for analyzing automobile exhaust gas.[59]

### 2.3.9 Thin-Layer Chromatography

Thin-layer chromatography has been used for identifying benzo[a]pyrene contained in samples of airborne particulate matter for a long time.[8,57,62,67] In this case the problem does not consist in the separation of all individual compounds, but in the separation of one PAH from the others. This of course does not necessitate a separation method with a very high separation power. When lining up the substance spots between initial and solvent front, about 10 to 15 substances can be arranged. Taking the more than 100 PAH detected in airborne particulate matter into account, this means a seven- to tenfold overlapping within this range.

The advantage of thin-layer chromatography consists in its easy technical handling. Beside cellulose acetate foils, mixed plates of alumina/acetylcellulose are mostly used as base material. The use of mixtures of low-polarity solvents such as hexane, benzene, methylene chloride, etc. has been recommended as well as the use of mixtures of pyridine, methanol, ether, water, etc. as mobile phase.

Two-dimensional processes for analyzing samples of airborne particulate matter were developed early to increase the separation capacity of thin-layer chromatography.[43,51,68] For identification of six selected PAH in water samples, a two-dimensional process has been proposed in which alumina and cellulose acetate are used as support material and n-hexane + benzene (9:1) or methanol + ether + water (4:4:1) as mobile phase.[45] In order to make a quantitative evaluation, the substance spots are scrubbed out with the base material, extracted, and evaluated in the PAH-typical range by spectrometric or fluorimetric methods. Differences between close maximal and minimal extinctions of characteristic wavelengths are used to eliminate background adsorption.

PAH dissolved in paraffin (n-pentane to n-decane) show spectra with a series of sharp individual bands at low temperatures (77.3 K), instead of the broad fluorescence bands observed at ambient temperature (Shpolsky effect). The lines are easily reproducible, although different types of spectra were found, for instance, for coronene. Eichhoff and Köhler[12] published a summary of this method.

According to another, more convenient method, fluorescent light radiated at an angle of 45° is measured. The exciting light (e.g., 365 nm) is emanated at the same angle to the developed thin-layer plate (chromatogram spectrophotometer). The disadvantage of this device is the high cost of acquisition, which is twice as high as that of a gas chromatograph.

### 2.3.10 High-Performance Liquid Chromatography (HPLC = High-Pressure Liquid Chromatography)

The separation capacity of liquid chromatography can be increased considerably by enlarging the surface of the adsorbing agent, i.e., by reducing the particle size to a few micrometers (high-performance liquid chromatography, a high-pressure liquid chromatography = HPLC). The dense packing of the fine material necessitates, however, a considerably increased pressure (up to 200 bar) in order to reach the rate of flow necessary in practice. In most cases UV spectrometers are used as detectors of PAH at the column end. In general they record the extinction at a certain wavelength; this can be done only if there is no interfering background, i.e., in case of optically very pure mixtures of PAH. Fluorescence measuring with excitation at a certain wavelength is also carried out because of its high detection sensitivity.[44,65] Silica gel or silica gels with chemically bound organic phases (reversed phase) have been found to be stationary phases which are suitable for the separation of PAH mixtures. In these silica gels hydrophobic alkyl residues, e.g., the octadecyl substituent, are bound chemically to the hydroxyl groups of silicic acid which the mobile phase cannot cleave off. In most cases water-alcohol mixtures serve as mobile phase.

HPLC was utilized for the separation of PAH mixtures isolated from samples of airborne particulate matter[9,13,55] and automobile exhaust gas.[60] So far a satisfactory separation of the complex mixture of airborne particulate matter has not been achieved due to the lower number of separation stages (about 20,000 HETP). According to comparative investigations carried out by Novotny et al.[55] the HPLC will "hardly be a final solution" for the separation of PAH mixtures obtained from environmental samples ("real PAH samples").

On the other hand, the separation power of HPLC seems to be adequate for investigating a few main components of water samples, if the evaluation of secondary components can be neglected and the quantitation be limited to the predominant base components.[35] In this case it is a genuine alternative to the two-dimensional thin-layer chromatography.

### 2.3.11 Final Evaluation of Methods Used for Separating Complex PAH Mixtures
*2.3.11.1 Gas Chromatography with Glass Capillaries*

Gas-chromatographic separation in glass capillaries is at present the most efficient method for separating complex PAH mixtures isolated from environmental samples and emissions. In routine operation, 70,000 and more HETP is achievable without difficulty. This number of separation stages means that about 200 completely separated peaks fit in between the retention time of fluoranthene and that of coronene. In case of "splitless" sample injection, a sample containing, for instance, 5 to 10 ng BaP is needed to record an amplitude of the FID signal extending over the whole recording width (detection limit: about 0.1 ng, defined as triple noise).

Beside its high sensitivity, the FID recording offers the same advantages as a mass detector. Known and unidentified PAH are recorded in proportion to their quantity and — without needing correcting factors — give a direct representation of the quantitative composition of the PAH mixture isolated from a sample (PAH profile).

*2.3.11.2 Gas Chromatography with Packed High-Performance Columns*

Packed high-performance columns are separation columns whose value of separation number exceeds 20,000. This separation capacity is achieved without difficulty when using column lengths of 7 to 10 m. About 100 separate peaks are obtained between the retention time of fluoranthene and that of coronene.

For routine operation, when about 1/8 of the maximum electronic amplification is used, the sample has to contain 100 to 1000 ng BaP to induce a 100% FID signal. This gas-chromatographic method is also suitable for PAH profile analysis of most types of sample.

*2.3.11.3 Gas Chromatography with Packed Columns (<5 m)*

The separation power of packed columns with a length below 5 m is not sufficient for a complete profile analysis of most sample types because the peak sequences are not separated adequately to enable an exact quantitative evaluation of individual signals. Only "peak mountain ranges" without valleys extending to the base line are recorded.

*2.3.11.4 Thin-Layer Chromatography*

In the case of most sample types, the unidimensional thin-layer chromatography (TC) is suitable for the identification of individual components only. Since the size of the substance spot does not depend upon the diameter of the application point alone, but also upon the ratio to the solvent front $R_f$ value), only 10 to 20 substances can be arranged between the starting point and the solvent front, even at complete separation of the spots.

A possible quantitation by measuring UV light fluorescence requires information on the characteristic properties of the substances. The concentration of an unidentified PAH, therefore, cannot be determined.

The two-dimensional TC considerably improves the separation of PAH but does not achieve the separation capacity of high-capacity columns used in gas chromatography.

*2.3.11.5 High-Performance Liquid Chromatography (HPLC)*

The presently used columns reach a maximum separation number of 20,000. This corresponds to the separation power accomplished with packed high-performance columns of a length of 7 to 10 m in gas chromatography. In practice this means that the predominating main components (some unsubstituted base components) can be separated completely from each other, whereas, for instance, mixtures of isomeric methyl derivatives of a base component cannot be separated. Since there is but a slight shift in the wavelengths of their maxima or minima, different measurements give no information when using specific wavelengths for such isomeric mixtures.

In case of mixtures such as samples of drinking water in which frequently occurring PAH (mostly unsubstituted base components) prevail, these main components may be quantified, but an assessment of secondary components is not feasible. Figure 11 shows the separation by HPLC of a sample of airborne particulate matter collected in Baltimore Harbor Tunnel.[13]

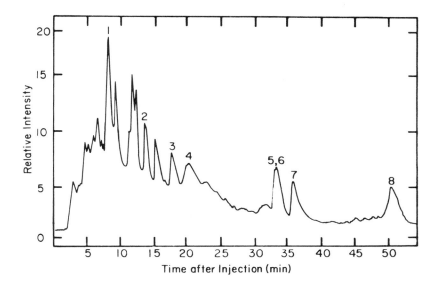

FIGURE 11. HPLC chromatogram of a benzene extract of airborne particulate matter (atmospheric particulates) collected in Baltimore Harbor Tunnel.[13]

Compared with this figure, the separation results obtained by means of gas chromatography with glass capillaries (Figure 9) or packed high-capacity columns (Figure 10) are considerably better.

## REFERENCES

1. **Bjørseth, A.**, Analysis of polycyclic aromatic hydrocarbons in particulate matter by glass capillary gas chromatography, *Anal. Chim. Acta,* 94, 21, 1977.
2. **Bolbertitiz, K.**, Water analysis using the CCE (carbon-chloroform-extract) method, *Fortschr. Wasserchem. Ihrer Grenzgeb.,* 15, 43, 1973.
3. **Borneff, J. and Kunte, H.**, Kanzerogene Substanzen in Wasser und Boden. XXVI. Routinemethode zur Bestimmung von polycyclischen Aromaten in Wasser, *Arch. Hyg. Bakteriol.,* 153, 220, 1969.
4. **Blumer, M. and Blumer, W.**, Polycyclic aromatic hydrocarbons in soil of a mountain valley: correlation with highway traffic and cancer incidence, *Environ. Sci. Technol.,* 11, 1082, 1977.
5. **Brune, H., Habs, M., and Schmähl, D.**, The tumor-producing effect of automobile exhaust condensate and fractions thereof, *J. Environ. Pathol. Toxicol.,* 1, 737, 1978.
6. **Cantreels, W. and van Cauwenberghe, K.**, Determination of organic compounds in airborne particulate matter by gas chromatography — mass spectrometry, *Atmos. Environ.,* 10, 447, 1976.
7. **Cantuti, V., Cartoni, G. P., Liberti, A., and Torri, A. G.**, Improved evaluation of polynuclear hydrocarbons in atmospheric dust by gas chromatography, *J. Chromatogr.,* 17, 60, 1965.
7a. **Clar, E.**, *Polycyclic Hydrocarbons,* Vol. 1 and 2, Academic Press, New York, 1964.
8. Coke Research Report 76, The Determination of Polynuclear Aromatic Hydrocarbons in Airborne Particulate Matter, British Coke Research Association, Chesterfield, England, 1973.
9. **Dong, M. and Locke, D. C.**, High pressure liquid chromatographic method for routine analysis of major parent polycyclic aromatic hydrocarbons in suspended particulate matter, *Anal. Chem.,* 48, 368, 1976.
10. **Dontenwill, W., Chevalier, H. J., and Harke, H. P., Klimisch, H. J., Brune, H., Fleischmann, B., and Keller, W.**, Experimentelle Untersuchungen über die tumorerzeugende Wirkung von Zigarettenrauch-Kondensat an der Mäusehaut, *Z. Krebsforsch.,* 85, 155, 1976.

11. Doran, T. and McTaggert, N. G., Combined use of high efficient liquid and capillary gas chromatography for the determination of polycyclic aromatic hydrocarbons in automotive exhaust condensate and other hydrocarbon mixtures, *J. Chromatogr. Sci.*, 12, 715, 1974.
12. Eichhoff, H. J. and Köhler, M., Identifizierung und Bestimmung polycyclischer aromatischer Kohlenwasserstoffe durch Fluoreszensspektren fester Lösungen bei tiefen Temperaturen, *Z. Anal. Chem.*, 197, 271, 1963.
13. Fox, M. A. and Staley, S. W., Determination of polycyclic aromatic hydrocarbons in atmospheric particulate matter by high pressure liquid chromatography coupled with fluorescence technique, *Anal. Chem.*, 48, 992, 1976.
14. Gouw, T. H., Wittemore, I. M., Jentoft, R. E., and Ralph, E., Versatile short capillary column in gas chromatography, *Anal. Chem.*, 42, 1394, 1970.
15. Grimmer, G., Eine Methode zur Bestimmung von 3,4-Benzpyren in Tabakrauchkondensaten, *Beitr. Tabakforsch.*, 1, 107, 1961.
16. Grimmer, G. and Hildebrandt, A., Kohlenwasserstoffe in der Umgebung des Menschen. I. Eine Methode zur simultanen Bestimmung von dreizehn polycyclischen Kohlenwasserstoffen, *J. Chromatogr.*, 20, 89, 1965.
16a. Grimmer, G., Glaser, A., and Wilhelm, G., Die Bildung von Benzo[a]pyren und Benzo[e]pyren beim Erhitzen von Tabak in Abhängigkeit von Temperatur und Strömungsgeschwindigkeit in Luft- und Stickstoffatmosphäre, *Beitr. Tabakforsch.*, 3, 415, 1966.
17. Grimmer, G. and Hildebrandt, A., Kohlenwasserstoffe in der Umgebung des Menschen. VI. Mitteilung: der Gehalt polycyclischer Kohlenwasserstoffe in rohen Pflanzenölen, *Arch. Hyg. Bakteriol.*, 152, 255, 1968.
18. Grimmer, G. and Wilhelm, G., Der Gehalt polycyclischer Kohlenwasserstoffen in europäischen Hefen, *Dtsch. Lebensm., Rundsch.*, 65, 229, 1969.
19. Grimmer, G. and Böhnke, H., Bestimmung des Gesamtgehaltes aller polycyclischen aromatischen Kohlenwasserstoffe in Luftstaub und Kraftfahrzeugabgas mit der Capillar-Gas-Chromatographie, *Z. Anal. Chem.*, 261, 310, 1972.
20. Grimmer, G., Die quantitative Bestimmung von polycyclischen Aromaten mit der Kapillargaschromatographie, *Erdöl Kohle Erdgas Petrochem.*, 25, 339, 1972.
21. Grimmer, G., Hildebrandt, A., and Böhnke, H., Probennahme und Analytik polycyclischer aromatischer Kohlenwasserstoffe in Kraftfahrzeugabgasen, *Erdöl Kohle Erdgas Petrochem.*, 25, 531, 1972.
22. Grimmer, G. and Hildebrandt, A., Concentration and estimation of 14 polycyclic aromatic hydrocarbons at low level in high-protein foods, oils, and fats, *J. Assoc. Off. Anal. Chem.*, 55, 631, 1972.
23. Grimmer, G., Hildebrandt, A., and Bohnke, H., Investigation on the carcinogenic burden by air pollution in man. II. Sampling and analysis of polycyclic aromatic hydrocarbons in automobile exhaust gas, *Zentralbl. Bacteriol. Parisitenkd. Infektionskr. Hyg., I. Abt. Orig. Reihe B*, 158, 22, 1973.
23a. Grimmer, G., et al., *J. Assoc. Off. Anal. Chem.*, 58, 725, 1975.
24. Grimmer, G. and Böhnke, H., Profile analysis of polycyclic aromatic hydrocarbons and metal content in sediment layers of a lake, *Cancer Lett.*, 1, 75, 1975.
25. Grimmer, G., Hildebrandt, A., and Böhnke, H., Profilanalyse der polycyclischen aromatischen Kohlenwasserstoffe in proteinreichen Nahrungsmitteln, Ölen und Fetten (gas-chromatographische Bestimmungsmethode), *Dtsch. Lebensm. Rundsch.*, 71, 93, 1975.
26. Grimmer, G. and Böhnke, H., Anreicherung und gaschromatographische Profil-Analyse der polycyclischen aromatischen Kohlenwasserstoffe in Schmieröl, *J. Chromatogr.*, 9, 30, 1976.
27. Grimmer, G., Böhnke, H., and Hildebrandt, A., Packed high-performance GC-columns (about 50,000 HETP) for profile analysis of carcinogenic polycyclic aromatic hydrocarbons in food, mineral oil products, vehicle exhaust, and cigarette smoke condensate etc., *Z. Anal. Chem.*, 279, 139, 1976.
28. Grimmer, G. and Böhnke, H., Untersuchungen von Sedimentkernen des Bodensees. I. Profile der polycyclischen aromatischen Kohlenwasserstoffe, *Z. Naturforsch.*, 32c, 703, 1977.
29. Grimmer, G., Böhnke, H., and Glaser, A., Investigation on the carcinogenic burden by air pollution in man. XV. Polycyclic aromatic hydrocarbons in automobile exhaust gas — an inventory, *Zentralbl. Bakteriol. Parisitenkd. Infektionskr. Hyg. Abt. 1, Orig. Reihe B*, 164, 218, 1977.
30. Grimmer, G., Böhnke, H., and Borwitzky, H., Gas-chromatographische Profilanalyse der polycyclischen aromatischen Kohlenwasserstoffe in Klärschlammproben, *Z. Anal. Chem.*, 289, 91, 1978.
31. Grimmer, G. and Böhnke, H., Polycyclische aromatische Kohlenwasserstoffe und Heterocyclen-Beziehung zum Reifegrad von Erdolen des Gifhorner Troges (Nordwestdeutschland), *Erdöl Kohle*, 31, 1978.
32. Grimmer, G., unpublished data.
33. Grob, K. and Grob, G., Splitless injection on capillary columns. I. The basic technique, steroid analysis as an example, *J. Chromatogr. Sci.*, 7, 584, 1969.
34. Grob, K. and Grob, G., Methodik der Kapillar-Gas-Chromatographie. Hinweise zur vollen Ausnutzung hochwertiger Säulen. I. Teil: Die Direkteinspritzung, *J. Chromatogr.*, 5, 3, 1972.

35. Hagenmaier, H., Feierabend, R., and Jäger, W., Bestimmung polycyclischer aromatischer Kohlenwasserstoffe in Wasser mittles Hochdruckflüssigkeitschromatographie, *Z. Wasser Abwasser Forsch.*, 99, 1977.
36. Herlan, A., Kanzerogene polycyclische Aromaten und Metaboliten als mögliche Bestandteile von Emissionen, *Zentralbl. Bakteriol. Parasitenkd. Infektionskr. Hyg., Abt. 1 Orig. Reihe B*, 165, 174, 1977.
37. Herlan, A. and Mayer, J., Polycyclische Aromaten und Benzol in den Abgasen von Hauschaltsfeuerungen. I. Ölofen, *Staub*, 38, 134, 1978.
38. Howard, J. W., Turicchi, E. W., White, R. H., and Fazio, T., Extraction and estimation of polycyclic aromatic hydrocarbons in vegetable oils, *J. Assoc. Off. Anal. Chem.*, 49, 1236, 1966.
38b. Howard, J. W., et al., *J. Assoc. Off. Anal. Chem.*, 51, 122, 1968.
39. Ikan, R., Baedecker, M. J., and Kaplan, I., Die thermische Veränderung organischer Substanz in jüngeren marinen Ablagerungen, *Erdöl Kohle*, 28, 489, 1975.
40. International Union of Pure and Applied Chemistry, *Nomenclature of Organic Chemistry*, Butterworths, London, 1966.
41. Janini, G. M., Johnston, K., and Zielinski, W., Use of nematic liquid crystal for gas-liquid chromatographic separation of polyaromatic hydrocarbons, *Anal. Chem.*, 47, 670, 1975.
42. Jager, J. and Kassowitzova, B., Determination of 3,4-benzopyrene in drinking water, *Chem. Listy*, 62, 216, 1968.
43. Köhler, M. and Eichhoff, H. J., Eine Schnellmethode zur Bestimmung von mehrkernigen aromatischen Kohlenwasserstoffen in Luftstaub, *Z. Anal. Chem.*, 401, 1967.
44. Krstulovic, A. M., Rosie, D. M., and Brown, Ph.R., Selective monitoring of polynuclear aromatic hydrocarbons by high pressure liquid chromatography with a variable wavelength detector, *Z. Anal. Chem.*, 48, 1383, 1976.
45. Kunte, H. and Borneff, J., Nachweisverfahren für polycyclische aromatische Kohlenwasserstoffe in Wasser, *Z. Wasser Abwasser Forsch.*, 9, 35, 1976.
46. Lang, K. F. and Eigen, I., Im Steinkohlenteer nachgewiesene organische Verbindungen, *Fortschr. Chem. Forsch.*, 8, 91, 1967.
47. Lao, R. C., Thomas, R. S., Oja, H., and Dubois, L., Application of gas-chromatography — mass-spectrometer data processor combination to the analysis of the polycyclic aromatic hydrocarbon content of airborne pollutants, *Z. Anal. Chem.*, 45, 908, 1973.
48. Lee, M. L., Novotny, M., and Bartle, K. D., Gas chromatography/mass spectrometric and nuclear magnetic resonance determination of polynuclear aromatic hydrocarbons in airborne particulates, *Z. Anal. Chem.*, 48, 1566, 1976.
49. Leoni, V., Puccetti, G., and Grella, A., Preliminary results on the use of TENAX for the extraction of pesticides and polynuclear aromatic hydrocarbons from surface and drinking waters for analytical purposes, *J. Chromatogr.*, 106, 119, 1975.
50. Liberti, A., Cartoni, G. P., and Cantuti, V., Gas chromatographic determination of polynuclear hydrocarbons in dust, *J. Chromatogr.*, 15, 141, 1964.
51. Matsushita, H. and Suzuki, Y., Two-dimensional dual-band thin-layer chromatographic separation of polynuclear hydrocarbons, *Bull. Chem. Soc. Jpn.*, 42, 460, 1969.
52. Matsushita, H. and Esumi, Y., Rapid determination method of benz[a]pyrene in air pollutants, *Bunseki Kagaku*, 21, 722, 1972.
53. Matsushita, H. and Hanya, T., Polynuclear aromatic hydrocarbons in the environment. I. Determination of polynuclear aromatic hydrocarbons in water by mass fragmentography, *Bunseki Kagaku*, 24, 505, 1975.
54. Monkman, J. L., Dubois, L., and Baker, C. J., Rapid measurement of polycyclic hydrocarbons in air by micro-sublimation, *Pure Appl. Chem.*, 24, 731, 1970.
55. Novotny, M., Lee, M. L., and Bartle, K. D., Methods for fractionation, analytical separation, and identification of polynuclear aromatic hydrocarbons in complex mixtures, *J. Chromatogr. Sci.*, 12, 606, 1974.
56. Patterson, A. M., Capell, L. T., and Walker, D. F., The ring index, in *A List of Ring Systems Used in Organic Chemistry*, 2nd ed., American Chemical Society, Washington, D.C., 1960, 1425.
57. Sawicki, E., Stanley, T. W., Elbert, W. C., and Pfaff, J. D., Application to thin-layer chromatography to the analysis of atmospheric pollutants and determination of benzo[a]pyrene, *Anal. Chem.*, 36, 497, 1964.
58. Sawicki, E., Stanley, T. W., Elbert, W. C., Meeker, J., and McPherson, S., Comparison of methods for determination of benzo[a]pyrene in particulates from urban and other atmospheres, *Atmos. Environ.*, 1, 131, 1967.
59. Sawicki, E., Atkins, P. R., Belsky, T., et al., Tentative method of analysis for polynuclear aromatic hydrocarbons in automobile exhaust, *Health Lab. Sci.*, 11, 228, 1974.
60. Schmit, J. A., Henry, R. A., Williams, R. C., and Dieckman, J. F., Application of high speed reversed-phase liquid chromatography, *J. Chromatogr. Sci.*, 9, 645, 1971.

61. **Scholz, L. and Altmann, H. J.**, Bestimmung von 3,4 Benzpyren in Wasser, *Z. Anal. Chem.*, 81, 1968.
62. **Seifert, B. and Steinbach, I.**, Dünnschicht-chromatographische Routinebestimmung von Benzo[a]pyren im Schwebestaub mit "in situ"-Auswertung, *Z. Anal. Chem.*, 287, 264, 1977.
63. **Shcherbak, N. P.**, Detection of 3,4-benzpyrene in soil, *Vopr. Onkol.*, 13, 77, 1967.
64. **Snook, M. E., Severson, R. F., Higman, H. C., Arrendale, R. F., and Chortyk, O. T.**, Polynuclear aromatic hydrocarbons of tobacco smoke: isolation and identification, *Beitr. Tabakforsch.*, 8, 250, 1976.
65. **Snyder, L. R. and Kirkland, J. J.**, *Introduction to Modern Liquid Chromatography*, John Wiley & Sons, New York, 1974.
66. **Stanley, T. W., Meeker, J. E., and Morgan, M. J.**, Extraction of organics from airborne particulates. Effects of various solvents and conditions on the recovery of benzo[a]pyrene, benz[c]acridine, and 7H-benz[d,e]anthracene-7-one, *Environ. Sci. Technol.*, 1, 927, 1967.
67. **Stanley, T. W., Morgan, M. J., and Meeker, J. E.**, Thin layer chromatographic separation and spectrophotometric determination of benzo[a]pyrene in organic extracts of airborne particulates, *Anal. Chem.*, 39, 1327, 1967.
68. **Stanley, T. W., Bender, D. F., and Elbert, W. C.**, Quantitative aspects of thin-layer chromatography in air pollution measurements, *Quant. Thin Layer Chromatogr.*, 305, 1973.
69. **Stepanova, M., Ilina, R. I., and Shaposhnikov, Y. K.**, Determination of polynuclear aromatic hydrocarbons in chemical and petrochemical waste water, *Zh. Anal. Khim.*, 27, 1201, 1972.
70. **Swallow, W. H.**, Survey of polycyclic aromatic hydrocarbons in selected foods and food additives available in New Zealand, *N. Z. J. Sci.*, 19, 407, 1976.
71. **Tomingas, R. and Brockhaus, A.**, Anwendung der Sublimationsmethode bei der Bestimmung von Benzo[a]pyren in Grossstadtaerosolen, *Staub Reinhalt. Luft*, 33, 481, 1973.
72. **Zoccolillo, L., Liberti, A., and Brocco, D.**, Determination of polycyclic hydrocarbons in air by gas chromatography with high efficient packed columns, *Atmos. Environ.*, 6, 715, 1972.

Chapter 3

# OCCURRENCE OF PAH

*G. Grimmer and F. Pott*

## TABLE OF CONTENTS

3.0  Processes During Which PAH Are Formed — G. Grimmer .............62

3.1—3.4  The Most Frequent Sources of Emission — G. Grimmer................64

    3.1  Domestic Heating.........................................................69
        3.1.1  Domestic Heating with Coal ......................................70
        3.1.2  Domestic Oil-Fired Heating (Extra-Light Oil) ..................73
            3.1.2.1  Type and Amount of PAH Emitted....................73

    3.2  Thermal Power Plants ..................................................73
        3.2.1  Hard Coal-Fired Power Plants ....................................73
        3.2.2  Lignite-Fired Installations........................................74
        3.2.3  Fuel Oil (Heavy)-Fired Installations..............................74
            3.2.3.1  Type and Amount of PAH Emitted....................74

    3.3  Motor Traffic ...........................................................74
        3.3.1  Gasoline-Engined Automobiles ...................................74
            3.3.1.1  Type and Amount of PAH Emitted....................74
            3.3.1.2  Variation of Automobile Emissions ..................74
        3.3.2  Diesel-Engined Automobiles......................................77
            3.3.2.1  Type and Amount of PAH Emitted....................77
            3.3.2.2  Carcinogenic Activity of Diesel Exhaust Condensate ....77

    3.4  Cigarettes...............................................................77
        3.4.1  Type and Amount of PAH Emitted .............................77
        3.4.2  Dependency of the PAH Amount on the Tobacco Brand .........77
        3.4.3  Intake of PAH by "Breathing-In" of Cigarette Smoke...........83
        3.4.4  Carcinogenic Effect of Cigarette Smoke .......................83

3.5—3.8  Environmental Contamination by PAH .............................84

    3.5  Air — F. Pott ...........................................................84
        3.5.1  Size Distribution of PAH-Containing Particles .................84
        3.5.2  Sampling of PAH-Containing Particles..........................84
        3.5.3  Types of PAH, PAH Profiles ....................................86
        3.5.4  An Index for the Carcinogenic Potency of Airborne PAH .......90
        3.5.5  The Range of PAH Concentration ...............................93

    3.6  Soil — Sediments — Crude Oil — G. Grimmer .....................101
        3.6.1  Soil Investigations ...............................................101
            3.6.1.1  Type of PAH ........................................101
            3.6.1.2  Range of PAH Concentration ......................102
        3.6.2  Sediment Investigations .........................................102
            3.6.2.1  Type and Amount of PAH..........................102
            3.6.2.2  Range of PAH Concentration ......................102

3.6.3 Investigation of Crude Oil Samples........................104
    3.6.3.1 Type and Amount of PAH......................104
    3.6.3.2 Range of PAH Concentration...................105

3.7 Water — Sewage Sludge — G. Grimmer............................105
  3.7.1 Water.................................................105
    3.7.1.1 Type and Amount of PAH......................105
    3.7.1.2 Range of PAH Concentration...................105
  3.7.2 Sewage Sludge .......................................105
    3.7.2.1 Type and Amount of PAH......................105
    3.7.2.2 Range of PAH Concentration...................106

3.8 Foodstuffs — G. Grimmer......................................109
  3.8.1 Smoked Foodstuffs....................................109
    3.8.1.1 Type and Amount of PAH......................109
    3.8.1.2 Range of PAH Concentration...................109
  3.8.2 Food Contaminated with Airborne Particulate Matter...111
    3.8.2.1 Type and Amount of PAH......................111
    3.8.2.2 Range of PAH Concentration...................111
  3.8.3 Smoke-Dried Foodstuffs...............................112

3.9 A Chemical Analytical Index for the Assessment of Carcinogenic Effects of Emissions and Environmental Samples — G. Grimmer ......113

3.10 Summary — G. Grimmer ........................................123
  3.10.1 Benzo[a]pyrene as an Index Substance of the Sum of All PAH.........................................123
  3.10.2 The Meaning of the Term "Guide Substance" ........123

References......................................................124

## 3.0 PROCESSES DURING WHICH PAH ARE FORMED

### G. Grimmer

As shown in Chapter 2, PAH are especially formed by the incomplete combustion or pyrolysis (with exclusion of air) of organic material containing carbon and hydrogen. During this thermal decomposition, in the presence of air or nitrogen, the same carcinogenic and noncarcinogenic PAH are formed. This is demonstrated, for instance, by comparing the burning of hard coal in a model incinerator at 1000°C with air (Figure 12) and with nitrogen (Figure 13).

The thermal decompositon of other types of material such as polystyrene, polypropylene, polyvinyl chloride (PVC), cellulose, cellulose acetate, polyamide, etc. yields the same PAH. When relating the PAH formed to 1 g of burned or pyrolyzed material, it becomes evident, however, that different types of material produce different amounts of PAH. In Figure 14 the standardized gas chromatograms of the formed PAH fraction (four-, five-, six-, and seven-ring compounds: chrysene to coronene) are compared with each other.

Polyvinyl chloride, which at low temperatures (about 200°C) cleaves off hydrochloric acid completely, also produces the usual PAH profile under the given conditions

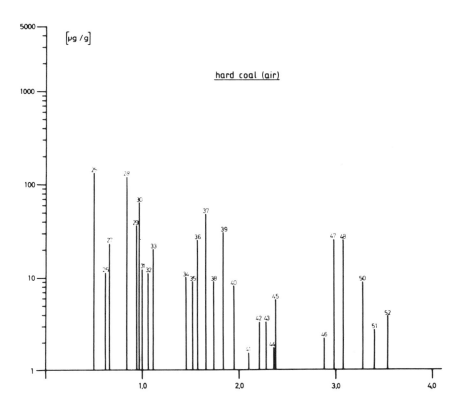

FIGURE 12. PAH profile obtained by combustion of hard coal at 1000°C in an air flow. Standardized gas chromatogram (semilogarithmic), abbreviations according to Table 13. (25) TRI + - CHR, (28) BF, (29) BeP, (30) BaP, (31) PER, (36) IF + DBa,hA (Dibenz[a,h]anthracene), (37) IP, (39) BghiP, (40) ANT, (47) M 302, (48) M 302, (52) COR + M302 (μg PAH per gram burned coal).

of decomposition. Consequently, chlorine-containing PAH are not formed during the combustion of PVC, only the characteristic PAH. On the other hand, metallurgical coke or polytetrafluoroethylene (PTFE) produce only small or nondetectable amounts of PAH. This is due to the fact that both materials do not contain any hydrogen. The PAH formed by thermal decomposition or incomplete combustion of different types of organic material are the same in each case and can be summed up in the following five groups:

1. M 228 (chrysene etc.), M 240 (methylene derivatives), M 242 (methyl derivatives of 228)
2. M 252 (benzofluoranthenes, BeP, BaP, perylene), M 266 (methyl derivatives of M 228)
3. M 276 (indenofluoranthene, indenopyrene, benzo[ghi]perylene, anthanthrene
4. M 278 (dibenzanthracene, picene), M 290 (methyl M 276)
5. M 302 (dibenzofluoranthenes, dibenzopyrenes), M 300 (coronene)

Even under different conditions of decomposition, such as in nitrogen at temperatures between 700 and 1000°C, or at a decomposition temperature of 700°C in air (incomplete combustion at 700°C), similar PAH profiles are formed by different types of material.

Nitrogen or sulfur compounds produce carbazole or thiophene derivatives, respectively, during thermal decomposition. Thus, for instance, hard coal (N and S) or poly-

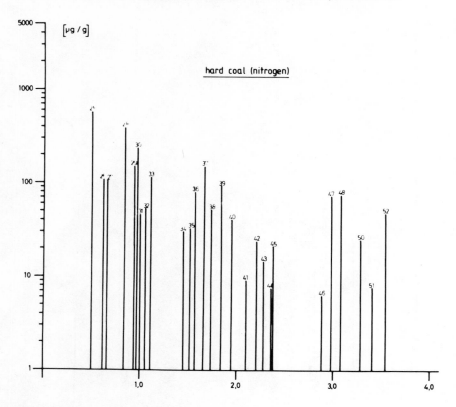

FIGURE 13. PAH profile obtained by pyrolysis of hard coal at 1000°C in a nitrogen flow, standardized GC (as in Figure 12).

amide (N) are materials which form heterocyclic compounds beside the usual PAH and can be used for identifying the source of emission.

The similarity of PAH profiles obtained during the thermal decomposition of different organic materials is probably due to a radical mechanism of formation (Chapter 2) which is characterized in that $C_2$ and, to a lesser degree, $C_3$ fragments produce thermodynamically stable compounds.

At lower combustion temperatures — for instance, when smoking cigarettes — other alkyl-substituted PAH are formed. These originate from precursors which are already present in unburned tobacco.

## 3.1—3.4 THE MOST FREQUENT SOURCES OF EMISSION

### G. Grimmer

At present it is not possible to list a total inventory of all PAH emitted by the most frequent sources of emission. Such an inventory exists for only few sources. So far estimates on the percentage of individual groups of emitters in the total emission have only been carried out for benzo[a]pyrene in the U.S. It is doubtful however, whether the importance attached to the percentage of BaP really corresponds to the carcinogenic hazard arising from these groups of emitters (Table 20).

These doubts are based on several findings. When investigating the carcinogenic effect of noxious substances from these condensates in appropriate model animal experiments, it can be shown that BaP accounts for only a small proportion of the total

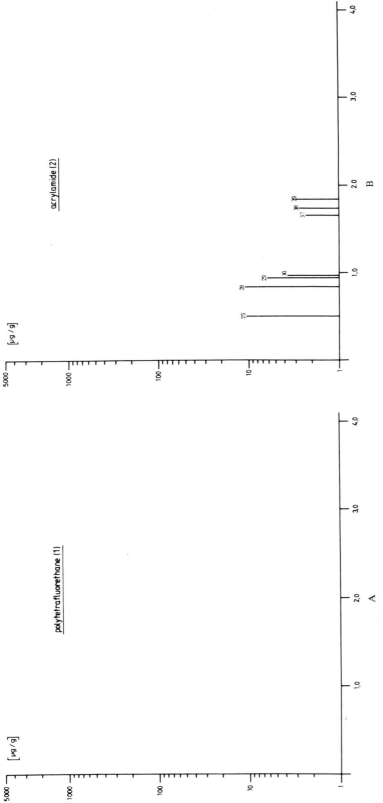

FIGURE 14. Comparison of PAH profiles obtained from polytetrafluorethane (A), acrylamide (B), polyamide (C), cellulose (D), cellulose acetate (E), polyvinylchloride (F), polypropylene (G), and polystyrene (H) burned at 1000°C in air (numbers as in Figures 12 and 13).

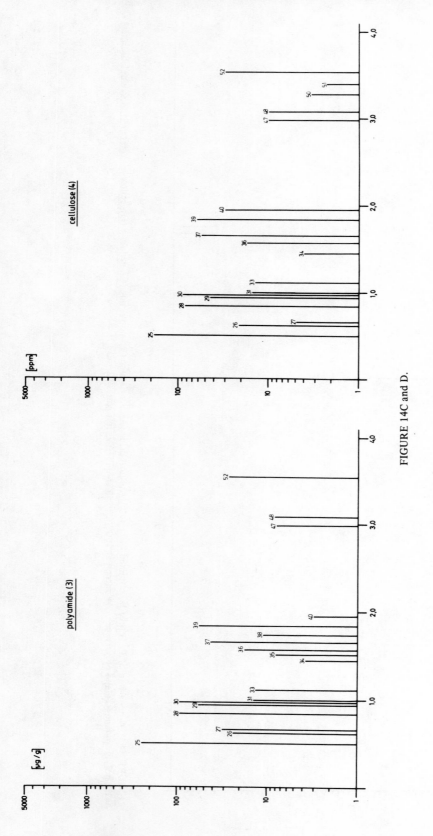

FIGURE 14C and D.

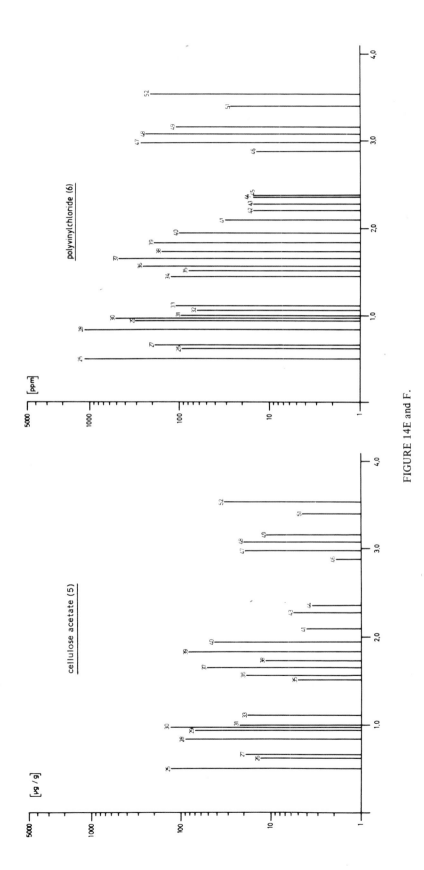

FIGURE 14E and F.

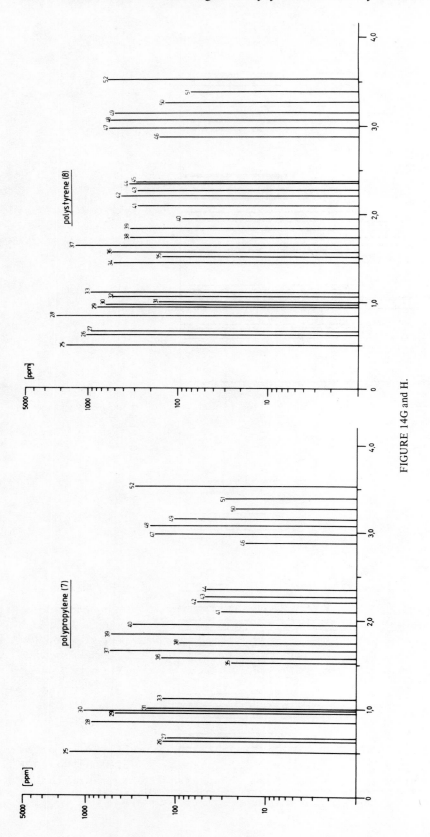

FIGURE 14G and H.

Table 20
ESTIMATED BENZO[a]PYRENE EMISSION IN THE U.S.

| | Tons/year | Ref. |
|---|---|---|
| Automotive exhaust | | 3b,51 |
| Gasoline-powered | | |
| Automobiles | 10.0 | |
| Trucks | 12.0 | |
| Diesel fuel-powered | | |
| Trucks and buses | 0.4 | |
| Total (percentage) | 22 (1.7%) | |
| Heating | | 3a,49a,51,57a,69a,70a, 83a,91 |
| Coal | | |
| Hand-stoked residential furnaces | 420 | |
| Intermediate units (chain-grate and spreader-stokers) | 10 | |
| Coal-fired steam power plants | 1 | |
| Oil | | |
| Low-pressure air-atomized and others | 2 | |
| Gas | 2 | |
| Wood | 40 | |
| Total (percentage) | 475 (38%) | |
| Refuse-burning | | |
| Commercial, industrial, institutional and apartment refuse-burning | 33 | 83a |
| Open burning | | 69a,70a,83a,91 |
| Forest and agricultural | 140 | |
| Vehicle disposal | 50 | 3a,49a |
| Coal refuse fires (10) | 340 | 67a |
| Total (percentage) | 563 (45%) | |
| Industrial plants | | |
| Petroleum | 6 | 51,75a |
| Asphalt/air-blowing | 1 | |
| Coke production (1.8 g BaP per ton) | 192 | 78a |
| Total (percentage) | 200 (16%) | |
| Total (tons/year) | 1260 | |

effect. Furthermore, the crucial point is not the total amount of PAH formed in West Germany, but only the amount of noxious material in the ambient air of a man. The local composition does not necessarily correspond to the global distribution of this noxious matter, but is determined by emission sources in the vicinity. The following sections describe the most frequent sources of emission, as far as their PAH emissions have been investigated.

## 3.1 DOMESTIC HEATING

In 1976 the consumption of energy in commercial establishments as well as in domestic heating in West Germany amounted to 73 million tons of coal equivalent (CE). Since 1960 the percentage in domestic heating of individual fuels has changed considerably (Table 21).[83]

Table 21
CONSUMPTION OF ENERGY IN DOMESTIC HEATING
($10^6$ CE)

|  | 1960 | 1965 | 1970 | 1975 | 1980 | 1985 | 1990 |
|---|---|---|---|---|---|---|---|
| Coal | 16.8 | 19.9 | 13.5 | 6.1 | 2.5 | 1.2 | — |
| Oil fuel EL | 4.4 | 13.7 | 34.5 | 42.3 | 46.9 | 48.7 | 48.6 |
| Gas | 1.3 | 2.4 | 4.5 | 8.8 | 11.9 | 14.7 | 16.6 |
| Electricity | 0.1 | 0.4 | 0.9 | 1.4 | 2.5 | 3.6 | 4.3 |
| Distant heating | 0.2 | 0.9 | 1.6 | 3.0 | 4.8 | 7.3 | 9.9 |
| Total | 22.8 | 37.3 | 55.0 | 61.6 | 68.6 | 75.5 | 79.4 |

*Note:* CE = coal equivalent. EL = extra-light.

Table 22
PERCENTAGE OF APARTMENTS HEATED BY
VARIOUS METHODS (END OF 1974)

|  | Coal | Fuel oil EL | Gas | Electricity | Distant heat |
|---|---|---|---|---|---|
| Schleswig-Holstein and Hamburg | 18 | 55 | 11 | 6 | 10 |
| Lower Saxony, Bremen | 19 | 54 | 17 | 3 | 7 |
| North Rhine-Westphalia | 29 | 41 | 17 | 9 | 4 |
| Hesse | 15 | 58 | 15 | 5 | 7 |
| Rhineland-Palatinate and Saarland | 20 | 63 | 12 | 4 | 1 |
| Baden-Württemberg | 20 | 59 | 15 | 5 | 1 |
| Bavaria | 19 | 63 | 10 | 3 | 5 |
| Berlin | 49 | 32 | 4 | 4 | 11 |
| West Germany | 23 | 53 | 15 | 5 | 5 |

At present coal-fired domestic heating accounts for about 8 to 9% while oil-fired heating installations produce about 70% of the total heat energy. Heating installations operated with gas, electric current, and distant energy emit practically no PAH at the site of heating, whereas coal- or briquet-fired stoves produce by far the largest proportion of the total PAH emission. Table 22 shows the regional distribution in percent of coal, fuel oil EL, natural gas, electric current, and distant heating for domestic use at the end of 1974.[83] The percentage of coal-heated apartments is surprisingly high in Berlin. Table 23 gives a survey of the consumption of fossil fuel types.

### 3.1.1 Domestic Heating with Coal Type and Amount of PAH Emitted

There are no data on the total PAH profile of flue gas particles emitted by different types of stoves. The emissions of various fuels such as anthracite nuts, lignite briquets, anthracite, broken coke, and other special coals (Anzit, Extrazit) in a universal slow-combustion stove (according to the German industrial standard DIN 18890) were compared with each other. Table 24 gives the concentrations of aerosol and 12 PAH in exhaust gas emitted by combustion of these fuels.[15]

The concentrations of PAH emitted per kilogram of fuel were not calculated in this paper; therefore it is not possible to give even a rough estimate of the amount of PAH produced annually by burning different fuels in different types of stoves in West Germany. As expected, the concentration of PAH emitted depends largely on the type of

Table 23
## CONSUMPTION OF LIQUID AND SOLID FUELS IN 1976 (IN 1000 t)

| | Public thermal power stations | Mining power plants | Other industrial thermal power plants | Distant heating plants | Refinery own consumption | Industry | Rail traffic | Road traffic | Air traffic | Inland waterway traffic | Households, small consumers | Military | Total consumption |
|---|---|---|---|---|---|---|---|---|---|---|---|---|---|
| Engine gasoline | | | | | | | | 20413 | | | 170 | 445 | 21020 |
| Aviation turbine fuel | | | | | | | | | 2099 | | | 46 | 2145 |
| Diesel oil | | | | | | | 556 | 7650 | | 890 | 1781 | 146 | 11023 |
| Fuel oil, light | 45 | 36 | 25 | 214 | | 5542 | 139 | | | | 42673 | 413 | 49087 |
| Fuel oil, heavy | 4597 | 11 | 2208 | 1074 | 3744 | 14987 | 77 | | | 7 | 66 | 224 | 26995 |
| Hard coal | 21455 | 5210 | 3905 | 2765 | | 3196 | 149 | | | 1 | 1383 | 1115 | 39169 |
| Hard coal coke | | | | 16 | | 12473 | 59 | | | | 2374 | 245 | 15160 |
| Hard coal briquets | | | | | | 4 | | | | | 1128 | 1 | 1133 |
| Crude lignite | 116767 | 1479 | 2880 | | | 1295 | | | | | 1 | 9 | 122431 |
| Lignite briquets | 748 | | 42 | | | 229 | 45 | | | | 3843 | 18 | 4925 |
| Woody lignite | 1263 | | 55 | | | 237 | | | | | 26 | | 1581 |
| Firing wood | | | 60 | | | 70 | | | | | 570 | | 700 |
| Firing peat | | | | | | | | | | | 238 | | 238 |
| Total | 144865 | 6736 | 9175 | 4069 | 3744 | 38033 | 1018 | 28063 | 2099 | 898 | 54253 | 2662 | 295615 |

Source: Arbeitsgemeinschaft Energiebilanzen, Düsseldorf, Steinstrasse 9-11.

Table 24
CONCENTRATION OF AIR SUSPENDED PARTICULATE MATTER AND
PAH IN FLUE GAS EMITTED BY COMBUSTION OF SOLID FUELS.
SAMPLING 10 m ABOVE THE CONNECTION BETWEEN STOVE AND
CHIMNEY[a]

| Fuel | ANB | BKB | ANT | BK | ANZ | EXT |
|---|---|---|---|---|---|---|
| Aerosol | $17.7 \times 10^6$ | $16.6 \times 10^6$ | $4.9 \times 10^6$ | $8.2 \times 10^6$ | $4.6 \times 10^6$ | $6.3 \times 10^6$ |
| Dilution % | 62.9 | 54.1 | 72.3 | 53.9 | 43.9 | 69.4 |
| FLU | $376.5 \times 10^3$ | $69.7 \times 10^3$ | 86.6 | a | 3904.0 | 28.3 |
| PYR | $86.6 \times 10^3$ | $15.4 \times 10^3$ | 24.9 | 1.7 | 129.7 | 4.4 |
| BaA | $159.0 \times 10^3$ | $20.9 \times 10^3$ | 69.3 | 8.7 | 1819.0 | 4.5 |
| CHR | $163.3 \times 10^3$ | $15.5 \times 10^3$ | 109.5 | 18.5 | 3490.0 | 17.7 |
| BbF | $205.1 \times 10^3$ | $14.5 \times 10^3$ | 163.5 | 71.2 | 1871.0 | 15.8 |
| BkF | $66.9 \times 10^3$ | $3.2 \times 10^3$ | 27.0 | 6.7 | 25.8 | 0.6 |
| BeP | $197.8 \times 10^3$ | $34.3 \times 10^3$ | 192.0 | 1152.0 | 677.2 | 32.1 |
| BaP | $64.9 \times 10^3$ | $5.0 \times 10^3$ | 11.8 | 10.2 | 2.2 | 0.5 |
| PER | $138.5 \times 10^3$ | $12.5 \times 10^3$ | 22.9 | 17.4 | 140.0 | a |
| DBahA | $22.6 \times 10^3$ | $10.2 \times 10^3$ | 12.5 | a | 53.8 | a |
| BghiP | $95.9 \times 10^3$ | $20.2 \times 10^3$ | 27.0 | 120.6 | 73.0 | 5.1 |
| COR | $40.5 \times 10^3$ | $6.3 \times 10^3$ | 7.6 | 5.8 | 11.1 | 1.6 |
| FLU-COR | $1618.0 \times 10^3$ | $227.4 \times 10^3$ | 754.6 | 1412 | $12.2 \times 10^3$ | 110.6 |

*Note:* Fuels: ANB = anthracite nuts, BKB = lignite briquets, ANT = anthracite, BK = broken coke, ANZ = "Anzit", EXT = "Extrazit"; a = not determined. PAH: FLU = fluoranthene, PYR = pyrene, BaA = benz[a]anthracene, CHR = chrysene, BbF = benzo[b]fluoranthene, BkF = benzo[k]fluoranthene, BeP = benzo[e]pyrene, BaP = benzo[a]pyrene, PER = perylene, DBahA = dibenz[a,h]anthracene, COR = Coronene.

[a] Concentration in exhaust gas in $ng/m^3$.

fuel. Thus, the combustion of "extrazit" yields 0.5 ng BaP per cubic meter, whereas anthracite nuts yield about 65 μg BaP per cubic meter (= 65000 ng BaP per cubic meter). These two extreme values of PAH emission differ from each other by a factor of more than 100,000. The interrelationship of the 12 investigated PAH also differs widely. Beine[4] also found a dependency of the amount of BaP emitted on the type of fuel used. He also gave evidence of a strong dependency of BaP emission of the type of stove. Thus the BaP emission from anthracite nuts is 0.15 mg BaP per gram of emitted matter in a universal slow-combustion stove, 10.3 in a down draft stove, and 8.5 in a continuous-combustion stove; this corresponds to about 2.2, 376, or 379 mg BaP per kilogram anthracite nuts, respectively.

Table 23 shows that in 1976 approximately 1,128,000 t of hard coal briquets were burned in domestic stoves. An extrapolation of the total amount of BaP formed, however, would be possible only if the amount of hard coal briquets burned in the stove types were known. Assuming the most favorable case, that the total amount of briquets were burned in low-emission, universal slow-combustion stoves, an annual BaP emission of 2000 kg would be derived for 1976. In the other two types of stove the amount of benzo[a]pyrene formed would be 170 times greater. Nowadays low smoke producing briquets are used, the BaP emission of which, according to random sampling, is considerably lower.

These figures are very rough estimates since the few experimental data obtained in only one stove do not permit such far-reaching conclusions. The domestic use of gasflame, gas, and fat coal is permitted nowadays only in universal slow-combustion stoves, whereas low smoke producing hard coal briquets may be burned in any type of stove.

### 3.1.2 Domestic Oil-Fired Heating (Extra-Light Oil)
*3.1.2.1 Type and Amount of PAH Emitted*

The amount of light fuel oil burned in households and by small consumers in West Germany was about 43 million tons in 1976. The consumption in 1979 of fuel oil combusted in heating installations with atomizing burners or in heating stoves with vaporizing burners (pot burners) is about 24,800 tons/year and 4,200,000 tons/year, respectively.

An inventory of PAH emitted by an oil-fired installation with a vaporizing pot burner (6500 kcal/hr) was compiled by Herlan et al.[54] The method used, high-resolution low-voltage mass spectrometry, does not permit the distinguishing of individual PAH in the mixture of isomers. Consequently, PAH emission profiles cannot be obtained because compounds of the same mass as, for example, mass 252 (BaP, BeP, perylene, benzo[b]fluoranthene, BjF, BkF, BaF, etc.) are recorded as a sum.

As expected, the emission is largely dependent on the efficiency factor ($\lambda$ = 5.9, 3.2, and 2.3). From previous experience, the collecting arrangement described by the authors for the quantitative collection of PAH-containing condensates on a hot glass-fiber filter (138°, 269°, and 419°C) cannot operate successfully.

Our own experiments (unpublished) with a commercially available burner-boiler combination (nominal capacity, 34 kW) produced low PAH emissions at the normal regulation fuel-air ratio. For instance, during intermittent operation (duration: 5 min), less than 0.05 µg BaP per kilogram of oil EL were emitted. The emitted amounts of coronene, indeno[1,2,3-cd]pyrene, benzofluoranthenes, or anthanthrene were of about the same magnitude, whereas two to three times as much chrysene and triphenylene were emitted. An oil-fired heating installation operating continuously without intervals emitted about half the amount of PAH. This relationship of individual PAH to BaP differs clearly from the relationships observed in emissions of different types of coal, as shown in Table 24.

There is no point, however in extrapolating the PAH emission data obtained in this burner-boiler combination to the total consumption of fuel oil EL in West Germany as long as these PAH amounts have not been confirmed by data of other burner-boiler combinations and different ratios of fuel and air.

## 3.2 THERMAL POWER PLANTS

### 3.2.1 Hard Coal-Fired Power Plants
*3.2.1.1 Type and Amount of PAH Emitted*

The majority of flue gas particles (flue dust) emitted by industrial power stations is precipitated in collectors (cyclones, electrostatic separators, etc.) The amount of flue dust actually emitted, therefore, amounts to only a few percent of the flue dust formed, unlike the emissions from domestic heating installations.

Investigations on the total PAH profile of flue ash emitted by thermal power stations are still lacking. Similarly, the content of individual PAH, such as BaP, has not yet been investigated in West Germany.

The amount of hard coal burned in coal power stations is about 21 million tons/year. In order to assess the environmental contamination by individual PAH such as BaP, we refer to earlier investigations carried out in the U.S.[17,25,50] Flue ash from a total of eight high-capacity power stations were investigated with respect to their content of BaP and, in some cases, also of ten other PAH.

These large power plants (electric generators, steam generators), which were fired predominantly with pulverized coal and achieved a capacity of 35 to 550 tons of steam per hour were equipped with fly ash precipitators (cyclones) or cyclones and electrostatic separators. According to the present state of technological development the 85 to 95% precipitation achieved with these collectors has to be considered as low.

In four of these power stations the BaP emission lay between 0.50 to 0.97 µg BaP per kilogram of hard coal.[50] Two installations emitted 0.5 µg BaP per kilogram at full load and 3.0 µg BaP per kilogram at partial load (75%).[17] Almost the same emission parameters, i.e., 3.5 µg BaP per kilogram and 0.5 µg BaP per kilogram of pulverized coal, were reported by Gerstle et al.[25] who investigated a tangential- and turbo-fired power plant boiler (430 tons and 68 tons of steam per hour, respectively).

The emission rates of other PAH are of the same magnitude. In comparison with domestic heating, the emitted amounts of PAH are small and can probably be neglected when assessing the total hazard arising from PAH emitted by different groups of emitters. An extrapolation of the above figures gives an annual emission of 10 to 70 kg BaP formed by combustion of $21 \times 10^9$ kg of hard coal per annum in coal-fired power plants in West Germany.

### 3.2.2 Lignite-Fired Installations

About 116 million tons of lignite are burned annually in thermal power stations in West Germany. Investigations on the PAH content of emitted flue dust have not yet been carried out in West Germany nor in other countries.

### 3.2.3 Fuel Oil (Heavy)-Fired Installations

In West Germany about 5 million tons of heavy oil are burned in thermal power plants.

*3.2.3.1 Type and Amount of PAH Emitted*

Investigations on the PAH profile of flue gas particles do not exist. Findings have been reported, however, on the emission of some PAH (BaP, pyrene, fluoranthene, and phenanthrene) by two oil-fired installations for the production of process heat (10 tons and 14 tons of steam per hour).[50] The amount of BaP emitted was <0.6 µg and 2.0 µg BaP per kilogram of heavy fuel oil, respectively.

## 3.3 MOTOR TRAFFIC

### 3.3.1 Gasoline-Engined Automobiles

At present the annual fuel consumption in West Germany is about 20 million tons (regular and premium grade gasoline).

*3.3.1.1 Type and Amount of PAH Emitted*

About 150 PAH were detected in the emissions of two series-produced automobiles, emitted during the so-called "Europa test" (ECE-Reglement 15, simulated urban traffic). By comparison with reference substances (GC retention time and mass spectra) 73 PAH could be identified. For 60 of the components the slightly different concentrations in the exhaust gases of the two automobiles were determined, and, similarly, the PAH content of the gasoline was analyzed. The gas-chromatographic profile analysis of PAH with four and more rings is shown in Figure 10, Chapter 2.[41]

*3.3.1.2 Variation of Automobile Emissions*

In collaboration with the Technical Control Board (TÜV, Essen), five automobiles each of the 20 most frequently registered automobile types were investigated.[38] The

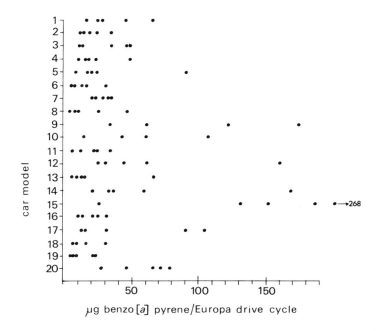

FIGURE 15. Amounts of BaP emitted during a so-called "Europa test". Comparison of 20 most frequently registered types of passenger cars in West Germany (5 automobiles per type).

range of 14 selected PAH emitted in the exhaust gases of these 20 automobile types was examined. Figure 15 shows a graphic presentation of the variation of BaP emission during a "Europa test".

**Amount of BaP emitted per annum by passenger cars** — On the basis of the average fuel consumption of 100 cars, (410 g/Europa test) and the simultaneously emitted BaP amount (41 μg BaP per Europa test), an annual emission of about 2 tons of BaP was calculated for West Geramny. This amount corresponds approximately to the 10 tons of BaP calculated for passenger cars in the U.S.[51]

**Carcinogenic components of condensate of automobile exhaust** — The cancer-inducing activity of the condensate from automobile emissions could be proven in various animal experiments (for details see Chapter 1).

**Influence of engine parameters, fuel, and lubricant on the amount of PAH in exhaust gas** — Summarizing, it can be stated that merely the fuel-air ratio and the engine temperature have a considerable influence on the amount of PAH emitted. There are partly contradictory experimental results on the impact of some other factors on the emitted amount of BaP and other PAH. A survey of papers published until 1971[71] and investigations carried out until 1974[38] show that single experiments do not suffice to clarify this complex problem. Reliable statements can be made only on the basis of a variety of experiments. According to our present knowledge it is, for instance, necessary to carry out at least 5 "Europa tests" under identical conditions in order to solve a single question.

**Ratio of gasoline-air mixture** — An excess of gasoline in the gasoline-air mixture above the stoichiometric ratio ($\lambda = 1.0$) considerably increases the rate of PAH emitted. When starting an engine, the working temperature has not yet been reached, and a high fuel concentration in the gasoline-air mixture is inevitable. Therefore, the PAH emission with respect to the combusted amount of gasoline is increased many times.

For example, a badly tuned standard engine (8% by volume of CO at idling) emits up to 0.4 mg BaP in the first Europa cycle (195 sec) during which the cold engine needs about 170 g of gasoline. This is a 40-fold increase in comparison with normal operation of a warm, well-adjusted engine.[46] The concentration of other PAH is also increased so that the profile of emitted PAH is hardly changed. In U.S.-built automobiles the emission of BaP was increased from 15 µg/A.G.* (0.9 to 1.4% by volume of CO) to 160 µg/A.G. (at 2.58% by volume of CO).[58] The authors also reported on the considerable effect of the working temperature of the engine. Thus at a "cold" 12-federal 7-mode test cycle about 1500 µg BaP per A.G. were emitted, whereas a "warm" cycle yielded only about 50 µg BaP per A.G. For analytical reasons no further PAH were investigated. The results obtainable with other PAH would probably correspond to those obtained with German automobiles.

**Percentage of aromatic compounds in gasoline** — The variation of the percentage of aromatic compounds in gasoline (0, 26, 32, 42, 48%) has a significant influence on the amount of PAH emitted (positive regression coefficient). To investigate this effect, a standard automobile (1.6 $\ell$), was driven in the "Europa test", using each of the five different gasolines four times. The average emission values of 18 most frequently occurring PAH (boiling range FLU-COR) were compared with each other. Increase in the content of aromatic compounds in gasoline was associated with a linearly increasing PAH emission which in the case of benzo[a]pyrene ranged from 4.7 to 28.4 µg BaP per Europa test (average gasoline consumption: 420 g).[48]

**Content of PAH in gasoline** — The effect of the PAH content on the profile of emitted PAH could not be detected in tests which were repeated several times with PAH-containing and PAH-free gasoline. This means that the PAH present in gasoline are combusted completely or retained in the lubricating oil. The PAH detected in exhaust gas are formed exclusively during combustion. Meyer et al.[68] who used gasoline prior to and after removal of PAH (bp >300°C) did not find a difference in the amount of the ten most frequent PAH emitted during combustion.[46]

**Dependency on operating period (age of engine)** — In an extensive study in which 140 "Europa tests" were evaluated ("Influences of operating mode and period of engine and lubricating oil on the emission of PAH by gasoline-engined automobiles", Project 110 sponsored by BMI-DGMK), two brand-new standard-type automobiles were driven in simulated urban traffic over 20,000 km and their exhaust gas pattern was measured. The emissions of both automobiles in the "Europa test" (6 tests with cold start performance and 12 subsequent tests with warm start performance) were first analyzed at the beginning of the test series, and then after 5,000, 10,000, and 20,000 km. The described 18 individual tests were carried out with fresh, as well as used, lubricating oil and the PAH emissions analyzed.

**Dependency on operating period: formation of deposits** — Under conditions of city traffic the PAH emission of a brand-new engine is only slightly increased after 10,000 km (e.g., by about 10% in automobile A). After high-speed driving (1000 km) following this long operating period in urban traffic, the PAH emission is reduced by about the same value.

**PAH concentration in lubricating oil** — In lubricating engine oil a considerable accumulation of PAH takes place with increasing time of operation. Under the described operating conditions a PAH profile is formed in used oil which is similar to the PAH profile of gasoline.

**PAH emission during operation with fresh/used oil** —The period of use and the corresponding accumulation of PAH in lubricating oil do not significantly affect PAH emission. A detailed presentation of experimental data is included in the research re-

---

* A.G. = American gallon.

port No. 110 of the German Society of Mineral Oil Science and Coal Chemistry (DGMK).[4a]

### 3.3.2 Diesel-Engined Automobiles

In West Germany about 8 million t of diesel fuel per annum are used at present by road traffic.

#### 3.3.2.1 Type and Amount of PAH Emitted

So far a systematic investigation of diesel exhaust condensate has not been reported. According to our own investigations[48] the profile of emitted PAH does not differ significantly from the PAH profile of gasoline engine exhaust. The concentration of PAH emitted during a "Europa test", however, is more than 50% lower in comparison with spark plug-ignited engines of the same capacity. This finding was obtained with the diesel engines of passenger cars, while large engines have not yet been investigated. It is therefore not possible to present data on the amount of BaP emitted annually by trucks.

Investigations carried out in the U.S. with trucks produced PAH emissions similar to those of comparable automobiles with gasoline engines.[71] Consequently, the amount of BaP emitted annually by diesel automobiles seems to be markedly lower than the amount produced by gasoline engines.

#### 3.3.2.2 Carcinogenic Activity of Diesel Exhaust Condensate

The carcinogenic effect of diesel exhaust condensate in an experimental animal model is described in Chapter 1. Investigations on interrelations between operating period, fuel composition, effect of lubricating oil and PAH emission of diesel engines have not been carried out so far.

## 3.4 CIGARETTES

In West Germany about 130 billion cigarettes are smoked per annum.

### 3.4.1 Type and Amount of PAH Emitted

The relative PAH composition (PAH profile) of the main stream smoke inhaled by smokers is largely similar to the profile of the side stream smoke formed by glowing between the smoking phases. This could be determined with machine-smoked cigarettes.[43] Figure 16 shows a comparison of both profiles.

In recent investigations, including gas chromatography and mass spectrometry, about 150 signals (peaks) could be characterized as PAH due to their typical fragmentation in mass spectrometry.[64,79] Also, 60 compounds could be identified by comparison with the corresponding reference substances. An additional UV spectrometric analysis of the resolved PAH mixture suggests that as well as the identified, frequently occurring base components, several hundred methyl and alkyl derivatives are contained in cigarette smoke.[56,79] The quantitative composition of this PAH profile is given in Table 25.[64]

### 3.4.2 Dependency of the PAH Amount on the Tobacco Brand

Comparative investigations of smoke condensate from different tobacco brands showed different concentrations of benzo[a]pyrene in condensates of these various cigarette brands. Thus condensate obtained from Maryland and Burley contained about 0.8 µg BaP per gram, whereas Orient yielded about 1.1 µg BaP per gram and Virginia 1.4 µg BaP per gram.[57] The content of benzo[a]pyrene nearly correlated to the carcinogenic effect of this condensate seen after topical application to mouse skin. Since the PAH profiles of these tobacco brands showed small but significant differences in their

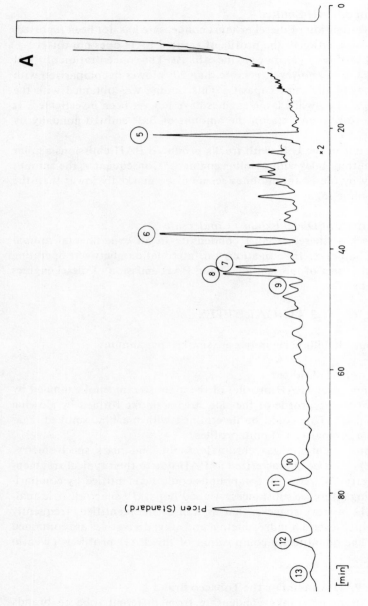

FIGURE 16A. Air sample collected in a smoking room (side stream smoke). Recording strip of the gas chromatogram of PAH (74 to 180 m$\ell$, chrysene to anthanthrene) in an air sample (7.6 m$^3$ obtained of 38 m$^3$, 30 cigarettes). Glass column (2 mm × 10 m) Silicone OV 101, 5% on Gas Chrom Q 100 to 120 mesh per square inch, isothermal 270°C, FID.

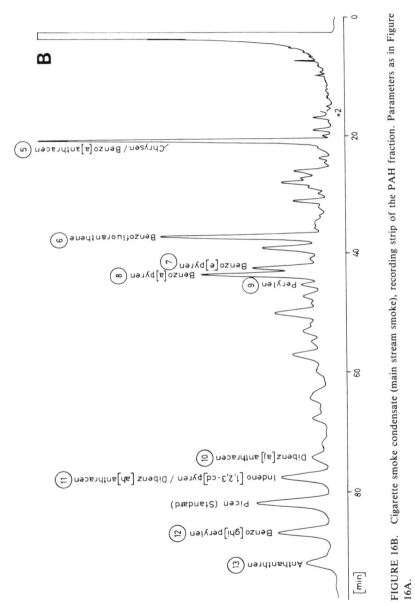

FIGURE 16B. Cigarette smoke condensate (main stream smoke), recording strip of the PAH fraction. Parameters as in Figure 16A.

## Table 25
## PAH IN TOBACCO SMOKE CONDENSATE

| Signal no. | Tobacco μg/100 cigarettes | Mol wt | Name |
|---|---|---|---|
| 1 | 0.3 | 131 | Methylindole |
| 2 | — | 145 | Ethylindole[a] |
| 3 | — | 168 | Dibenzofurane |
| 4 | 0.5 | 166 | Methylacenaphthylene |
| 5 | 0.3 | 180 | 2-Methylfluorene |
| 6 | 0.3 | 180 | 1-Methylfluorene |
| 7 | 8.5 | 178 | Phenanthrene |
| 8 | 2.3 | 178 | Anthracene |
| 9 | 0.1 | 196 | Ethylmethylbiphenyl[b] |
| 10 | — | 167 | Carbazole |
| 11 | 0.4 | 192 | — |
| 12 | 2.0 | 192 | 3-Methylphenanthrene |
| 13 | 5.6 | 192 | 2-Methylphenanthrene |
| 14 | 2.4 | 192 | 2-Methylanthracene |
| 15 | 2.4 | 190 | 4H-Cyclopenta[def]phenanthrene |
| 16 | 2.7 | 192 | 9-Methylphenanthrene |
| 17 | 3.2 | 192 | 1-Methylphenanthrene |
| 18 | — | 181 | Methylcarbazole |
| 19 | — | 181 | Methylcarbazole |
| 20 | — | 181 | Methylcarbazole |
| 21 | 1.6 | 204 | Methyl-4H-cyclopenta[def]phenanthrene |
| 22 | — | 181 | Methylcarbazole |
| 23 | 0.4 | 206 | Ethylphenanthrene or ethylanthracene[a] |
| 24 | 0.6 | 206 | Ethylphenanthrene or ethylanthracene[a] |
| 25 | 0.5 | 206 | Ethylphenanthrene or ethylanthracene[a] |
| 26 | 0.5 | 206 | Ethylphenanthrene or ethylanthracene[a] |
| 27 | 0.8 | 206 | Ethylphenanthrene or ethylanthracene[a] |
| 28 | 0.6 | 206 | Ethylphenanthrene or ethylanthracene[a] |
| 29 | 0.7 | 206 | Ethylphananthrene or ethylanthracene[a] |
| 30 | 1.6 | 206 | Ethylphenanthrene or ethylanthracene[a] |
| 31 | 1.8 | 206 | Ethylphenanthrene or ethylanthracene[a] |
| 32 | 1.9 | 206 | Ethylphenanthrene or ethylanthracene[a] |
| 33 | 8.3 | 202 | Fluoranthene |
| 34 | 1.6 | 206 | Ethylphenanthrene or ethylanthracene[a] |
| 35 | 1.2 | 202 | Benzacenaphthylene |
| 36 | 3.4 | 206 | Ethylphenanthrene or ethylanthracene[a] |
| 37 | 6.8 | 202 | Pyrene |
| 38 | 0.7 | 218 | Ethyl-4H-cyclopenta[def]phenanthrene[a] |
| 39 | 0.7 | 218 | Ethyl-4H-cyclopenta[def]phenanthrene[a] |
| 40 | — | 282 | p,p´-TDEE |
| 41 | 1.4 | 218 | Ethyl-4H-cyclopenta[def]phenanthrene[a] |
| 42 | 0.8 | 218 | Ethyl-4H-cyclopenta[def]phenanthrene[a] |
| 43 | 0.5 | 220 | Ethylmethylphenanthrene or ethylmethylanthracene[b] |
| 44 | 0.7 | 220, 218 | Ethylmethylphenanthrene or ethylmethylanthracene[b], ethyl-4H-cyclopenta[def]phenanthrene[a] |
| 45 | 1.6 | 218 | Ethyl-4H-cyclopenta[def]phenanthrene[a] |
| 46 | 4.6 | 216 | Methylfluoranthene |
| 47 | 1.8 | 216 | Methylfluoranthene |
| 48 | 3.6 | 216 | Methylfluoranthene |
| 49 | 4.9 | 216 | Benzo[a]fluorene |
| 50 | 5.5 | 216 | 2-Methylpyrene and benzo[b]fluorene |
| 51 | — | 318 | o,p´-TDE |
| 52 | 1.2 | 220 | Ethylmethylphenanthrene or ethylmethylanthracene[b] |
| 53 | 4.4 | 216 | 4-Methylpyrene |

## Table 25 (continued)
## PAH IN TOBACCO SMOKE CONDENSATE

| Signal no. | Tobacco μg/100 cigarettes | Mol wt | Name |
|---|---|---|---|
| 54 | 5.6 | 216 | 1-Methylpyrene |
| 55 | 0.9 | 216 | Methylfluoranthene |
| 56 | 0.3 | 216 | Methylfluoranthene |
| 57 | — | 318 | p,p'-TDE |
| 58 | 1.5 | 230 | Ethylfluoranthene or ethylpyrene[a] |
| 59 | 0.5 | 230 | Ethylflouranthene or ethylpyrene[a] |
| 60 | 0.9 | 230 | Ethylfluoranthene or ethylpyrene[a] |
| 61 | 1.0 | 230 | Ethylfluoranthene or ethylpyrene[a] |
| 62 | 2.4 | 230 | Ethylfluoranthene or ethylpyrene[a] |
| 63 | 2.4 | 230 | Ethylfluoranthene or ethylpyrene[a] |
| 64 | 2.7 | 230 | Ethylfluoranthene or ethylpyrene[a] |
| 65 | 1.8 | 230 | Ethylfluoranthene or ethylpyrene[a] |
| 66 | 1.6 | 230, 226 | Ethylfluoranthene or ethylpyrene[a], acefluoranthylene |
| 67 | 3.0 | 230 | Ethylfluoranthene or ethylpyrene[a] |
| 68 | 2.6 | 230, 226 | Ethylfluoranthene or ethylpyrene[a], acepyrylene |
| 69 | 1.4 | 230 | Ethylfluoranthene or ethylpyrene[a] |
| 70 | 1.7 | 230 | Ethylfluoranthene or ethylpyrene[a] |
| 71 | 1.3 | 230 | Ethylfluoranthene or ethylpyrene[a] |
| 72 | 0.4 | 226, 230 | Benzo[ghi]fluoranthene, ethylfluoranthene or ethylpyrene[a] |
| 73 | 2.6 | 228 | Benz[a]anthracene |
| 74 | 5.1 | 228 | Chrysene |
| 75 | 0.8 | 244 | Ethylmethylfluoranthene or ethylmethylpyrene[b] |
| 76 | 0.6 | 244 | Ethylmethylfluoranthene or ethylmethylpyrene[b] |
| 77 | 0.6 | 244 | Ethylmethylfluoranthene or ethylmethylpyrene[b] |
| 78 | 0.7 | 244 | Ethylmethylfluoranthene or ethylmethylpyrene[b] |
| 79 | 0.6 | 244 | Ethylmethylfluoranthene or ethylmethylpyrene[b] |
| 80 | 0.7 | 244 | Ethylmethylfluoranthene or ethylmethylpyrene[b] |
| 81 | 0.7 | 244 | Ethylmethylfluoranthene or ethylmethylpyrene[b] |
| 82 | 0.6 | 242 | Methylchrysene or methylbenz[a]anthracene |
| 83 | 0.5 | 242 | Methylchrysene or methylbenz[a]anthracene |
| 84 | 2.2 | 242 | Methylchrysene or methylbenz[a]anthracene |
| 85 | 2.2 | 242 | Methylchrysene or methylbenz[a]anthracene |
| 86 | 1.1 | 242 | Methylchrysene or methylbenz[a]anthracene |
| 87 | 0.7 | 242 | Methylchrysene or methylbenz[a]anthracene |
| 88 | 1.9 | 242 | Methylchrysene or methylbenz[a]anthracene |
| 89 | 2.9 | 242 | Methylchrysene or methylbenz[a]anthracene |
| 90 | — | — | — |
| 91 | 0.5 | 254 | Binaphthyl |
| 92 | 0.3 | 254 | Binaphthyl |
| 93 | 0.7 | 256 | Ethylchrysene or ethylbenz[a]anthracene[a] |
| 94 | 0.6 | 256 | Ethylchrysene or ethylbenz[a]anthracene[a] |
| 95 | 0.7 | 256 | Ethylchrysene or ethylbenz[a]anthracene[a] |
| 96 | 0.6 | 256 | Ethylchrysene or ethylbenz[a]anthracene[a] |
| 97 | 0.7 | 256 | Ethylchrysene or ethylbenz[a]anthracene[a] |
| 98 | 0.7 | 256 | Ethylchrysene or ethylbenz[a]anthracene[a] |
| 99 | 0.3 | 256 | Ethylchrysene or ethylbenz[a]anthracene[a] |
| 100 | 0.7 | 256 | Ethylchrysene or ethylbenz[a]anthracene[a] |
| 101 | 0.6 | 268 | Methylbinaphthyl |
| 102 | 0.4 | 268 | Methylbinaphthyl |
| 103 | 0.3 | 268 | Methylbinaphthyl |
| 104 | 0.3 | 268 | Methylbinaphthyl |
| 105 | 0.3 | 268 | Methylbinaphthyl |
| 106 | 0.6 | 270 | Ethylmethylchrysene or ethylmethylbenz[a]anthracene[b] |
| 107 | 0.4 | 270 | Ethylmethylchrysene or ethylmethylbenz[a]anthracene[b] |
| 108 | 0.4 | 282 | Ethylbinaphthyl[a] |

## Table 25 (continued)
## PAH IN TOBACCO SMOKE CONDENSATE

| Signal no. | Tobacco µg/100 cigarettes | Mol wt | Name |
|---|---|---|---|
| 109 | 0.3 | 282 | Ethylbinaphthyl[a] |
| 110 | 2.1 | 252 | Benzo[j]fluoranthene |
| 111 | 1.2 | 252 | Benzo[k]fluoranthene |
| 112 | 0.7 | 252 | Benzofluoranthene |
| 113 | 0.5 | 252 | Benzofluoranthene |
| 114 | 1.3 | 252 | Benzo[e]pyrene |
| 115 | 1.7 | 252 | Benzo[a]pyrene |
| 116 | — | 252 | Perylene |
| 117 | 0.2 | 266 | Methylbenzopyrene or methylbenzofluoranthene |
| 118 | 0.6 | 266 | Methylbenzopyrene or methylbenzofluoranthene |
| 119 | 0.5 | 266 | Methylbenzopyrene or methylbenzofluoranthene |
| 120 | 0.6 | 266 | Methylbenzopyrene or methylbenzofluoranthene |
| 121 | 0.6 | 266 | Methylbenzopyrene or methylbenzofluoranthene |
| 122 | 0.6 | 266 | Methylbenzopyrene or methylbenzofluoranthene |
| 123 | 0.7 | 266 | Methylbenzopyrene or methylbenzofluoranthene |
| 124 | 0.6 | 266 | Methylbenzopyrene or methylbenzofluoranthene |
| 125 | 0.5 | 266 | Methylbenzopyrene or methylbenzofluoranthene |
| 126 | 0.5 | 266 | Methylbenzopyrene or methylbenzofluoranthene |
| 127 | 0.3 | 266 | Methylbenzopyrene or methylbenzofluoranthene |
| 128 | 0.2 | 266 | Methylbenzopyrene or methylbenzofluoranthene |
| 129 | 0.4 | 266, 280 | Methylbenzopyrene, ethylbenzopyrene, or ethylbenzofluoranthene[a] |
| 130 | 0.5 | 280 | Ethylbenzopyrene or ethylbenzofluoranthene[a] |
| 131 | 0.3 | 280 | Ethylbenzopyrene or ethylbenzofluoranthene[a] |
| 132 | 0.3 | 280 | Ethylbenzopyrene or ethylbenzofluoranthene[a] |
| 133 | 0.3 | 280 | Ethylbenzopyrene or ethylbenzofluoranthene[a] |
| 134 | — | 276[c] | |
| 135 | — | 276,[c] 278 | Dibenz[a,j]anthracene |
| 136 | — | 276[c] | |
| 137 | 0.3 | 276[c] | |
| 138 | — | 276[c] | |
| 139 | 0.6 | 278 | Dibenz[a,h]anthracene or dibenz[a,c]anthracene |
| 140 | 0.2 | 276[c] | |
| 141 | 0.3 | 276 | Benzo[ghi]perylene |
| 142 | — | 276[c] | |
| 143 | — | 276 | Anthanthrene |
| 144 | — | 290[d] | |
| 145 | — | 290[d] | |
| 146 | — | 290[d] | |
| 147 | — | 290[d] | |
| 148 | — | 290[d] | |
| 149 | — | 290[d], 302 | Dibenzopyrene |
| 150 | — | 290[c], 302[c] | Dibenzopyrene |
| 151 | — | 290[d] | |
| 152 | — | 304, 306 | Diphenylacenaphthylene, quaterphenyl |
| 153 | — | 306 | Quaterphenyl |

[a] May also be dimethyl.
[b] May also be trimethyl or propyl.
[c] Compounds with a molecular weight of 276 may be: indeno[1,2,3-cd]pyrene, indeno[1,2,3-cd]fluoranthene, aceperylene, phenanthro[10,1,2,3-cdef]fluorene, acenaphth[1,2-a]acenaphthylene, dibenzo[b,mno]fluoranthene. Further possibilities are benzo derivatives of acepyrylene and acefluoranthylene.
[d] Compounds with a molecular weight of 290 are methyl derivatives of compounds with the molecular weight of 276.

Table 26
CONCENTRATION OF POLYCYCLIC AROMATIC HYDROCARBONS AFTER SMOKING OF 5 CIGARETTES PER HOUR IN A ROOM OF 36 m³ WITH SINGLE AIR REVERSAL

|  | Individual tests (ng/m³) | | | | | Medium (ng/m³) |
|---|---|---|---|---|---|---|
| Number of cigarettes | 41 | 41 | 41 | 42 | 0 | 41 |
| Size of air sample (m³) | 2 | 2 | 6 | 6 | 6 | |
| Fluoranthene | 109 | 116 | 80 | 89 | 50 | 99 |
| Pyrene | 62 | 84 | 53 | 65 | 26 | 66 |
| Benzo[a]fluorene | 34 | 35 | 42 | 44 | 6 | 39 |
| Benz[a]anthracene/chrysene | 107 | 101 | 103 | 90 | 31 | 100 |
| Benzo[b/j/k]fluoranthene | 35 | 29 | 39 | 35 | 5 | 35 |
| Benzo[e]pyrene | 17 | 16 | 21 | 18 | <2 | 18 |
| Benzo[a]pyrene | 18 | 16 | 28 | 25 | <3 | 22 |
| Perylene | 18 | 10 | 8 | 8 | 6 | 11 |
| Dibenz[a,j]anthracene | 0 | 0 | 6 | 6 | 0 | 6 |
| Indeno[1,2,3-cd]pyrene/dibenz[a,h]anthracene | 12 | 10 | 14 | 14 | 3 | 13 |
| Benzo[ghi]perylene | 18 | 18 | 15 | 15 | 5 | 17 |
| Anthranthrene | 0 | 0 | 3 | 3 | 0 | 3 |

composition, the quantitative assessment of further known carcinogenic PAH might bring about a better correlation to the biological activity.

### 3.4.3 Intake of PAH by "Breathing-In" of Cigarette Smoke

Greatly varying data exist on the additional hazard for man by benzo[a]pyrene, arising from breathing in cigarette smoke, i.e., not by inhalation of smoke but by passive intake via respiration of breathing air. To assess this hazard the concentration of BaP was measured directly in restaurants and lounges. The PAH detected in these rooms were of course derived from different emitters. It was not possible to assess the percentage of PAH formed by smoking tobacco products. Excluding this basic concentration produced by other emitters, maximum values of 22 ng BaP per cubic meter of air were measured.[42] The maximum tolerable concentration in the side stream smoke of cigarettes was reached when smoke from 2 to 3 cigarettes per hour strongly irritate the eyes of smokers in a workroom of 7.3 m²/person (= 18 m³/person). An air sample of 2 or 6 m³ was collected during a period of 8 hr in order to simulate passive smoking. Since only filtered outdoor air was sucked into the room, the collected air sample showed the typical PAH profile of cigarette smoke. A comparison with this profile of the same room after ventilation yielded the differences shown in Table 26. Only main components are evaluated.

The BaP concentration measured on the "nonsmoking day" lay below 3 ng/m³, whereas 22 ng BaP per cubic meter were calculated as the average value for the four experimental days during which two smokers smoked five cigarettes per hour for 8 hr. The room size was 36 m³ (single air reversal per hour). In general, the extreme conditions of the experiment under which smoker's eyes are irritated to such an extent that they water, are not reached in closed rooms. The measured concentration of 22 ng BaP per cubic meter has to be considered as a maximum BaP concentration attainable by smoking. In practice nobody would tolerate this concentration.

### 3.4.4 Carcinogenic Effect of Cigarette Smoke

The carcinogenic effect of cigarette smoke and the effect of PAH-containing and PAH-free fractions are described in Chapter 1.

The carcinogenic hazard arising for man as a result of air pollution in industrial areas and big cities, or by side-stream smoke of glowing cigarettes in closed rooms, can certainly not be described by comparing merely the BaP concentrations. In both "types" of air there are further carcinogenic substances. The percentage of BaP in the total carcinogenic activity is probably not identical for these two "types" of air.

## 3.5—3.8 ENVIRONMENTAL CONTAMINATION BY PAH

## 3.5 AIR

### F. Pott

### 3.5.1 Size Distribution of PAH-Containing Particles

PAH formed during incomplete combustion are mostly adsorbed on soot particles and emitted with them. In the atmosphere, however, PAH are hardly present as gases. This amount increases parallel with temperature rise and is higher for PAH with a high vapor pressure (PAH with three to four rings) than for PAH with five or more rings.

Only particles with an aerodynamic diameter below 10 μm can be deposited in bronchi and alveoli (see Chapter 4.1). Consequently, it is important to know the size of particles on which PAH are predominantly adsorbed. A survey of presently available data is given in Table 27. Generally speaking, these data suggest that almost the total amount of BaP adsorbed on particles in the atmosphere is respirable. Numerous PAH were also found on deposited particulate matter.[53] On closer examination, it can be said that there are essential differences in the distribution of PAH on particles of different sizes. They depend on the site of sampling, influences of certain emitters, the time of the year type of PAH, and possibly the composition of particle nuclei. These largely unknown conditions can produce different results in different sampling methods. They are, furthermore, of importance for the assessment of PAH deposition in the different sections of the respiratory tract (see Chapter 4.1).

### 3.5.2 Sampling of PAH-Containing Particles

An optimal, quantitatively reproducible method for the collection of airborne PAH meets with methodical difficulties due, especially, to three factors:

1. Their different vapor pressures
2. Their different stability
3. Their still insufficiently known distribution on particles of varying size

In routine sampling it is only possible to collect PAH adsorbed on particles. This should be carried out by means of filters with a very high rate of precipitation (filters for atmospheric particulate matter, special grade S), since a large part of the PAH is adsorbed onto very small particles. But even when using extremely fine-pored filters, the problem arises that the air passes over the growing number of deposited particles and blows off gaseous PAH which are in equilibrium with the solid matter on the filter.

This "blowing-off effect" does not seem to be relevant in the case of 24-hr sampling of benzo[a]pyrene in winter. Ester-impregnated filters did not have a higher collecting capacity than simple glass-fiber filters. When arranging several filters in series, the

## Table 27
## PAH CONTENT OF AIRBORNE PARTICLES OF DIFFERENT SIZE

| Origin of dust sample | Particle size (µm) | PAH content in particle size fraction (%) | PAH | Ref. |
|---|---|---|---|---|
| Pittsburgh | "respirable" | >75 | 6 PAH | 19 |
|  |  | 95 | BaP |  |
| Green Bay | 0.01—1.1 | 29 | Bap |  |
| Wisconsin | 1.1—2.0 | 11 | BaP |  |
| (downtown) | 2.0—3.3 | 12 | BaP |  |
|  | 3.3—7.0 | 20 | BaP |  |
|  | >7.0 | 27 | BaP |  |
| Budapest | <0.14 | su. 42 | BaP | 59 |
|  |  | wi. 55 | BaP |  |
|  | <0.47 | su. 65 | BaP |  |
|  |  | wi. 83 | BaP |  |
|  | <1.2 | su. 79 | BaP |  |
|  |  | wi. 92 | BaP |  |
|  | <3.8 | su. 93 | BaP |  |
|  |  | wi. 98 | BaP |  |
| Town in | 0.01—1.1 | 60 | BaP + BkF | 2 |
| Canada | 1.1—7.0 | 25 | BaP + BkF |  |
|  | >7.0 | 15 | BaP + BkF |  |
| Fine coke dust | "respirable" | <50 | BaP | 66 |
| Toronto, Canada |  |  |  | 72 |
| York, summer | <1.0; <3.0 | 14; 62 | BaP |  |
|  |  | 6; 56 | BkF |  |
|  |  | 8; 56 | Per |  |
| York, winter | <1.0<3.0 | 30; 76 | BaP |  |
|  |  | 22; 72 | BkF |  |
|  |  | 28; 78 | Per |  |
| Evans, summer | <1.0; <3.0 | 40; 70 | BaP |  |
|  |  | 40; 71 | BkF |  |
|  |  | 29; 68 | Per |  |
| MacDonald-Cartier, summer | <1.0; <3.0 | 56; 78 | BaP |  |
|  |  | 58; 79 | BkF |  |
|  |  | 54; 74 | Per |  |
| Bombay, India |  |  |  | 69 |
| near coking plant | <1.0 | 64 | BaP |  |
| heavy road traffic | <1.0 | 79 | BaP |  |

*Note:* BaP = benzo[a]pyrene, BkF = benzo[k]fluoranthene, Per = perylene.

second impregnated filter contained only 1% — and in one case 5% — of the total BaP amount of both filters, taking three measurements.[87]

During longer-lasting sampling, for 4 and 12 weeks without exchange of filters, e.g., when collecting particulate matter needed for animal experiments, the loss of PAH, especially of PAH with three or four rings, becomes more evident (Figure 17). The filter used for 12 weeks still contained about 65 to 80% of the sum of each PAH, ranging from benz[a]anthracene to coronene; this sum being calculated by analyzing 3 filters used at the same time for 4 weeks each (Figure 17B). Filter arrangements with impregnated second and third filters increased and collected amount only of those PAH which have a lower molecular weight than benz[a]anthracene (Figure 18).

Besides the "blowing-off effect", chemical reactions of PAH may also take place on filters. It is not known to what extent these transformations are also effected in clean air by the time factor alone, as well as by other air pollutants such as $SO_3$, $O_3$, or $NO_2$.

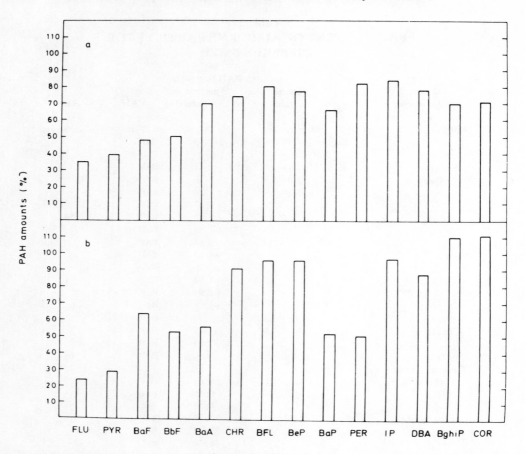

FIGURE 17. PAH amounts on nonimpregnated filters during a collection period of (a) 12 weeks and (b) 4 weeks as a percentage of the total PAH amount [= 100%] during a collection period of (a) 3 × 4 weeks and (b) 4 × 1 weeks.[60] Abbreviations: FLU = fluoranthene, PYR = pyrene, BaF = benzo[a]fluorene, BbF = benzo[b]fluorene, BaA = benz[a]anthracene, CHR = chrysene, BFL = benzofluoranthenes, BeP = benzo[e]pyrene, BaP = benzo[a]pyrene, PER = perylene, IP = indeno[1,2,3-cd]pyrene, DBA = dibenzanthracenes, BghiP = benzo[ghi]perylene, COR = coronene.

Further efforts will be needed to clarify these problems in order to create the necessary preconditions for an optimal comparative PAH sampling, and a correct evaluation of PAH data obtained from air samples.

### 3.5.3 Types of PAH, PAH Profiles

During qualitative analysis of the PAH content of airborne particulate matter extracts, the masses of 124 peaks in the gas chromatogram were identified by Lao et al.;[62] 122 peaks were identified by Lee et al.[65] They determined the complete structure of 29 of these substances. An inventory of PAH in some samples of the atmosphere of West Germany was carried out by Grimmer et al.[47] (Figure 19; see also Figure 24) and König[61a] (Table 28).

PAH, especially those with a higher molecular weight, i.e., starting with benz[a]anthracene, have to be considered as potential carcinogens, unless the contrary has been proved.

Since 1974, atmospheric particulate matter has been collected on glass-fiber filters at several sampling sites in the western part of West Germany, and 8 PAH were deter-

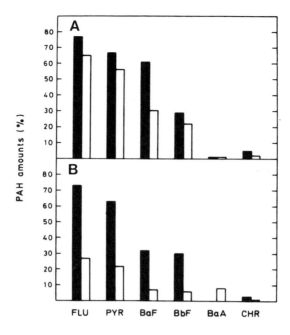

FIGURE 18. PAH amounts on impregnated no. 2 and no. 3 filters as a percentage of the total PAH (no. 1, no. 2 and no. 3 filter; ≅ 100%) in the case of nonimpregnated no. 1 filters during varying collection periods.[60] 18A. ■ = 12 week filters; □ = sum of three 4 week filters. 18B. ■ = 4 week filters; □ = sum of four 1 week filters. For abbreviations, see Figure 17.

mined by means of thin-layer chromatography and direct fluorimetric measurement.[86,88] The sampling sites cannot be considered as representative for all types of residential areas; however, they were located in areas with different types of emission sources: industrial complexes, urban residential districts adjacent to major traffic arteries, and rural districts. Figure 20 demonstrates an example of five locations for one collecting period. A total of 54 profiles were measured; 32 of which were determined after a sampling period of 24 hr, while 22 profiles were obtained after a sampling period of more than 2 months. Figure 21 shows the relative amount of each of the 8 PAH based on the total amount of all of them. The rather small standard deviation of the average values indicates a similarity of the PAH profiles. This similarity contrasts with the variations of the PAH profiles of the emissions from burning different species of coal, from an oil stove, and from an automobile engine (Figure 22). Obviously, the different PAH emission profiles mix and change in the atmosphere to form similar PAH profiles.

The thin-layer chromatography method permits a simultaneous quantitative PAH determination for only a small number of PAH. To obtain information on other PAH, particulate matter was collected in five cities (Bremen, Duisburg, Frankfurt, Karlsruhe, Muenchen) during the winter of 1978/79 and was analyzed by means of the gas chromatography/mass spectrometric techniques.[60a] The amounts of 134 PAH, relative to chrysene, are shown in Figure 23; it illustrates the similarity of the five comprehensive PAH profiles.

Even after the short collecting time of only 1 hr, Grimmer[47] found similar PAH profiles in airborne particles analyzed by gas chromatography (Figure 24). The three

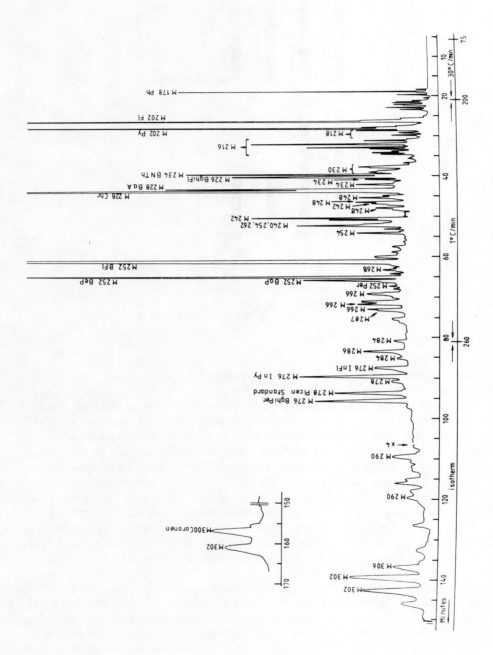

FIGURE 19. PAH profile of airborne particulate matter collected in a German city (Düsseldorf). Recording strip of a glass capillary gas chromatogram (0.27 × 25 m), Silicone OV 101. Charging at 100°C, temperature program, from 260°C, isothermal.[47]

Table 28
PAH CONCENTRATIONS
IN THE ATMOSPHERE OF
DUISBURG CENTER;
PARTICLES COLLECTED
ON THREE FILTERS FOR 4
WEEKS FROM DECEMBER
1977 TO MARCH 1978

| PAH | Concentration (ng/m³) |
|---|---|
| Fluoranthene | 40.3 |
| Pyrene | 28.2 |
| Benzo[a]fluorene | 3.6 |
| Benzo[b]fluorene | 6.2 |
| Benz[a]anthracene | 10.1 |
| Chrysene | 27.5 |
| Benzofluoranthenes | 34.4 |
| Benzo[e]pyrene | 12.6 |
| Benzo[a]pyrene | 4.7 |
| Perylene | 5.1 |
| Indenopyrene | 7.3 |
| Dibenzanthracenes | 2.3 |
| Benzo[ghi]perylene | 8.3 |
| Coronene | 1.9 |

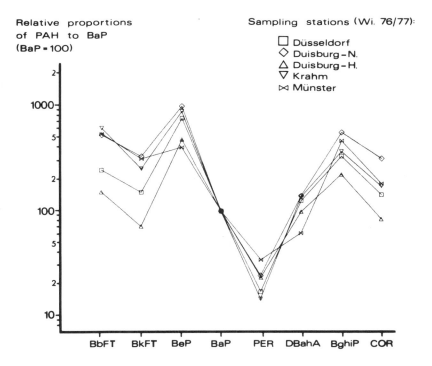

FIGURE 20. Profiles of PAH contained in atmospheric particulate matter collected at 5 sampling stations. Period: winter 1976/77.[86] Abbreviations: BbFT = benzo[b]fluoranthene, BkFT = benzo[k]fluoranthene, BeP = benzo[e]pyrene, BaP = benzo[a]pyrene, PER = perylene, DBahA = dibenz[a,h]anthracene, BghiP = benzo[ghi]perylene, COR = coronene.

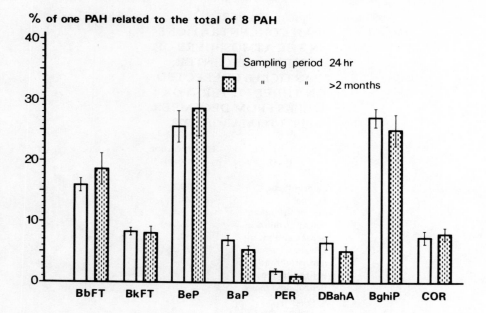

FIGURE 21. Percentage of one PAH related to the total of 8 PAH; average of 32 sampling periods of 24 hr and 22 of 2 to 4 months. (The average values characterize the "standard profile").[74] For abbreviations see Figure 20.

sampling sites were characterized by different emission sources. The differences between the results of the three working groups can be explained by differences in the methods used.

It is of interest to compare data from West Germany with results obtained in the U.S. Whereas Hoffmann and Wynder[56a] indicate in their compilation (see Table 29) that dibenz[a,h]anthracene could not be detected in the atmosphere in the U.S., the particulate matter collected in the western part of Germany contained this PAH at about the same level as BaP. This difference can be explained by the fact that in American cities exhaust gases originate mainly from combustion of oil and gasoline, which contain hardly any dibenz[a,h]anthracene (Figure 22).

### 3.5.4 An Index for the Carcinogenic Potency of Airborne PAH

As mentioned before, exhausts from incomplete combustion, as well as atmospheric particulate matter, contain much more than 100 PAH. The question arises to what extent the composition of a complex of PAH in a dust sample — the PAH profile — should be identified to give all necessary information for establishing a scale or index for the carcinogenic potency of airborne PAH. Hitherto, the prevailing idea was to assess the carcinogenicity of a PAH mixture in relation to the content of a single carcinogenic component, namely benzo[a]pyrene. However, the concentrations of benzo[a]pyrene in the air are frequently considered as insufficient for serving as an index for the carcinogenic potency of the total group of PAH. Such an index is one of the preconditions for setting an exposure limit for PAH.[73]

As shown by the three working groups, on the whole there are no great differences in the PAH profiles in various locations of West Germany. The rather different emission profiles are mixed and changed in the atmosphere to relatively uniform profiles of PAH. This statement should not be generalized for all locations and for all countries. However, from the view of preventive medicine, it seems to present a sufficiently

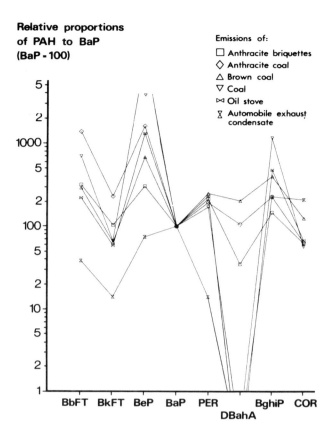

FIGURE 22. PAH profiles in exhaust from several species of coal, from oil stove[15] and from automobiles.[41] The values are expressed as a ratio to BaP (BaP = 100). For abbreviations see Figure 20.

solid base for developing a scale or index for the carcinogenic potency of airborne PAH.

Such an index has been elaborated as a base for discussion.[74] PAH units for characterizing the carcinogenic potency of airborne PAH are derived from the concentrations of five PAH in the atmosphere in combination with five evaluation factors. The evaluation factors were developed according to the average PAH values shown in Figure 22, which are based on the analytical results of a period of several years. The determination of five PAH will probably suffice to average the analytical errors of the individual PAH measurements or the deviations in the PAH profiles. Besides the traditional benzo[a]pyrene, four further PAH among those with the highest concentration were selected: benzo[b]fluoranthene, benzo[k]fluoranthene, benzo[e]pyrene, and benzo[ghi]perylene. The average concentrations of these five PAH constitute the standard profile. Since each PAH in this standard profile is in a fixed relation to the other four PAH, each PAH is of the same significance with respect to the evaluation. By the same token, the individual carcinogenic potency of any one of these PAH is immaterial, because the concentration of a noncarcinogenic PAH indicates the concentration of the others, and thus, also, that of a carcinogenic PAH. To establish the same statistical weight for each of the five PAH in the standard profile, special evaluation factors must be defined. They ensure that the high concentration of

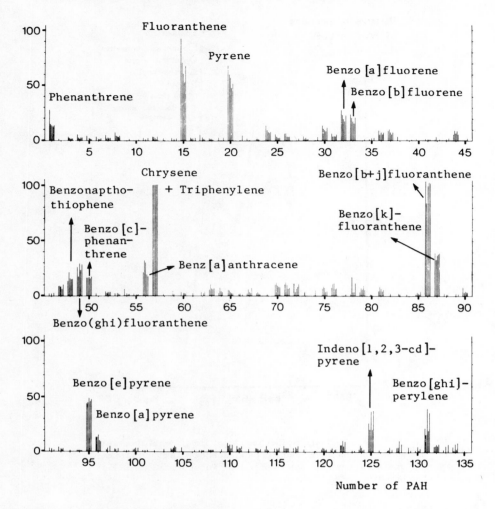

FIGURE 23. Relative proportions of 135 PAH (chrysene = 100) isolated from suspended particulate matter collected in five cities of the Federal Republic of Germany. The five peaks in each group are in the sequence Bremen, Duisburg, Frankfurt, Karlsruhe, München.[60a]

benzo[e]pyrene will have the same weight in the evaluation as the relatively low concentration of benzo[a]pyrene. The values of the five evaluation factors are derived from the percentage of the five PAH in the standard profiles (see Figure 25). The factors are defined in such a way that each of the five selected PAH in a standard profile contributes 20% to the PAH index.

As Figure 21 shows, the different periods of sampling have little influence on the profile. A distinct difference is seen in the case of benzo[a]pyrene due to its well-known instability. However, even this deficiency is averaged out in the evaluation, when other PAH are included.

In the case of data obtained by gas chromatography as given in Figure 23 and 24, the PAH selected would be different. Even the evaluation factors would change; however, the principle of establishing an index for the PAH level and the carcinogenic potency of PAH by averaging numerous similar PAH profiles may still be retained.

Contemporaneous with the discussion on adopting an index for the carcinogenic potency of airborne PAH further analytical work should be done in order to correct the evaluation factors.

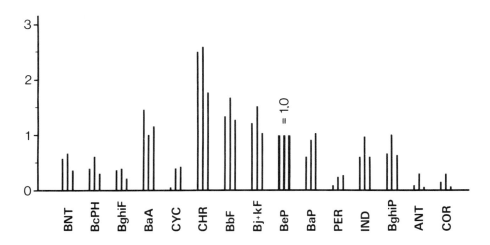

FIGURE 24.  PAH profiles of three residential areas; sampling of airborne particles in 1 hr. The three columns in each group are in the sequence (1) predominantly coal heated area, (2) predominantly oil heated area, (3) area close to coking plant.[47] Abbreviations: BNT = benzo[b]naphtho[2,1-d]thiophene, BcPH = benzo[c]phenanthrene, BghiF = benzo[ghi]fluoranthene, BaA = benz[a]anthracene, CYC = cyclopenta[cd]pyrene, CHR = chrysene, BbF = benzo[b]fluoranthene, Bj+kF = benzo[j]fluoranthene + Benzo[k]fluoranthene, BeP = benzo[e]pyrene, BaP = benzo[a]pyrene, PER = perylene, IND = indeno[1,2,3-cd]pyrene, BghiP = benzo[ghi]perylene, ANT = anthanthrene, COR = coronene.[47]

### 3.5.5 The Range of PAH-Concentration

Most published data on PAH concentrations in the atmosphere are restricted to BaP, and even these are incomplete. Sawicki[76] compiled a comprehensive survey of measured values of BaP in the air (Table 30). BaP values taken in Germany are reported in Table 31.

In addition to these data concerning the ambient air, there are informative data on particularly heavy pollution of the atmosphere (partly at the workplace) and of emissions[76] (Table 32).

A comparison of BaP concentrations, described by various authors, can be carried out only with great reservations because a large number of different dust-sampling procedures and analytical methods were used. As mentioned before, this means that the results may be modified considerably. Also of significance is the site at which the airborne particulate matter is collected: a single sampling station in a big city cannot provide representative data for the whole city. Therefore an apparent comparison of data from two cities is actually only an observation made at two sites in these cities. Thus the data compiled in Tables 27—29 are not comparable if the different sampling methods are not taken into account. Nevertheless, the present data on BaP concentrations permit the formation of some general principal conclusions: The BaP concentration in the atmosphere is

- Much higher in winter than in summer
- Much higher in urban communities than in rural areas
- Much higher in many European cities than in most North American cities

These three facts are dependent on different causes which can only be mentioned briefly in this chapter:

- Type of emitters (see Chapter 3.1)

## Table 29
## PAH CONCENTRATION (EXCLUDING BaP) IN THE ATMOSPHERE OF URBAN AREAS[56a]

| PAH | Concentration (ng/m³) | Urban area |
|---|---|---|
| Benzo[b]fluoranthene | 2.3—7.4 | U.S. cities (1958) |
| | 0.5—1.5 | Sidney (1962) |
| Benzo[j]fluoranthene | 0.8—4.4 | Detroit (1962-63) |
| Dibenzo[a,h]pyrene | | Presence questionable |
| Tribenzo[a,e,i]pyrene | | Presence questionable |
| Dibenzo[a,h]anthracene | 3.2—32.0 | German cities (1966); (not detected in U.S. towns) |
| Benzo[a]anthracene | 0.1—21.6 | U.S. cities (1958) |
| | 4.0 | Average of 100 U.S. cities (1958) |
| Chrysene | 1.5—13.3 | Cincinnati (1965) |
| | 1.3—11.6 | Detroit (1962—63) |
| Benzo[e]pyrene | 1.0—25.0 | 12 U.S. cities (1958) |
| | 5.0 | Average of U.S. cities (1958) |
| | ≤37.0 | London (1967—68) |
| | 5.0—8.0 | Oslo (1962—63) |
| Indeno[1,2,3-cd]pyrene | 1.5—8.2 | Detroit (1962—63) |
| Benzo[ghi]perylene | 2.0—35.0 | 12 U.S. cities (1958) |
| | 18.0—26.0 | Belfast (1961—62) |
| | 12.0—46.0 | London (1962) |
| Anthanthrene | 0.26 | Average of U.S. cities (1958) |
| | 2.0—6.0 | London (1962) |
| Coronene | 2.0 | Average of U.S. cities (1958) |
| | 4.0—20.0 | London (1962) |
| Perylene | 0.7 | Average of U.S. cities (1958) |
| Pyrene | traces - 35.0 | 12 U.S. cities (1958) |
| | 1.3—19.3 | Detroit (1962—63) |
| | 4.0 | Average in the U.S. (1958) |
| Fluoranthene | 0.9—15.0 | Detroit (1962—63) |

- Size of emitters
- Density of emitter localizations
- Height of exhaust emission
- Meteorology

The role of photolysis has not yet been elucidated. Although individual details have to be assessed cautiously because of the imponderable factors mentioned earlier, the high BaP concentration in the atmosphere of cities and at certain workplaces (see Chapter 3.1) can be correlated with the prevalence of coal heating as well as with the type of combustion. Changes in heating techniques resulted in a distinct reduction of BaP concentrations in London (Table 33) and the Ruhr district (Figure 26).

In London this development has begun after enactment of the Clean Air Act in 1956, i.e., much earlier than in the Ruhr district (West Germany). Results of recent measurements in Duisburg[60a,84] permit the anticipation that here too the annual average value has fallen below 10 ng/m³. Between 1958 and 1970 the average BaP concentration in the atmosphere of U.S. cities fell from 6.6 to 2 ng/m³ (Table 19). Only in rare cases were values exceeding 10 ng/m³ measured.[1]

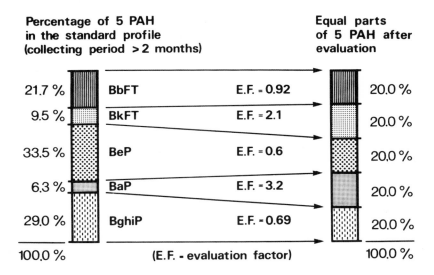

FIGURE 25. Definition of evaluation factors providing equal weight of five selected PAH in the standard profiles. The standard profile is derived from 22 determinations of PAH after a sampling period longer than 2 months (compare with Figure 21; for abbreviations see Figure 20). On account of these factors each of the selected PAH contributes 20% to the PAH index.[74]

Table 30
BENZO[a]PYRENE CONCENTRATIONS IN URBAN ATMOSPHERES[76] (ABRIDGED)

|  |  | Concentration[a] (ng/m$^3$) | |
| --- | --- | --- | --- |
| Country — city | Year | Winter | Summer |
| Australia |  |  |  |
| Sydney | 1962/63 | 8.0 | 0.8 |
| Belgium |  |  |  |
| Liege | 1958—62 | 110.0 | 15.0 |
| Canada, Ontario | 1961—62 |  |  |
| Sarnia |  | 3.5 | 1.6 |
| Windsor |  | 15.0 | 7.8 |
| Windsor |  | 15.0 | 7.8 |
| Chatham |  | 5.0 | 2.3 |
| London |  | 3.2 | 1.7 |
| Kitchener |  | 2.7 | 1.2 |
| Brantford |  | 5.2 | 2.2 |
| Hamilton |  | 9.4 | 5.7 |
| St. Catharines |  | 9.1 | 3.8 |
| Toronto |  | 5.4 | 6.4 |
| Oshawa |  | 2.6 | 1.7 |
| Peterborough |  | 10.0 | 1.8 |
| Belleville |  | 2.0 | 1.7 |
| Kingston |  | 11.0 | 4.0 |
| Brockville |  | 1.5 | 1.6 |
| Cornwall |  | 20.0 | 18.5 |
| Ottawa |  | 2.6 | 0.6 |
| Orillia |  | 14.0 | 1.3 |
| North Bay |  | 4.9 | 2.9 |
| Sudbury |  | 11.0 | 1.2 |
| Sault St. Marie |  | 3.9 | 4.0 |

Table 30 (continued)
BENZO[a]PYRENE CONCENTRATIONS IN URBAN ATMOSPHERES[76]
(ABRIDGED)

| | | Concentration$^a$ (ng/m$^3$) | |
|---|---|---|---|
| Country — city | Year | Winter | Summer |
| Port Arthur | | | 1.3 |
| Czechoslovakia | | | |
|   Prague | | 122(high) | 19(low) |
|   100 m distance from pitch battery at Orlova—Lazy Coke Kilns | 1964 | | 1800—3000 |
|   400 m distance | | | 1100—1900 |
| Denmark | | | |
|   Copenhagen | 1956 | 17 | 5 |
| England | 1956 | | |
|   Bilston | | | 27 |
|   Bristol | | | 13 |
|   Burnley | | | 27 |
|   Cannock | | | 19 |
|   Hull | | | 18 |
|   Leicester | | | 29 |
|   Sheffield | | | 42 |
|   Salford | | (Feb. 1953) | 210 |
|   Salford | | | 110 |
|   Burnley | | | 32 |
|   Darwen | | | 35 |
|   Gateshead | | | 62 |
|   Lancaster | | | 20 |
|   Merseyside (St. George's Dock) | | | 31 |
|   Ripon | | | 15 |
|   Salford | | | 108 |
|   Warrington | | | 31 |
|   York | | | 24 |
|   Northern England and Wales | | | 11—108 |
| Finland | | | |
|   Helsinki | 1962/63 | 5 | 2 |
| France | | | |
|   Paris | 1958 | 300—500 | — |
|   Lyon | March 1972 | | 1.3 |
| Hungary | | | |
|   Budapest | 1968 | 1000$^b$ | 32 |
| | 1971—72 | | 27$^c$ |
| Iceland | | | |
|   Reykjavik | 1955 | | 3 |
| India | | | |
|   Bombay | 1973 | | |
|   Near gas plant (coal used as fuel) | | | 170—860 |
|   Street (traffic density: 60 vehicles min$^{-1}$) | | | 15—36 |
|   Near street and kiln for firing pottery | | | 17—230 |
|   Residential suburbs | | | 0.8—3.9 |
| Iran | | | |
|   Teheran | 1971 | 6 | 0.6 |
| Ireland | | | |
|   Belfast | 1961/62 | 51 | 9 |
|   Dublin | 1961/62 | 23 | 3 |
| Italy | | | |
|   Bologna | | 212(high) | 6(low) |

## Table 30 (continued)
## BENZO[a]PYRENE CONCENTRATIONS IN URBAN ATMOSPHERES[76]

| Country — city | Year | Concentration[a] (ng/m³) | |
| --- | --- | --- | --- |
| | | Winter | Summer |
| Genoa | | 37 | 1 |
| Milan | 1958—60 | 610(high) | 3(low) |
| Rome | 1963—66 winter | | 20—147 |
| Japan | | | |
| Muroran | | | 110—160 |
| Osaka | 1965 | | 50 |
| | 1970 | | 15 |
| | 1971 | | 11 |
| Sapporo | Feb. 1961 | | 200 |
| Tokyo | Feb. 1964 | | 15 |
| Netherlands | | | |
| Amsterdam | 1968 | 22 | — |
| | 1969 | 18 | 2 |
| | 1970 | 5 | 2 |
| | 1971 | 8 | — |
| Delft | 1968 | 20 | 3 |
| | 1969 | 18 | 1 |
| | 1970 | 12 | 3 |
| | 1971 | 6 | 3 |
| Rotterdam | 1968 | 15 | 3 |
| | 1969 | 23 | 1 |
| | 1970 | 19 | 3 |
| | 1971 | 23 | 2 |
| Vlaardingen | 1968 | 35 | 3 |
| | 1969 | 32 | 4 |
| | 1970 | 13 | 5 |
| | 1971 | 16 | 9 |
| The Hague | 1968 | 23 | — |
| | 1969 | 12 | 4 |
| | 1970 | — | 4 |
| | 1971 | 13 | — |
| Norway | | | |
| Oslo | 1956 | 15 | 1 |
| | 1962/63 | 14 | 0.5 |
| Poland (average for 10 large cities) | 1966/67 | 130 | 30 |
| Gdansk | 1966/67 | | 84, 64 |
| Katowice | | | 76, 75 |
| Krakow | | | 63, 63 |
| Lodz | | | 45, 45 |
| Opole | | | 54, 47 |
| Poznan | | | 48, 49 |
| Szczecin | | | 44, 62 |
| Warszawa | | | 29, 29 |
| Wroclaw | | | 57, 64 |
| Zabrze | | | 130, 100 |
| South Africa | | | |
| Durban | June 1964 | | 5, 14, 28 |
| Johannesburg | May 1964 | | 49 |
| | May 1964 | | 1100[d] |
| Pretoria | 1963/64 | 10 | 22 |
| Spain | | | |
| Madrid | 1969/70 | 120 | 0 |
| | 1969 | 9 | 0 |

## Table 30 (continued)
## BENZO[a]PYRENE CONCENTRATIONS IN URBAN ATMOSPHERES[76] (ABRIDGED)

| | | Concentration[a] (ng/m³) | |
|---|---|---|---|
| Country — city | Year | Winter | Summer |
| Sweden | | | |
|   Stockholm | 1960 | 10 | 1 |
| | 1967 | 27(high) | 2(low) |
| Switzerland | | | |
|   Basle | 1963—64 | 5—80 | |
| U.S. — 100 large urban | 1958—59 | 6.6 | |
|   communities | 1962 | 5 | |
|   32 large urban stations | 1966—67 | 3.3 | |
| | 1968 | 2.7 | |
| | 1969 | 2.9 | |
| | 1970 | 2 | |
|   Los Angeles | June 1971—June 1972 | 1.1, 0.5, 3.5, 0.03 | |
| U.S.S.R. | | | |
|   Leningrad | 1965 | 15 | |
|   Kiev | 1965 | 9 | |
|   Tashkent | 1965 | 110 | |
|   100 meters distance from a pitch boiling plant of a cardboard factory in Tashkent | | 129 | |
|   Near coke furnaces | | 570 | |
|   500 meters to the south of the furnaces | | 120 | |

[a] Average values. Where one value is reported it is an annual average value, unless otherwise stated. In some cases high and low values of the year are reported.
[b] Taken during a period of heavy smog.
[c] Winter values.
[d] Near a road tarring operation.

## Table 31
## BENZO[a]PYRENE CONCENTRATION IN THE ATMOSPHERE IN GERMANY

| | | Time of sampling | | | |
|---|---|---|---|---|---|
| Measuring station | Year | Month | Duration | BaP (ng/m³) | Ref. |
| Bonn | 1965 | Feb. | Per station: | 133 | 55 |
| Düsseldorf | | Feb. | 20 days | 125 | |
| Bochum | | Feb. | | 144 | |
| Bonn | 1965 | July | Per station: | 4 | |
| Düsseldorf | | July | 20 days | 5 | |
| Bochum | | July | | 19 | |
| Ruhr district (12 stations) | 1966 | Feb. or March | Per station: day | 133 | |
| Ruhr district (12 stations) | 1967 | Feb. or March | Per station: 1 day | 103 | |
| Essen (3 stations) | 1967 | Jan. | Per station: 4 × 1 day | 333 | |
| Ruhr district (13 stations) | 1968 | Feb.—Dec. | Per station: 28 days all year round | 110 | 20 |

## Table 31 (continued)
## BENZO[a]PYRENE CONCENTRATION IN THE ATMOSPHERE IN GERMANY

| Measuring station | Year | Time of sampling Month | Duration | BaP (ng/m³) | Ref. |
|---|---|---|---|---|---|
| East Berlin (2 stations) | 1970 | July (stn. 1) Sept. (stn.2) | Average of 6 hr: for 11 days for 9 days | 18 18 | 75 |
| Duisburg Düsseldorf Krahm | 1969 until 1973 | Per year: April—Sept. | Per station: 24 hr daily; analysis of collected filters of 1 month | 23 6 1 | 13 (and additional data) |
| Duisburg Düsseldorf Krahm | 1969 until 1974 | Per year: Oct—March | Per station: 24 hr daily: analysis of collected filters of 1 month | 121 50 7 | |
| Duisburg Düsseldorf Krahm | 1969 until 1974 | | All monthly values combined (average of (5 years) | 72 28 4 | |
| Cologne | 1970 until 1973 | Nov.— Feb. total of 31 days | 5 Sampling periods | 60 | 18 |
| Gelsenkirchen Mannheim (Rhine shore) Westerland (Sylt) Waldhof (Lüneburg Heath) Deuselbach (Hunsrück) Brotjacklriegel (Bavarian Forest) Schauinsland (Black Forest) | 1970 until 1973 | Jan.—Dec. | Per station: 24 hr daily; analysis of collected filters of 1 month   Given are average values over 4 years. Monthly average values are reported in publication. | 91 11 3.4 2.3 1.5 0.7 0.4 | 14 |
| Duisburg | 1973 | Dec. | On 6 days | 93 | 85 |
| Düsseldorf | 1974 | Dec. | On 13 days | 40 | |
| Duisburg (15 stations) | 1974/75 | Dec.—March | Per station: once or twice 1—2 hr | 5 737/x = 190 | 72 |
| Düsseldorf (3 stations) | 1975 | Jan. | Per station: for 28 days | 19 44 16 | 87 |
| | 1975 | Feb. | Per station: for 28 days | 3 31 11 | |
| Karlsruhe Nuclear research center Park Subway Park Subway | 1974/75 1975 1975 1975/76 1975/76 | Nov.—March May—June May—June Oct.—March Oct.—March | Per station: 3 times weekly | 1.8 0.1 2.9 4.8 9.5 | 52 |
| Duisburg (center) | 1977/78 | Dec.—March | 3 times weekly for 4 weeks | 5 | 61a |

## Table 32
## BENZO[a]PYRENE CONCENTRATIONS IN HIGHLY POLLUTED ATMOSPHERES AND EMISSIONS[76] (ABRIDGED)

| Area | Concentration (ng/m³) |
|---|---|
| Coal-fired residential furnaces | 2200—1,500,000 |
| Coal-fired power plants | 30—930 |
| Coal-fired unit (intermediate size) | 49—7900 |
| Coke oven, above gasworks retorts | 216,000 |
| Coke oven battery | |
|   On battery locations | 172—15,900 |
|   Off battery locations | 21—1200 |
|   Battery roof | 6700 |
|   Larry car | 6300 |
|   Pusher | 960 |
|   Pump house | 260 |
|   Brick shed | 380 |
|   Cortez van | 150 |
| Garage air (Cincinnati downtown) | 33 |
| Roof-tarring operations | 90, 870, 14,000 |
| Gas works retort houses | 3000 (average) |
|   Above the retorts | 220,000 |
| Gas-fired heat generation units | 20—350 |
| Incinerators, municipal[a] | 17, 19, 2700 |
| Incinerators, commercial | 11,000, 52,000 |
| Oil-fired heat generation units | 20—1900 |
| Open burning | 2800, 4200, 173,000 |
| Retort houses, maximal results | 2,300,000 |
|   Above horizontal retorts | 220,000 |
| Sidewalk tarring operations | 52, 110, 78,000 |
| Silicon carbide (carborundum) plant | |
|   Air in crusher shop | 300—900 |
|   Air from coke ovens | 400—730 |
|   100 m from coke ovens | 200—410 |
|   500 m from coke ovens | 72—180 |
|   100 m from plant | 28—56 |
| Smoky atmosphere | |
|   Beer hall in Prague | 28—144[b] |
|   Arena | 0.7—22 |
| Tar paper plant[c] | |
|   Mass boiling shop | 1100—1500 |
|   Plant territory air | 230—290 |
|   100 m from plant | 125—135 |
|   500 m from plant | 38—61 |
| Tunnel, Blackwell | 350 |
| Tunnel, Sumner | 690 |
| Wall-tarring operations | $520 \times 10^3$, $640 \times 10^3$ |
| | $1600 \times 10^3$, $6000 \times 10^{3\,d}$ |

[a] In the combustion of municipal refuse in a continuous feed incinerator it was found that about 9.3450 and 3.5 mg of BaP and BeP were emitted per day in the stack gases, the residues, and the water effluents, respectively.

[b] Dependent on number of people smoking and ventilation. At same time urban air in Prague contained 2.8 to 4.6 ng BaP per cubic meter.

[c] The coal tar, coal tar pitch, and the boiling mass utilized in the plant contained 0.35 to 1.0, 0.4 to 2.0, and 0.3 to 1.4% BaP, respectively.

[d] Within 1 hr a worker could inhale ~3 mg of BaP, the amount of BaP he would get from smoking approximately 300,000 cigarettes or breathing polluted air containing 10 ng BaP per cubic meter for approximately 70 years.

## Table 33
### BENZO(a)PYRENE CONCENTRATIONS IN THE ATMOSPHERE OF CENTRAL LONDON IN 1949—1973[63]

| Period of time | Sampling station | BaP (ng/m³) |
|---|---|---|
| 1949—51 | County Hall | 46 |
| 1953—56 | St. Bartholomew's Hospital | 17 |
| 1957—64 | County Hall | 14 |
| 1972—73 | St. Bartholomew's Medical College | 4 |

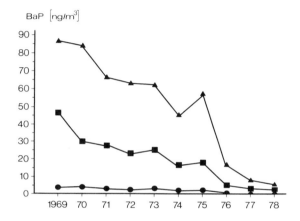

FIGURE 26. Benzo[a]pyrene concentrations in the atmosphere (annual averages 1969—78) at three sampling sites.[13,88] ▲ = Duisberg, ■ = Düsseldorf, ● = Krahm (Berg Land).

The BaP concentrations in residential buildings seem to be 25 to 70% lower than the atmospheric values[82] as long as no PAH are formed by cooking or heating in these buildings. In badly aired huts of primitive tribes in Kenya, values between 24 and 291 ng BaP per cubic meter were measured during cooking.[56a] Data on PAH other than BaP in the ambient air are very few. Even a more recent survey by Hoffmann and Wynder[56a] is almost exclusively comprised of references which are 15- to 20-years old (Table 29). Table 28 gives data on concentrations of 14 PAH in Duisburg calculated as an average value for the three months during the winter of 1977/78.

A considerably increased measurement of PAH concentrations in the atmosphere is necessary to obtain data which, unlike the presently available results, will permit an accurate account to be given of the hazard accruing to man from PAH in the whole of West Germany and to evaluate long-term effects of changes in emission.

### 3.6 SOIL — SEDIMENTS — CRUDE OIL

G. Grimmer

#### 3.6.1 Soil Investigations
##### 3.6.1.1 Type of PAH

In examinations of soil samples by means of capillary gas chromatography, known

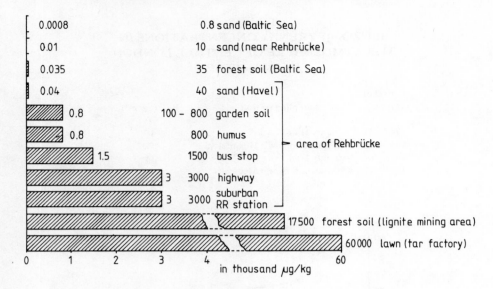

FIGURE 27. Content of benzo[a]pyrene in soil samples (µg/kg).

PAH (three-ring to seven-ring compounds) could be identified by their UV and mass spectra. The presence of alkyl-substituted compounds (up to $C_5$) was also indicated. Concentrations of PAH were not reported.[6]

*3.6.1.2 Range of PAH Concentration*

The majority of investigations were carried out between 1967 and 1977 by Russian authors (29 papers). However, only the BaP content of samples was reported. The concentrations of BaP measured in the U.S.S.R. ranged from 0.0008 mg to 200 mg/kg, the maximum value being found in the vicinity of an oil refinery.[78] Similarly high concentrations (650 mg/kg) were measured in the area of a carbon black factory.[21] In samples of sandy and forest soil collected in West Germany, considerably lower concentrations of BaP, ranging from 0.001 to 0.0004 mg/kg, were found.[8] The contamination of soil can be attributed almost exclusively to emissions from combustion processes. In the majority of surface soil samples taken in Iceland, where hardly any fossil fuels are burnt, the PAH most commonly found (12 PAH) were not detected (detection limit for BaP, e.g., 0.02 µg/kg of soil). Soil samples taken at Reykjavik airport, however, were extremely contaminated, the BaP concentration reaching 0.785 mg/kg.[36] In Figure 27 the BaP content of different soil samples are compared.[21] As shown by several assays with cabbages, spinach, radishes, carrots, etc., roots in higher plants do not absorb PAH at all, or only in small amounts.[70]

**3.6.2 Sediment Investigations**
*3.6.2.1 Type and Amount of PAH*

In drilling cores taken from the bottom of Grosser Plöner See in West Germany, 64 PAH were detected by mass spectrometry. About 40 of these PAH were identified by comparison with reference compounds. The PAH profiles were evaluated with respect to 25 main components in annual sediment layers for the period from 1913 to 1969.[39] An example PAH profile is shown in Figure 28.

*3.6.2.2 Range of PAH Concentration*

The examined sediment samples from drilling cores taken near the banks of the more

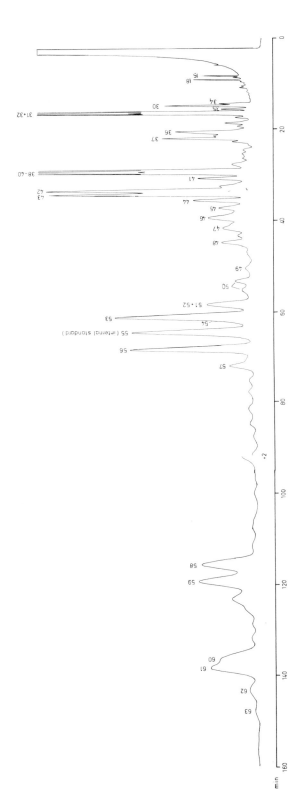

FIGURE 28. Profile analysis of a sediment layer (drilling core from the bottom of the northern part of lake "Grosser Plöner See"). Gas chromatogram of PAH fraction CHR-COR, glass: 2 mm × 10 m, 5% Silicone OV 101 on Gas Chrom Q 100 to 120 mesh per square inch isothermal 250°C, (15) FLU, (18) PYR, (34) benzo[b]naphtho[2,1-d]thiophene, (30) BghiF, (31 + 32) BaA+TRi+CHR, (42) BeP, (43) BaP, (44) PER, (51) IF, (53) IP, (56) BghiP, (57) ANT, (60), COR, (61) dibenzo[fg,op]tetracene.

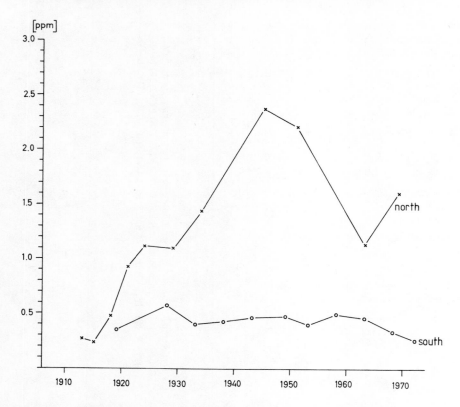

FIGURE 29. BaP concentration in the annual sediment layers of a drilling core taken near the banks of the northern (built-up, much traffic) and the southern (forested) part of lake "Grosser Plöner See" (1913-1972).[39]

densely populated part of Grosser Plöner See revealed an increase in PAH concentration (dry matter), an average of five times from 1913 to 1951, whereas the concentration decreased from 1951 to 1969 (e.g., BaP: 0.27 mg/kg to 2.2 mg/kg to 1.6 mg/kg). Contrary to this, no significant differences were detected in samples from the forested southern part of the lake over the whole period of time (about 0.4 mg/kg) (Figure 29).

Similar results were obtained by examination of several drilling cores taken from Lake Constance. In this case, too, the PAH concentration (dry matter) increased from the surface of the lake bottom to a layer of maximal contamination, and subsequently the contamination was reduced considerably.[40] In the heavily polluted areas of the lake the PAH concentration, in layers of maximal contamination (e.g., 1.6 mg BaP per kilogram) was five times greater than in the less contaminated parts (0.3 mg BaP per kilogram). Both results indicate that in these areas the contamination by carcinogenic PAH had increased five times over the period from the turn of the century until 1965.

Further examinations of sediment samples were carried out by Giger et al.[26] In this case only surface samples were analyzed (0.2 to 0.6 mg BaP per kilogram).

### 3.6.3 Investigation of Crude Oil Samples

Crude oil and its mineral oil products such as lubricating oil, light fuel oil, diesel fuel, etc. show PAH profiles which clearly differ from profiles formed by incomplete combustion (see also Chapter 2.1.1).

*3.6.3.1 Type and Amount of PAH*

In samples of crude oil found in northern Germany, 120 PAH, as well as about 60 carbazole and benzocarbazole derivatives, were detected by gas-chromatographic profile analysis and identified by their mass or UV spectra. By comparison with reference compounds, 50 PAH could be identified. The majority of these compounds were methyl derivatives of the following basic components: Naphthalene, diphenyl, fluorene, dibenzothiophene, phenanthrene, fluoranthene, pyrene, benzo[b]naphtho[2,1-d]thiophene,triphenylene, chrysene, benzo[b]fluoranthene, and benzo[e]pyrene. The nitrogen-containing hetercycles can be arranged in a $CH_2$-homologous series of carbazole and benzocarbazoles.[24]

*3.6.3.2 Range of PAH Concentration*

Surprisingly, in crude oil samples taken in the Gifhorn Trough (Gifhorner Trog), West Germany, the amount of BaP is relatively small in relation to other PAH. In eight samples examined, the BaP fell below 0.1 mg/kg of crude oil (maximum: 2.1 mg BaP per kilogram). A comparison of 30 additional PAH contained in these samples revealed a dependence of the amount of PAH detected on the degree of maturity of oil. Since the degree of maturity can be correlated to the period of exposure of deposits to maximum temperatures, the demonstrated dependence of PAH concentration on the depth of the deposit suggests an abiological mode of formation of PAH and nitrogen-containing heterocycles in this area.[44]

### 3.7 WATER — SEWAGE SLUDGE

### G. Grimmer

**3.7.1 Water**

*3.7.1.1 Type and Amount of PAH*

Up to now a comprehensive survey of PAH in drinking water, groundwater, river water, or other surface waters has not yet been published. Existing data are mostly restricted to BaP or six frequently occurring PAH.

*3.7.1.2 Range of PAH Concentration*

BaP concentrations of 2.5 to 9 ng/$\ell$ were detected in samples of tap water (14 samples). Samples of ground water (15 samples) taken in West Germany also contained comparable amounts of BaP (1 to 10 ng/$\ell$).[10] The investigations (about 70) carried out in the last few years (1967 to 1977) mostly comprised only data on BaP concentrations in water. Most of these data were presented by Soviet investigators. A special topic is the elimination of PAH by ozonization, chlorination, treatment with chlorine dioxide, flocculation, and adsorption to activated carbon.[11] A study covering international data on the BaP content of waters was published by Andelmann et al.[3]

**3.7.2 Sewage Sludge**

The colloidal, dispersed particle phase of waste water is a collector of those anthropogenic wastes which are neither collected as garbage nor emitted into the air. Analogous to airborne particles, sewage sludge contains all the noxious substances which are emitted into the environment by technological processes as well as by domestic and agricultural activities.

*3.7.2.1 Type and Amount of PAH*

About 70 PAH were detected in samples of sewage sludge by their mass-spectrometric fragmentation. Some thiophene derivatives, such as dibenzothiophene and benzonaphthothiophene and their methyl derivatives, were also found in comparable amounts.[12] In their description of a routine method of determination, Grimmer and

## Table 34
## PROFILE ANALYSIS OF PAH IN SAMPLES OF SEWAGE SLUDGE

| PAH (mg/kg of dried sample) | MÜN | NÜR | ING | SCH | NEU | MEM | LAN | MUR | UFF[a] |
|---|---|---|---|---|---|---|---|---|---|
| Fluoranthene | 3.40 | 4.29 | 6.67 | 4.18 | 2.43 | 3.92 | 4.11 | 3.95 | 1.94 |
| Pyrene | 3.37 | 4.01 | 6.94 | 4.87 | 1.91 | 3.24 | 3.51 | 4.16 | 1.67 |
| Benzo[a]fluorene | 1.05 | 1.15 | 1.95 | 1.99 | 0.57 | 1.02 | 1.06 | 1.36 | 0.64 |
| Benzo[b/c]fluorene | 0.82 | 1.02 | 1.31 | 2.14 | 0.44 | 0.76 | 0.80 | 1.06 | 0.54 |
| Benzo[b]naphtho[2,1-d]thiophene | 1.10 | 1.20 | 1.79 | 2.53 | 0.43 | 0.78 | 1.01 | 1.41 | 0.68 |
| Chrysene | 2.71 | 3.69 | 4.60 | 2.30 | 1.32 | 2.23 | 3.34 | 2.90 | 1.81 |
| Benzo[a]anthracene | 1.60 | 3.20 | 2.97 | 1.45 | 0.92 | 1.84 | 2.02 | 1.96 | 1.05 |
| Methyl derivative of M 228A | 0.31 | 0.23 | 1.51 | 2.19 | 0.16 | 0.34 | 0.73 | 1.10 | 0.22 |
| Methyl derivative of M 228B | 0.85 | 0.84 | 1.14 | 1.73 | 0.27 | 0.45 | 0.51 | 0.89 | 0.56 |
| Benzofluoranthenes (3 isomers) | 2.46 | 3.54 | 8.01 | 3.26 | 1.16 | 2.24 | 3.26 | 5.05 | 1.83 |
| Benzo[e]pyrene | 1.24 | 1.54 | 3.46 | 1.78 | 0.55 | 1.11 | 1.84 | 2.23 | 0.99 |
| Benzo[a]pyrene | 1.09 | 1.90 | 4.81 | 1.78 | 0.55 | 1.28 | 1.61 | 2.65 | 0.96 |
| Perylene | 0.44 | 0.58 | 1.37 | 0.94 | 0.19 | 0.39 | 0.70 | 0.69 | 0.37 |
| Dibenz[a,j]anthracene | 0.25 | 0.36 | 1.24 | 0.28 | 0.10 | 0.25 | 0.93 | 0.50 | 0.18 |
| Indeno[1,2,3-cd]pyrene | 0.81 | 1.26 | 3.42 | 0.76 | 0.33 | 0.77 | 1.22 | 1.81 | 0.59 |
| Benzo[ghi]perylene | 0.97 | 1.11 | 3.24 | 0.99 | 0.40 | 0.91 | 1.32 | 1.79 | 0.71 |
| Anthanthrene | 0.10 | 0.19 | 0.58 | 0.10 | 0.08 | 0.21 | 0.08 | 0.24 | 0.15 |
| M 302A | 0.24 | 0.36 | 1.26 | 0.20 | 0.14 | 0.38 | 0.50 | 0.70 | 0.25 |
| M 302B | 0.21 | 0.38 | 1.71 | 0.23 | 0.13 | 0.41 | 0.57 | 0.79 | 0.23 |
| M 302C + coronene | 0.26 | 0.41 | 1.60 | 0.33 | 0.12 | 0.38 | 0.68 | 0.83 | 0.21 |

*Note:* Time of sampling: June 14—21, 1976.

[a] MÜN = München, NÜR = Nürnberg, ING = Ingolstadt, SCH = Schwabach, NEU = Neumarkt, MEM = Memmingen, LAN = Landsberg, MUR = Murnau, UFF = Uffenheim.

co-workers[45] presented a typical PAH profile that may be used to analyze the 20 most frequent PAH.

### 3.7.2.2 Range of PAH Concentration

An examination of samples of sewage sludge collected in 25 different sewage treatment plants in West Germany showed that the PAH profile (number of individual components and their quantitative relation to each other) is very similar for all samples examined. This finding suggests that the main source of contamination is similar for all plants which purify waste water from industrial, as well as residential, areas. Table 34 shows a comparison of the PAH of sewage sludge collected in purification plants of rural, urban, and industrial catchment areas.

The concentration of individual PAH in the sewage sludge of these 25 purification plants varied within a decimal power. Thus the BaP content in the freeze-dried samples ranged from 0.54 to 5.77 mg/kg (average: 1.8 mg/kg of dried sewage sludge). Unlike PAH concentrations in the air which are subject to distinct variations in summer and winter months (about 1:20), the samples of sewage sludge collected in 3-month intervals from 3 plants did not produce any measurable changes in the PAH concentration in the course of the year.

The comparison of relative PAH profiles compiled in Table 34 shows that the composition of sewage sludge collected in the highly different catchment areas of these nine sewage treatment plants is rather similar (Figure 30).[48]

Comparative examinations of fresh sewage sludge (taken in front of the digestion tower) and activated sludge (taken from behind the accelerator) did not reveal any differences in the PAH content.[9] This investigation which was carried out in four sewage treatment plants yielded BaP concentrations between 0.12 and 1.33 mg/kg, i.e., a little below the average value (1.8 mg/kg) of the 25 plants mentioned earlier.

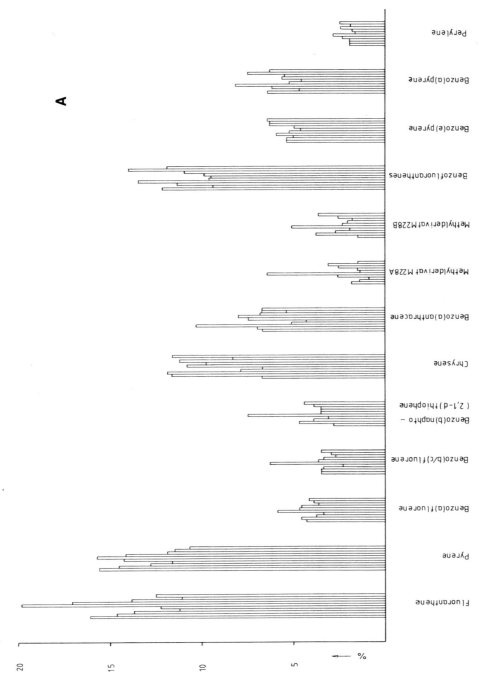

FIGURE 30 A and B. Percentage of 20 polycyclic aromatic hydrocarbons (PAH) based on the sum of the 20 compounds (relative composition). Bars represent, left to right, Geiselbullach, München, Nürnberg, Ingolstadt, Schwabach, Neumarkt, Memmingen, Landsberg, Murnau, Uffenheim.

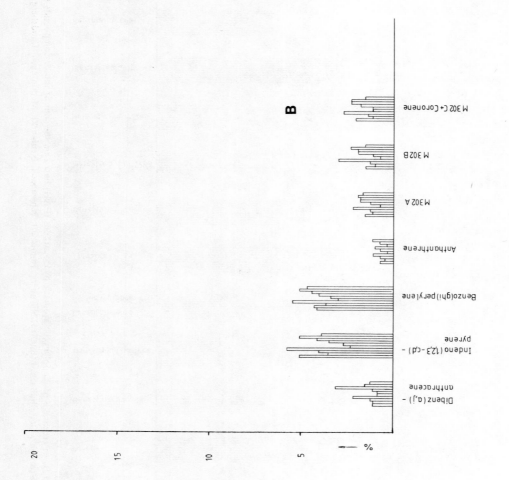

FIGURE 30 B.

Samples taken over a period of 24 hr from a sewage treatment plant in a residential area (20,000 inhabitants) did not give the slightest indication of biological decomposition of PAH during this time (without digestion tower). In this study the waste water entering into the plant, the precipitated coarse sand, the sewage sludge, and the purified water (output) were taken into consideration. Since the sewage sludge was deposited in the settling basin, the discharged water contained — as expected — much lower amounts of PAH than the original waste water. Thus the BaP concentration was reduced from 308 mg/day in the waste water to 39 mg/day in the discharged purified water. Each day, 4280 m³ of waste water were purified.[48]

## 3.8 FOODSTUFFS

### G. Grimmer

Foodstuffs may be contaminated with PAH via airborne particulate matter, by direct drying with smoke, or by absorbtion during the smoking process. Consequently the PAH profile obtained from smoked food differs from that of food contaminated with airborne particulate matter in that it lacks thiophene derivatives, such as benzonaphthothiophene. Surprisingly, large-leaved varieties of vegetables, e.g., kale contaminated with atmospheric particulate matter, in general contained much higher amounts of PAH than smoked fish or ham.[30]

In the course of baking[23] or frying,[31] i.e., flameless heating (no open fire), only very small amounts of PAH are formed, since practically no PAH are formed at the usual cooking temperatures of 200 to 230°C and in the time period required. Detectable amounts of PAH are formed from proteins, carbohydrates, and fats only at considerably higher temperatures. When frying in deep fat, which is occasionally used for several weeks, the content of BaP in the fat sometimes even decreases.[23,35] On the other hand, meat products which are broiled above a smoking open fire get contaminated with the PAH adsorbed to the particles of smoke.[24,90]

The amounts of PAH taken in with food are not known because only a few groups of foodstuffs have been analyzed with respect to their PAH profile.[37] Experiments in which condensates from smoked food and their PAH-free fractions are fed to animals have not yet been carried out. This would be necessary to assess the carcinogenic hazard for man arising from orally ingested PAH. It must be accepted that as well as PAH, other carcinogens, such as nitrogen-containing aromatic compounds, nitrosamines, and mycotoxins also constitute a possible carcinogenic risk to the alimentary tract. Also, a quantitative assessment of the carcinogenic effect of smoke particles has not yet been carried out.

Fritz[22] estimated the amount of BaP taken in annually with food. According to this estimate, based on an average life expectancy of 70 years, an inhabitant of East Germany would therefore ingest a total of 24 to 85 mg of BaP. The distribution of PAH among individual groups of foodstuffs is outlined in Figure 31.

A calculation of the BaP intake in West Germany would yield similar data.

### 3.8.1 Smoked Foodstuffs
#### 3.8.1.1 Type and Amount of PAH
The PAH profile of smoked fish was analyzed by gas chromatography on high-performance columns. More than 100 individual compounds were identified as PAH.[37]

#### 3.8.1.2 Range of PAH Concentration
In the last few years (1967 to 1977) about 100 investigations were carried out which dealt with the PAH content of smoked food. With a few exceptions, only the BaP

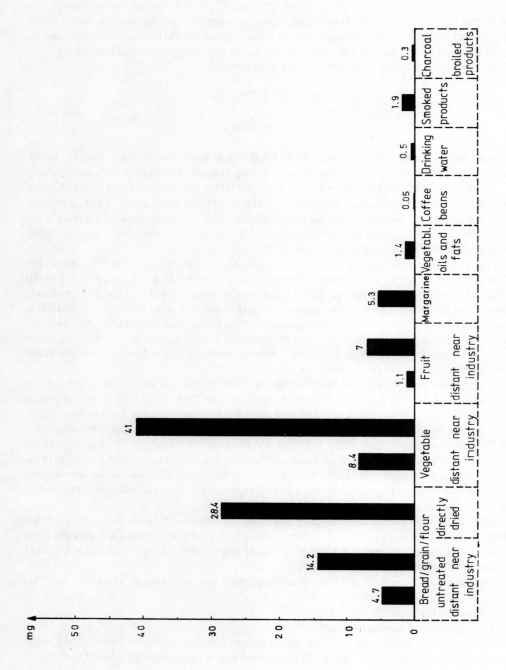

FIGURE 31. Estimated amount of BaP taken in with food per person in the course of 70 years.

content was recorded. In about 25 publications, American, Belgian, German, Yugoslavian, and Russian authors reported average values ranging from 0.2 to 0.9 μg BaP per kilogram for smoked meat products such as sausages, ham, bacon, etc. There are some reports of extremely high values (23 to 55 μg/kg) in intensely smoked products (black-smoked).

Similar BaP concentrations were measured in smoked fish (range from 0.1 to 9.8 μg BaP per kilogram; nine examinations). Maximum values were found in Japan in smoked fish such as Katsuobushi (up to 37 μg BaP per kilogram).[67]

PAH do not normally penetrate sausage skin; only natural sausage skin is permeable to traces of PAH.

After feeding provender containing 6.2 μg to 43 mg BaP per kilogram for about 1 year to cattle, rabbits, pigs, chickens, fowls, and ducks, the analyzed meat of these animals contained a maximum of 0.26 μg BaP per kilogram.[27] Long-term feeding of considerable amounts of BaP (e.g., 1.5 μg/day to pigs for 92 days or 0.8 μg/day to ducklings for 46 days) did not significantly increase the BaP content in meat, fat, or in the liver of these animals. The BaP content of milk or eggs also remained unchanged.[16]

When determining the average content of 18 different PAH in 51 samples collected in West Germany (ham, bacon, sausage), an average BaP concentration of 0.6 μg/kg was found. The average BaP content in black-smoked products, however, was 9 μg/kg (maximum: 55 μg/kg).[89] In West Germany, the maximum allowable amount of BaP in smoked meat products has been limited to 1 μg/kg.

### 3.8.2 Food Contaminated with Airborne Particulate Matter
#### 3.8.2.1 Type and Amount of PAH

Food, such as kale, lettuce, grain, yeast, etc., contaminated with airborne particulate matter, has the same PAH profile as airborne particulate matter. This can be seen, for instance, by the characteristic content of benzonaphthothiophenes.[48]

#### 3.8.2.2 Range of PAH Concentration

Kale is often very much contaminated with PAH due to the large surface of its leaves and the late harvesting during the winter months. Table 35 gives the main components of the PAH profile of 16 samples collected in several German cities.

The high content of benzo[b]naphtho[2,1-d]thiophene, for instance, is characteristic of air pollution. Even by frequent rinsing, the PAH cannot be removed from the wax layer on the leaf surface.[30] Lettuce which was contaminated to a similar degree contained PAH predominantly in the outer leaves.

PAH could also be detected in crude vegetable oils such as sunflower, rape-seed, palm-kernel, palm, peanut, cottonseed, soybean, and linseed oil. In about 60 samples the BaP concentration ranged from 1.2 to 15.3 μg/kg of oil. As expected, crude coconut oil obtained from smoke-dried copra contained particularly high amounts of PAH. Twelve main components were taken into consideration. The range of BaP concentration lay between 17.8 and 48.4 μg/kg.[33] By refining crude oils (vaporization and adsorptive processing) the PAH contained in crude oil can completely be removed.[5] Therefore, margarine and other similar products produced in West Germany do not contain any detectable amounts of PAH.

When examining different sorts of fruit, the dependence of the PAH content on the amount of air pollution at the site of growth was clearly demonstrated. Thus, for instance, apples grown in residential areas contained 0.2 to 0.5 μg BaP per kilogram whereas 30 to 60 μg BaP per kilogram were measured in apples grown near industrial plants. The PAH accumulated predominantly in the apple skin[22] while the apple core contained only 5 to 6 μg BaP per kilogram.

## Table 35
## PAH IN KALE SAMPLES COLLECTED IN BREMEN, OSNABRÜCK, MÜNSTER, DORTMUND, AND HANNOVER

| PAH | Range | Average |
|---|---|---|
| Fluoranthene | 43.0—488.0 | 134.96 |
| Pyrene | 28.1—289.0 | 92.72 |
| Benzo[a]fluorene | 4.4— 51.7 | 22.55 |
| Benzo[b/c]fluorene | 2.3— 28.5 | 11.59 |
| Benzonaphthothiophene | 5.5— 74.9 | 24.11 |
| Benzo[ghi]fluorene | 3.9— 57.7 | 17.31 |
| Chrysene + benzo[a]anthracene (5 + 1) | 21.1—209.0 | 74.84 |
| 3-Methylchrysene | 1.7—20.2 | 6.80 |
| 2-Methylchrysene | 0.9— 6.2 | 2.91 |
| 6-/4-Methylchrysene | 0.9— 2.6 | 1.68 |
| Benzo[b + k + j]fluoranthene (5 + 2 + 3) | 26.0—65.1 | 35.13 |
| Benzo[e]pyrene | 6.1—18.1 | 11.93 |
| Benzo[a]pyrene | 4.2—15.6 | 7.93 |
| Perylene | 1.3— 4.0 | 2.33 |
| Dibenz[a,j]anthracene | 1.6— 4.9 | 2.81 |
| Indeno[1,2,3-cd]pyrene | 5.8—15.2 | 10.60 |
| Benzo[ghi]perylene | 7.6—16.0 | 9.21 |
| Anthanthrene | 0.4— 2.1 | 0.92 |
| Coronene | 1.0—10.1 | 3.01 |

*Note:* 16 samples, content given in µg/kg of fresh weight.

Similar results are described in a study in which the contents of 13 different PAH in barley, rye, and wheat harvested in rural and industrial areas of West Germany were measured. Thus wheat samples from rural areas contained 0.19 to 0.34 µg BaP per kilogram, whereas 0.72 to 3.52 µg BaP per kilogram was measured in wheat grown near industrial plants. These differences in concentration were also observed with respect to the other PAH.[29]

When grinding grain, the PAH adsorbed on the surface of the grains pass predominantly over into the flour.

An investigation of baking yeasts sampled in different European countries also showed that the PAH content depended on the site of origin. These findings led to the conclusion that the air used for ventilation was in some cases not filtered (range 3.1 to 55.0 µg BaP per kilogram of dried yeast). On the other hand, yeasts grown on mineral oil or n-paraffins aerated with filtered air were hardly contaminated with PAH (range 0.2 to 2.6 µg BaP per kilogram of dried yeast.)[34]

### 3.8.3 Smoke-Dried Foodstuffs

Direct smoke drying is used for coconut meat (copra) or fodder cereals. The high content of PAH in crude coconut oil obtained from smoke-dried copra has already been mentioned. There have also been several investigations on the contamination of grain by direct smoke drying. Thus the BaP content may increase from 0.34 to 1.44 µg/kg in cases in which particulate matter is in the gas stream. In these cases the grain acts as a particle filter absorbing the gasborne particles.[29] Since the amount of soot and flue ash derived from different fossil fuels differs widely, only particle-free smoke should be used for direct drying. Figure 32 demonstrates the dependence of the amount of BaP formed during drying on the type of fuel used.[22]

Souci[80] published a report on the influence of combustion gases on foodstuff and

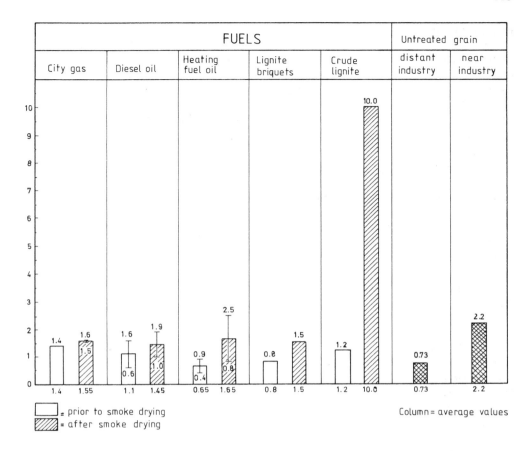

FIGURE 32. Contents of benzo[a]pyrene in grain dried by various means (μg/kg).

fodder. General surveys on the PAH content of foodstuff were published by several authors.[22,32,49]

## 3.9 A CHEMICAL ANALYTICAL INDEX FOR THE ASSESSMENT OF CARCINOGENIC EFFECTS OF EMISSIONS AND ENVIRONMENTAL SAMPLES

### G. Grimmer

Every organic substance which contains carbon and hydrogen atoms produces a large number of polycyclic aromatic hydrocarbons (PAH) during incomplete combustion, pyrolysis, or carbonization. Thus, PAH isolated from an environmental sample always constitute a complex mixture of more than 100 individual compounds, which are present in different quantities. By the term PAH profile of a sample it is understood that reference has been made to the number and relative amount of individual PAH compounds in the sample. A graphic presentation of such a PAH profile is, for instance, the recording strip of a highly resolved gas chromatogram when the area integrals of FID signals are linearly proportional to the combusted carbon mass. A "standardized gas chromatogram" is derived from the presentation mentioned earlier by transforming the (two-dimensional) peak areas into (one-dimensional) lines. This type of presentation is common in mass spectra. In order to compare several gas chromatograms, the retention time of each compound is identical in all "standardized gas chromatograms" (see Figures 33—40). The more completely a fossil fuel can be burned with the stoichiometric, or more than the stoichiometric, amount of oxygen to form carbon dioxide and water, the smaller will be the amount of each individual PAH generated in a side reaction. Since the burned material, as well as the conditions of combustion are generally different it cannot be expected that the number and amount of PAH formed will always be the same. Differences in the PAH profiles of flue gas are to be expected all the more so, since fuels frequently contain further heteroatoms such as nitrogen or sulfur. During their combustion or thermal decomposition in a protective atmosphere (e.g., nitrogen), corresponding heterocycles are formed.

Figures 33 to 40 show standardized PAH profiles from different emission sources and environmental samples. The basic capillary gas chromatograms were taken under the same conditions of temperature programming on Silicone OV 17. The same conditions were used in the GC-MS combination. In the case of emitters, the quantitative collection of PAH ranging from FLU to COR was verified by adding a control filter behind the collecting filter. Extraction yield, PAH enrichment, and FID detection were also checked by adding an internal standard. The evaluation of FID signals was checked for linear proportionality and substance independence by determining the quotient of a known weight of added mixture of ten reference compounds (FLU-COR)/FID signal area. Due to this control of conditions, the standardized PAH profiles are therefore comparable to each other.

The PAH profile analysis of environmental samples have shown that the PAH mixtures of these samples have a completely different composition. For instance, the percentage of BaP in the sum of all the recorded PAH is not constant. All the sample tapes investigated contained, as well as BaP, other PAH, which also showed carcinogenic activity in model animal experiments. In all these types of sample, BaP accounted for only a part of the total carcinogenic effect, and this proportion varied in amount.

If a single carcinogenic substance, such as BaP, is to be used as a relative measure of the total carcinogenic effect of environmental samples, this would have to be based on the presupposition — among others — that all the other carcinogens in the samples would always have to be present in the same quantitative relation to BaP, i.e., the profile of all carcinogenic PAH would have to be identical for all samples. In this case it would not even be necessary to know the percentage of BaP in the sample, because interest would be only in the statement that one sample is more active than another when the PAH profiles of the samples are identical.

The different PAH profiles of these types of samples raise the question whether PAH profiles are characteristic of their emitters. Characteristic compounds such as, e.g., benzonaphthothiophenes, occur predominantly in emissions from coal-fired heating installations or in crude oil. Consequently, these sulfur-containing heterocycles have to be regarded as substances characteristic of these emitters or as typical components of mineral oil products. Similarly, alkyl-substituted PAH may be considered to be characteristic constituents in some matrices (e.g., crude oil).

A dependence of the PAH profile on the sample type can also be proven for unsubstituted base components (FLU-COR).

Frequently the ratio of isomers to one another are clearly different. Thus, for example, the quantitative ratio of the two benzopyrenes and the benzofluoranthenes varies in different sample types. Only an investigation of a large number of samples will permit a statement as to the significance of these findings.

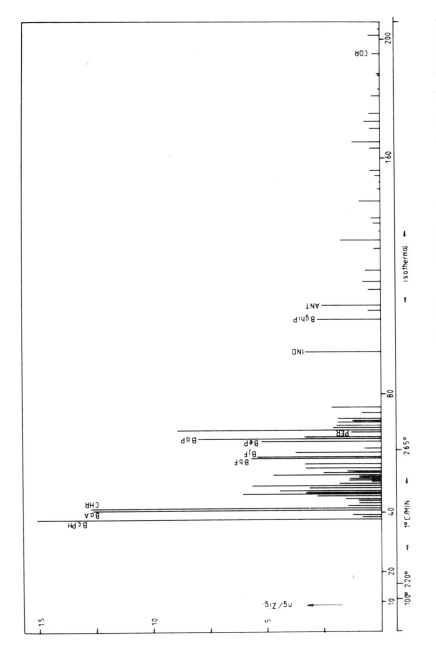

FIGURE 33. Condensate of cigarette smoke (main stream), PAH profile as standardized gas chromatogram (height of signal corresponds to area of FID signal and thus to quantity). The evaluated capillary gas chromatograms of Figures 33—40 were recorded under equal conditions in a temperature program on a glass capillary column (0.27 mm × 25m) OV 17. Identification by comparison with PAH reference substances and GC-MS combination (VARIAN MAT 112 S).

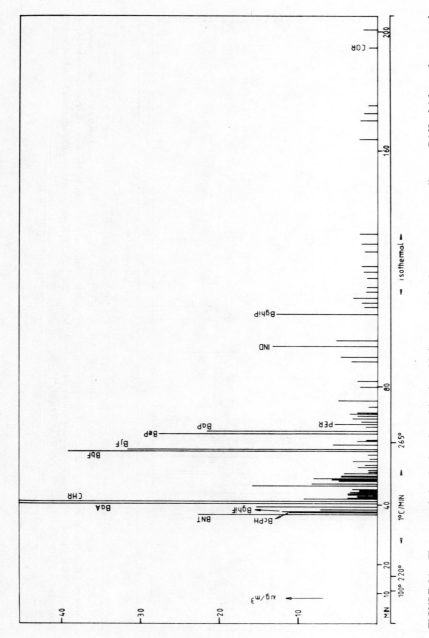

FIGURE 34. Flue gas emission by burning of hard coal briquets, μg/m³ exhaust gas corresponding mg PAH of 4 kg coal consumed. PAH-profile as in Figure 33.

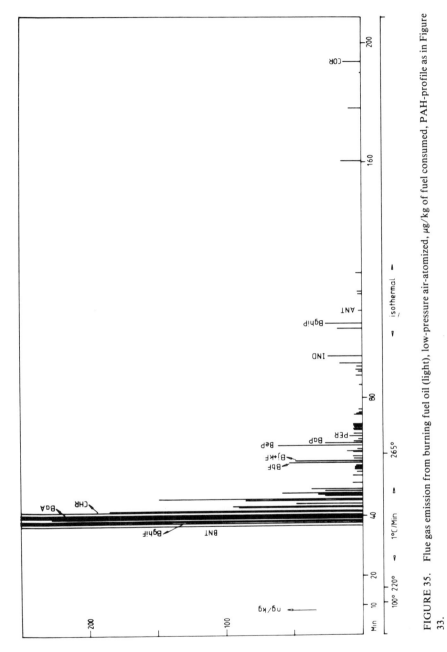

FIGURE 35. Flue gas emission from burning fuel oil (light), low-pressure air-atomized, µg/kg of fuel consumed, PAH-profile as in Figure 33.

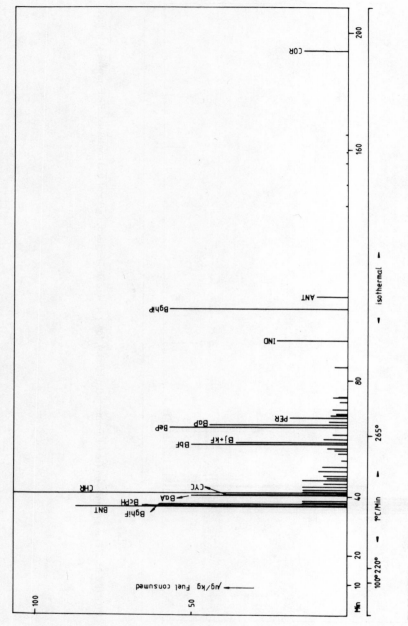

FIGURE 36. Automobile exhaust emission from gasoline engine (spark plug ignited), μg/kg of fuel consumed PAH profile as in Figure 33.

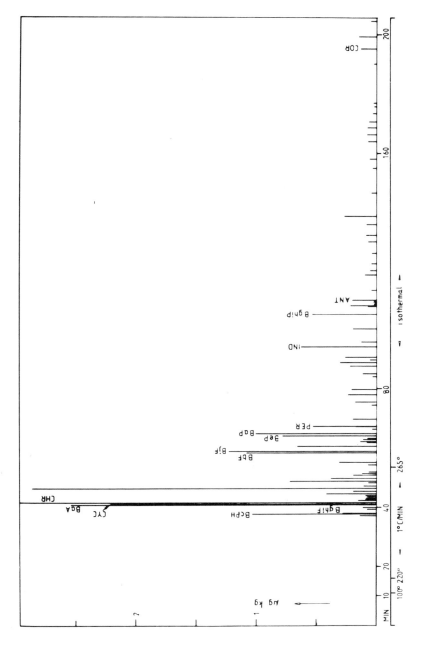

FIGURE 37. Smoked herring, μg/kg. PAH profile as in Figure 33.

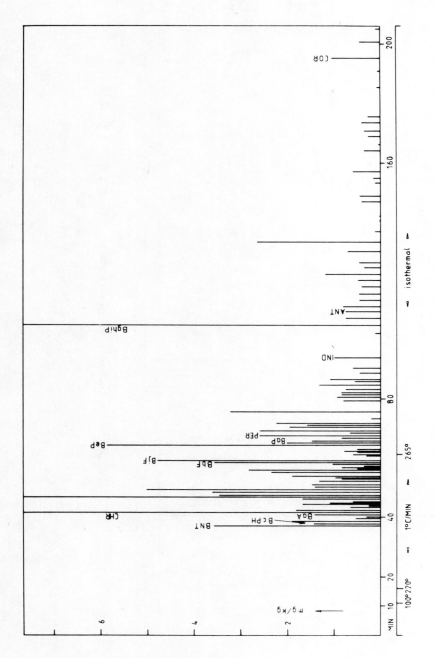

FIGURE 38. Crude oil (from "Gifhorner Trog"). PAH profile as in Figure 33.

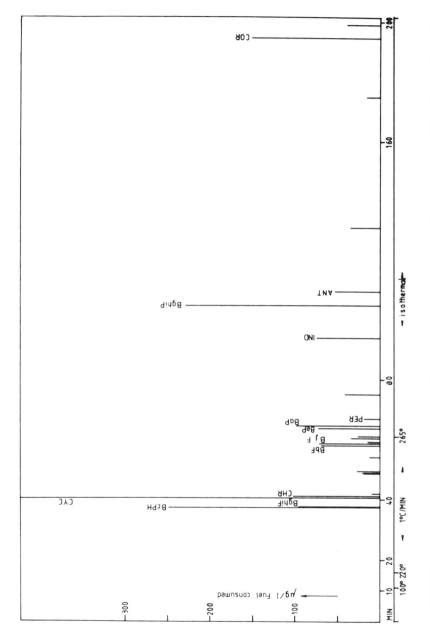

FIGURE 39. Automobile exhaust emission from diesel engine (passenger car), µg/kg of fuel consumed. PAH-profile as in Figure 33.

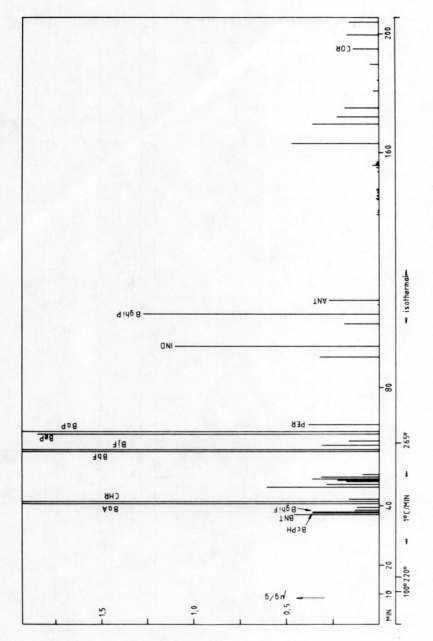

FIGURE 40. Sewage sludge collected in a sewage treatment plant, μg/g freeze-dried sludge. PAH-profile as in Figure 33.

## 3.10 SUMMARY

### G. Grimmer

### 3.10.1 Benzo[a]pyrene as an Index Substance of the Sum of all PAH

Correlation of the BaP concentration to the carcinogenic effect of a sample (or extract) in experimental animal models is as follows:

(1) During every incomplete combustion, pyrolysis, or carbonization, different amounts of the various PAH (BaP, among others) are generally formed. Like any other PAH, BaP can therefore be considered as an index substance indicating the presence of further PAH in the sample.
(2) In general, the percentage of BaP in the sum of all PAH formed is not constant. Therefore, BaP is not a quantitative index of the total amount of all PAH contained in a sample and cannot serve as a quantitative index of the carcinogenic effect of a sample in a certain experimental animal model.
(3) Within a certain type of sample (e.g., smoke from different brands of cigarettes) the PAH profile is often very similar. Only if this prerequisite is met may BaP serve as an index substance of the total amount of PAH and thus as a numerical index within the scope of this type of sample. If no further carcinogens or carcinogenesis-affecting substances in the sample can be identified in animal experiments, then the BaP concentration may be used as a comparative figure representing the carcinogenic effect in the case of samples of the same type.

### 3.10.2 The Meaning of the Term "Guide Substance"

So far the occasionally used term "guide substance" has not been precisely defined. Therefore, it is often used in completely different senses as, for example

1. Typical PAH for an emitter, i.e., a characteristic or typical compound or compounds (e.g., thiophene derivatives) for a certain type of emitter (e.g., hard coal, etc.). (Referring to benzo[a]pyrene as a sort of "guide substance")
2. Index substance of the presence of PAH: a certain PAH such as, e.g., BaP, serves as a representative for the occurrence of other PAH contained in the sample, but the amount of BaP is not considered a quantitative index to be used for the assessment of the amount of other PAH
3. Index of the contents of PAH in a sample: BaP is used as a quantitative index — as a sort of inspection glass — indicating the concentration of other PAH. As shown in the tables, however, the percentage of BaP in the sum of all PAH (FLU-COR ) is not constant. As demonstrated in Figures 33 to 40, the ratios of BaP to other carcinogenic PAH varies, so that this interpretation of the term is not acceptable or can be accepted only under special conditions.
4. Index of the carcinogenicity of a sample (carcinogenicity index, effect index): incorrectly, BaP serves as a quantitative index of carcinogenicity although it is known that the percentage of BaP in the carcinogenic activity of the total sample may vary from 1 to 20%; that means that other PAH than BaP contributes considerably to the total carcinogenicity (see Chapter 1.1.2).

It would therefore be expedient to bestow the term of "guide substance" — which is mostly used without a definition — with a definite meaning restricted to one of the interpretations outlined under points 1 to 4.

## REFERENCES

1. Air Quality Data for 1968, Publ. No. APTD-0978, National Surveillance Networks and Contributing State and Local Networks, Division of Atmospheric Surveillance Research, Environmental Protection Agency Publication, Triangle Park, N.C., 1972, 231.
2. Albagli, A., Eagan, L., Oja, H., and Dubois, L., Size-distribution measurements of airborne particulates, *Atmosph. Environ.*, 8, 201, 1974.
3. Andelmann, J. B. and Suess, M. J., Polynuclear aromatic hydrocarbons in the water environment, *Bull. WHO*, 43, 479, 1970.
3a. Automobile disposal, a national problem, U.S. Department of the Interior, U.S. Government Printing Office, Washington, D.C., 1967, 569.
3b. Begeman, C. R. and Burgan, J. G., Polynuclear hydrocarbon emission from automotive engines, Soc. Automot. Eng., SAE Paper 700469, 1970, 13.
4. Beine, H., Über den Gehalt an 3,4-Benzpyren in den Abgasen von Hausbrandöfen mit festen Brennstoffen, *Staub Reinhalt. Luft*, 30, 334, 1970.
4a. Behn, U., Meyer, J.-P., and Grimmer, G., PAH-Kumulierung im Motorenschmieröl und PAH-Emission aus Ottomotoren, Einfluss korrespondierender PAH-armer und PAH-reicher Kraftstoffe, *Erdöl Kohle Erdgas Petrochem.*, 33, 135, 1980.
4b. Behn, U., Meyer, J.-P., Grimmer, G., and Misfeld, J., Untersuchungen über den Einfluss korrespondierender PAH-armer und PAH-reicher Kraftstoffe auf die PAH-Kumulierung in Motorenschmieröl und auf die PAH-Emission aus O Homotoren, Research Report 228, Deutsche Gesellschaft für Mineralöl wissenschaft und Kohlechemie e.V., Hamburg, 1979.
5. Biernoth, G. and Rost, H. E., The occurrence of polycyclic aromatic hydrocarbons in coconut oil and their removal, *Chem. Ind. (London)*, 2002, 1967.
6. Blumer, M., Blumer, W., and Reich, T., Polycyclic aromatic hydrocarbons in soils of a mountain valley. Correlation with highway traffic and cancer incidence, *Environ. Sci. Technol.*, 11, 1082, 1977.
7. BMI-UBA, *Medizinische, biologische und ökologische Grundlagen zur Bewertung schädlicher Luftverunreinigungen*, Umweltbundesamt, Berlin, 1978, 225.
8. Borneff, J. and Kunte, H., Kanzerogene Substanzen in Wasser und Boden. XIV. Weitere Untersuchungen über polycyclische aromatische Kohlenwasserstoffe in Erdproben, *Arch. Hyg. Bakteriol.*, 147, 401, 1963.
9. Borneff, J. and Kunte, H., Kanzerogene Substanzen in Wasser und Boden. XIX. Wirkung der Abwasserreinigung auf polycyclische Aromaten, *Arch. Hyg. Bakteriol.*, 151, 202, 1967.
10. Borneff, J. and Kunte, H., Kanzerogene Substanzen in Wasser und Boden. XXVI. Routinemethode zur Bestimmung von polycyclischen Aromaten in Wasser, *Arch. Hyg. Bakteriol.*, 153, 220, 1969.
11. Borneff, J. and Kunte, H., Kanzerogene Substanzen in Wasser und Boden. XXVII. Weitere Untersuchungen zur Eliminierung kanzerogener, polycyclischer Aromaten aus Abwasser, *Zentralbl. Bakteriol. Paristenkd. Infektionskr. Hyg., Abt. I Orig. Reihe B*, 155, 18, 1971.
12. Borwitzky, H., Zur Kenntnis von polycyclischen aromatischen Kohlenwasserstoffen im Faulschlamm einer Grosskläranlage, Thesis, University of Hamburg, West Germany, 1975.
13. Brockhaus, A., Weisz, H., Friedrichs, K. H., and Krämer, U., Das Vorkommen von Benzo[a]pyren und partikulärem Blei bei unterschiedlichen Immissionssituationen, in Immissionssituation durch den Kraftverkehr in der Bundesrepublik Deutschland, *Schriftenr. Ver. Wasser. Boden, Lufthyg. Berlin Dahlem*, 42, 183, 1974.
14. Bockhaus, A., Rönicke, G., and Weisz, H., *Benzpyrenpegel des Luftstaubs In der Bundesrepublik Deutschland*, Deutsche Forschungs-Gemeinschaft, Bonn, 1975, 26.
15. Brockhaus, A. and Tomingas, R., Emission polyzyklischer Kohlenwasserstoffe bei Verbrennungsprozessen in kleinen Heizungsanlagen und ihre Konzentration in der Atmosphäre, *Staub Reinhalt. Luft*, 36, 96, 1976.
16. Cherepanova, A. I., Level of polycyclic hydrocarbons in feeds and mineral supplements, and their possible build-up in tissues, organs, eggs, and milk, *Zap Leningr. Skh. Inst.*, 141, 97, 1971.
17. Cuffe, S. T., Gerstle, R. W., Orning, A. A., and Schwartz, C. H., Air pollutant emissions from coal-fired power plants, Report No. 1, *J. Air Pollut. Control Assoc.*, 14, 353, 1964.
18. Deimel, M., Kohlenmonoxid-, Blei-, Stickoxid- und Benzo[a]pyrenbelastung in Kölner Strassen, in *Immissionssituation durch den Kraftverkehr in der Bundesrepublik Deutschland*, Meinck, V. F., Ed., Gustav Fischer Verlag, Stuttgart, 1974, 149.
19. Demaio, L. and Corn, M., Polynuclear aromatic hydrocarbons associated with particulates in Pittsburgh air, *J. Air Pollut. Control Assoc.*, 16, 67, 1966.
20. Friedrichs, K.-H., Stuke, J., Brockhaus, A., and Steiger, H., Luftverunreinigung im Ruhrgebiet — Beurteilung eines Messprogramms unter lufthygeinischen Gesichtspunkten, *Staub Reinhalt. Luft*, 31, 323, 1971.
21. Fritz, W. and Engst, R., Zur umweltbedingten Kontamination von Lebensmitteln mit krebserzeugenden Kohlenwasserstoffen, *Z. Gesamte Hyg. Ihre Grenzgeb.*, 17, 271, 1971.

22. Fritz, W., Umfang und Quellen der Kontamination unserer Lebensmittel mit krebserzeugenden Kohlenwasserstoffen, *Ernährungsforschung,* 16(4), 547, 1971.
23. Fritz, W., Zur technologisch und zubereitungsbedingten Verunreinigung von Lebensmitteln mit kanzerogenen Kohlenwasserstoffen, *Arch. Geschwulstforsch.,* 40, 81, 1972.
24. Fritz, W., Zur Bildung kanzerogener Kohlenwasserstoffe bei der thermischen Behandlung von Lebensmitteln. V. Mitteilung: Untersuchungen zur Kontamonation beim Grillen über Holzkohle, *Dtsch. Lebensm. Rundsch.,* 69, 119, 1973.
25. Gerstle, R. W., Cuffe, S. T., Orning, A. A., and Schwartz, C. H., Air pollutant emissions from coal-fired power plants, Report No. 2, *J. Air Pollut. Control Assoc.,* 15, 59, 1965.
26. Giger, W. and Schaffner, C., Determination of polycyclic aromatic hydrocarbons in the environment by glass capillary gas chromatography, *Anal. Chem.,* 50, 243, 1978.
27. Gorelova, N. D., Dikun, P. P., Dimitrochenko, A. P., Krsnitskaya, N. D., Cherepanova, A. I., and Shendrikova, I. A., Correlation between the content of polycyclic carcinogens in animal food products and in fodder for farm animals, *Vopr. Pitan.,* 29, 61, 1970.
28. Gorelova, N. D., Dikun, P. P., Kostenko, L. D., Gretskaya, O. P., and Emshanova, A. V., Detection of possible presence of 3,4-benzopyrene in fresh fish, *Nov. Onkol.,* 8, 1971.
29. Grimmer, G. and Hildebrandt, A., Kohlenwasserstoffe in der Umgebung des Menschen. II. Mitt.: der Gehalt polycyclischer Kohlenwasserstoffe in Brotgetreide verschiedener Standorte, *Z. Krebsforsch.,* 67, 272, 1965.
30. Grimmer, G. and Hildebrandt, A., Der Gehalt polycyclischer Kohlenwasserstoffe in verschiedenen Gemüsesorten. III. Mitt., *Dtsch. Lebensm. Rundsch.,* 61, 237, 1965.
31. Grimmer, G. and Hildebrandt, A., Kohlenwasserstoffe in der Umgebung des Menschen. V. Mitt.: der Gehalt polycyclischer Kohlenwasserstoffe in Fleisch und Räucherwaren, *Z. Krebsforsch.,* 69, 223, 1967.
32. Grimmer, G., Cancerogene Kohlenwasserstoffe in der Umgebung des Menschen, *Dtsch. Apoth. Ztg.,* 108, 529, 1968.
33. Grimmer, G. and Hildebrandt, A., Kohlenwasserstoffe in der Umgebung des Menschen. VI. Mitt.: der Gehalt polycyclischer Kohlenwasserstoffe in rohen Pflanzenölen, *Arch. Hyg. Bakteriol.,* 152, 255, 1968.
34. Grimmer, G. and Wilhelm, G., Der Gehalt polycyclischer Kohlenwasserstoffe in europäischen Hefen. VII. Mitt., *Dtsch. Lebensm. Rundsch.,* 65, 229, 1969.
35. Grimmer, G., Entstehen bei der Hitzesterilisierung von Lebensmitteln carcinogene Stoffe? in *Hitzesterilisierung und Werterhaltung von Lebensmitteln,* Steinkopff Verlag, Darmstadt, West Germany, 1971, 97.
36. Grimmer, G., Jacob, J., and Hildebrandt, A., Kohlenwasserstoffe in der Umgebung des Menschen. IX. Mitt.: der Gehalt polycyclischer Kohlenwasserstoffe in isländischen Bodenproben, *Z. Krebsforsch.,* 78, 65, 1972.
37. Grimmer, G. and Böhnke, H., Polycyclic aromatic hydrocarbon profile analysis of high-protein foods, oils, and fats by gas chromatography, *J. Assoc. Off. Anal. Chem.,* 58, 725, 1975.
38. Grimmer, G. and Hildebrandt, A., Investigation on carcinogenic burden by air pollution in man. XIII. Assessment on the contribution of passenger cars to air pollution by carcinogenic polycyclic aromatic hydrocarbons, *Zentralbl. Bakteriol. Parasitenkd. Infektionskr. Hyg., Abt. 1, Orig. Reihe B,* 161, 104, 1975.
39. Grimmer, G. and Böhnke, H., Profile analysis of polycyclic aromatic hydrocarbons and metal content in sediment layers of a lake, *Cancer Lett.,* 1, 75, 1975.
40. Grimmer, G. and Böhnke, H., Untersuchungen von Sedimentkernen des Bodensees. I. Profile der polycyclischen aromatischen Kohlenwasserstoffe, *Z. Naturforsch.,* 32, 703, 1977.
41. Grimmer, G., Böhnke, H., and Glaser, A., Investigation on the carcinogenic burden by air pollution in man. XV. PAH in automobile exhaust gas — an inventory, *Zentralbl. Bakteriol. Parisitenkd. Infektionskr. Hyg. Abt. 1 Orig. Reihe B,* 164, 218, 1977; Polycyclische aromatische Kohlenwasserstoffe im Abgas von Kraftfahrzeugen, *Erdöl Kohle,* 30, 411, 1977.
42. Grimmer, G., Böhnke, H., and Harke, H. P., Zum Problem des Passivrauchens. Aufnahme von polycyclischen aromatischen Kohlenwasserstoffen durch Einatmen zigarettenrauchhaltiger Luft, *Int. Arch. Occup. Environ. Health,* 40, 93, 1977.
43. Grimmer, G., Böhnke, H., and Harke, H. P., Zum Problem des Passivrauchens. Konzentrationsmessungen von polycyclischen aromatischen Kohlenwasserstoffen in Innenräumen nach dem maschinellen Abrauchen von Zigaretten, *Int. Arch. Occup. Environ. Health,* 40, 83, 1977.
44. Grimmer, G. and Böhnke, H., Polycyclische aromatische Kohlenwasserstoffe und Heterocyclen — Beziehung zum Reifegrad von Erdölen des Gifhorner Troges (Nordwestdeutschland), *Erdöl Kohle,* 31, 272, 1978.
45. Grimmer, G., Böhnke, H., and Borwitzky, H., Gas-chromatographische Profilanalyse der polycyclischen aromatischen Kohlenwasserstoffe in Klärschlammproben, *Z. Anal. Chem.,* 289, 91, 1978.

46. Grimmer, G. and Böhnke, H., The tumor-producing effect of automobile exhaust condensate and fractions thereof. I. Chemical studies, *J. Environ. Pathol. Toxicol.*, 1, 661, 1978.
47. Grimmer, G., Analytik und Vergleich der PAH-Profile aus Umweltproben, in *Luftverunreinigung durch Polycyclische Aromatische Kohlenwasserstoffe — Erfassung und Bewertung*, VDI-Verlag, Dusseldorf, 1980, 39.
48. Grimmer, G., unpublished data.
49. Haenni, E. O., Analytical control of polycyclic aromatic hydrocarbons in food and food additives, *Residue Rev.*, 24, 41, 1968.
49a. Hamburg, F. C., Economically feasible alternatives to open burning in railroad freight car dismantling, *J. Air Pollut. Control Assoc.*, 19, 477, 1969.
50. Hangebrauck, R. P., von Lehmden, D. J., and Meeker, J. E., Emission of polynuclear hydrocarbons and other pollutants from heat-generation and incineration processes, *J. Air Pollut. Control Assoc.*, 14, 267, 1964.
51. Hangebrauck, R. P., von Lehmden, D. J., and Meeker, J. E., Sources of Polynuclear Hydrocarbons in the Atmosphere, Public Health Service Publ. NO. 999-AP-33, U.S. Department of Health, Education, and Welfare, Cincinnati, 1967, 48.
52. Heinrich, G. and Güsten, H., Belastung der Atmosphäre durch polycyclische aromatische Kohlenwasserstoffe und Blei im Raume Karlsruhe, *Staub Reinhalt. Luft*, 38, 94, 1978.
53. Herlan, A., Kohlenwasserstoffe in Sedimentationsstäuben — eine massenspektrometrische Untersuchung, *Staub*, 35, 45, 1975.
54. Herlan, A. and Mayer, J., Polycyclische Aromaten und Benzol in den Abgasen von Haushaltsfeuerungen. I. Ölofen, *Staub Reinhalt. Luft*, 38, 134, 1978.
55. Hettche, H. O. and Grimmer, G., Die Belastung der Atmosphäre durch polycyclische Aromaten im Grossraum eines Industriegebietes, *Schriftenr. Landesanst. Immissions Bodennutzung. Landes Nordrhein Westfalen*, 12, 92, 1968.
56. Hoffman, D. and Wynder, E. L., A study of tobacco carcinogenesis. XI. Tumor initiators, tumor accelerators, and tumor promoting activity of condensate fractions, *Cancer (Philadelphia)*, 27, 848, 1971.
56a. Hoffmann, D. and Wynder, E. L., Environmental respiratory carcinogenesis, *ACS Monogr.*, 173, 324, 1976.
57. Hoffmann, D., Rathkamp, G., Brunnemann, K. D., and Wynder, E. L., Chemical studies on tobacco smoke. XXII. On the profile analysis of tobacco smoke, *Sci. Total Environ.*, 2, 157, 1973.
57a. Inventory of air pollutant emissions, 1968, Natl Air Pollut. Control Adm. Publ. No. AP-73, U.S. Department of Health, Education, and Welfare, Washington, D.C., 1970, 36.
58. Jacobs, E. S., Brandt, P. J., Hoffmann, C. S., Patterson, G. H., and Willis, R. L., PAH emissions from vehicles, paper presented to the American Chemical Society, Washington, D.C., March 31, 1971.
59. Kertesz-Saringer, M., Meszaros, E., Varkonyi, T., Experimentelle Untersuchungen über den 3,4-Benzpyren-Gehalt im Schwebstaub, *Z. Gesamte Hyg. Ihre Grenzgeb.*, 17, 571, 1971.
60. König, J., Funcke, W., Balfanz, E., Grosch, B., and Pott, F., Testing a high volume air sampler for quantitative collection of polycyclic aromatic hydrocarbons, *Atmos. Environ.*, 14, 609, 1980.
60a. König, J., Funcke, W., Balfanz, E., Grosch, B., Romanowski, T., and Pott, F., Untersuchung von 135 polyzyklischen aromatischen Kohlenwasserstoffen in atmospharischen Schwebstoffen aus 5 Stadten der Bundesrepublik Deutschland, *Staub Reinhalt. Luft*, 41, 73, 1981.
61. König, J., Funcke, W., Balfanz, E., Grosch, B., Romanowski, T., and Pott, F., Vergleichende Untersuchung von atmospharischen Schwebstoffproben aus 5 Städten der Bundesrepublik Deutschland auf ihren Gehalt an 135 polycyclischen aromatischen Kohlenwasserstoffen, *Staub Reinhalt. Luft*, in press, 1980.
61a. König, J., Unpublished data.
62. Lao, R. C., Thomas, R. S., Oja, H., and Dubois, J., Application of a gas chromatograph — mass spectrometer — data processor combination to the analysis of the polycyclic aromatic hydrocarbon content of airborne pollutants, *Anal. Chem.*, 45, 908, 1973.
63. Lawther, P. J. and Waller, R. E., Coal fires, industrial emissions of motor vehicles as sources of environmental carcinogens, *Inserm*, 52, 27, 1976.
64. Lee, M. L. and Novotny, M., Gas chromatography/mass spectrometric and nuclear magnetic resonance spectrometric studies of carcinogenic polynuclear aromatic hydrocarbons in tobacco and marijuana smoke condensates, *Anal. Chem.*, 48, 405, 1976.
65. Lee, M. L., Novotny, M., and Bartle, K. D., Gas chromatography/mass spectrometric and nuclear magnetic resonance determination of polycyclic aromatic hydrocarbons in airborne particulates, *Anal. Chem.*, 48, 1566, 1976.
66. Mašek, V., 3,4-Benzpyren in lungengängigen und nicht lungengängigen Teilen des Flugstaubes der Kokereien, *Zentralbl. Arbeitsmed. Arbeitsschultz*, 24, 213, 1974.

67. Masuda, Y. and Karatsune, M., Polycyclic aromatic hydrocarbons in smoked fish "Katsuobushi", *Gann*, 62, 27, 1971.
67a. McNab, L., in *Particulate Polycyclic Organic Matter*, National Academy of Science, Washington, D.C., 1972.
68. Meyer, J. P. and Grimmer, G., Einflüsse PAH-haltiger und PAH-freier Kraftstoffe auf die Emission von polycyclischen aromatischen Kohlenwasserstoffen eines Kraftfahrzeuges mit Ottomotor im Europa-Test, *German Soc. Mineral Oil Sci. Coal Chem.*, Research Report No. 4547, II, Hamburg, 1974.
69. Mohan Rao, A. M. and Vohra, K. G., The concentrations of benzo[a]pyrene in Bombay, *Atmos. Environ.*, 9, 403, 1975.
69a. Muhich, A. J., Klee, A. J., and Britton, P. W., 1968 National Survey of Community Solid Waste Practice, Public Health Service Publ. No. 1866, U.S. Department of Health, Education, and Welfare, Cincinnati, 1968.
70. Müller, H., Aufnahme von 3,4 Benzpyren durch Nahrungspflanzen aus künstlich angereicherten Substraten, *Z. Pflanzenernähr. Bodenkd.*, 685, 1976.
70a. 1968 Wildfire Statistics, Division of Cooperative Forest Fire Control, Forest Service, U.S. Department of Agriculture, Washington, D.C., 1969, 48.
71. *Particulate Polycyclic Organic Matter*, National Academy of Science, Washington, D.C., 1972.
72. Pierce, R. C. and Katz, M., Dependency of polynuclear aromatic hydrocarbon content on size distribution of atmospheric aerosols, *Environ. Sci. Technol.*, 9, 347, 1975.
73. Pott, F. and Dolgner, R., Problems involved in finding an exposure limit for polycyclic aromatic hydrocarbons, in *Luftverunreinigung durch polycyclische aromatische Kohlenwasserstoffe — Erfassung und Bewertung*, VDI-Verlag, Dusseldorf, 1980, 375.
74. Pott, F., Tomingas, R., and König, J., Problems and possibilities of determining the carcinogenic potency of inhalable fibrous dusts and polycyclic aromatic hydrocarbons, in *13th Rochester Int. Conf. Environ. Toxicity "Measurements of Risks"*, in press, 1980.
75. Prietsch, W., Wettich, K., and Kahl, H., 3,4-Benzpyren in der Stadtluft an zwei Berliner Verkehrsschwerpunkten, *Z. Gesamte Hyg. Ihre Grenzgeb.*, 17, 573, 1971.
75a. Refining survey, *Oil Gas J.*, 67, 115, 1969.
76. Sawicki, E., Analysis of atmospheric carcinogens and their cofactors, in *Environmental Pollution and Carcinogenic Risks*, Rosenfeld, C. and Davis, W., Eds., Institut National de la Sante et de la Recherche Medicale, Paris, 1976, 297.
77. Schneider, W., Matter, L., and Jerrmann, E., Benzpyrengehalte der Luft eines Ballungsgebietes, *Umwelthygiene*, 9, 273, 1975.
78. Shabad, L. M., Keesina, A. Y., and Khitrova, S. S., Carcinogenic hydrocarbons in automobile exhaust gases and their possible elimination, *Vestn. Akad. Med. Nauk SSSR*, 23, 6, 1968.
78a. Smith, W. M., Evaluation of coke oven emissions, 78th Gen. Meet. Am. Iron Steel Inst., New York, March 28, 1970.
79. Snook, M. E., Severson, R. E., Arrendale, R. F., Higman, H. C., and Chortyk, O. T., Multi-alkylated polynuclear aromatic hydrocarbons of tobacco smoke: separation and identification, *Beitr. Tabakforsch. Int.*, 9, 222, 1978.
80. Souci, S. W., Über den Einfluss von Verbrennungsgasen auf Lebensmittel und Futtermittel, *Dtsch. Lebensm. Rundsch.*, 64, 235, 1968.
81. Starkey, R. and Warpinski, J., Size distribution of particulate benzo[a]pyrene, *J. Environ. Health*, 36, 503, 1974.
82. Stocks, P., Commins, B. T., and Aubry, K. V., *Int. J. Air Water Pollut.*, 4, 141, 1961; as cited in Hoffman, D. und Wynder, E. L., *ACS Monogr.*, 173, 324, 1976.
83. *Systemanalyse, M.A.N. — Neue Technologie*, A.W. Genter Verlag, Stuttgart, 1976, 44.
83a. Technical-Economic Study of Solid Waste Disposal Needs and Practices, Public Health Serv. Publ. No. 1886, Bureau of Solid Waste Management, Environmental Control Administration, Public Health Service, U.S. Department of Health, Education, and Welfare, Washington, D.C., 1969, 44.
84. Tomingas, R., unpublished data 1978.
85. Tomingas, R. and Brockhaus, A., Immissionsmessungen an Sonntagen mit Fahrverbot an einer Messtelle in Düsseldorf, *Staub Reinhatt. Luft*, 34, 87, 1974.
86. Tomingas, R., Pott, F., and Voltmer, G., Profile von polycyclischen aromatischen Kohlenwasserstoffen in Immissionen verschiedener Schwebstoffsammelstationen in der westlichen Bundesrepublik Deutschland, *Zentralbl. Bakteriol. Parisitenkd. Infektionskr. Hyg., Abt. Orig. Reihe B*, 166, 322, 1978.
87. Tomingas, R. and Voltmer, G., Abscheidung von Benzo[a]pyren aus der Atmosphäre auf Glasfilter, *Staub Reinhalt. Luft*, 38, 216, 1978.
88. Tomingas, R., Untersuchung der PAH-Belastung im Ruhrgebiet — Vergleich mit einer Reinluftstation, in *Luftverunreinigung durch Polycyclische Aromatische Kohlenwasserstoffe — Erfassung und Bewertung*, VDI-Verlag, Dusseldorf, 1980, 147.

89. Toth, L. and Blaas, W., Effect of smoking technology on the content of carcinogenic hydrocarbons in smoked meat products. I. Effect of various smoking methods, *Fleischwirtschaft,* 52, 1121, 1972.
90. Toth, L. and Blaas, W., Cancerogene Kohlenwasserstoffe in gegrillten Fleischwaren, *Fleischwirtschaft,* 53, 1456, 1973.
91. Wadleigh, C. H., Wastes in relation to agriculture and forestry, *U.S. Dept. Agric. Misc. Publ.,* No. 1065, 112, 1968.

Chapter 4

# BEHAVIOR OF PAH IN THE ORGANISM

## TABLE OF CONTENTS

4.1 Intake and Distribution of PAH — F. Pott and G. Oberdörster ........... 130
    4.1.1 Deposition of Particles in the Respiratory Tract ................ 130
    4.1.2 Retention and Clearance of Deposited Substances .............. 133
    4.1.3 Retention and Clearance of Deposited PAH ................... 134
        4.1.3.1 Upper Respiratory Tract Compartment ........... 135
        4.1.3.2 Tracheobronchial Compartment .................. 135
        4.1.3.3 Lower Respiratory Tract Compartment ........... 136
    4.1.4 Experimental Results on the Activity of PAH in the Respiratory Tract ................................................. 136

4.2 Metabolism of Polycyclic Aromatic Hydrocarbons — J. Jacob and G. Grimmer ................................................. 137
    4.2.1 Metabolism of PAH ........................................ 137
        4.2.1.1 Formation of Epoxides Catalyzed by the Monooxygenase System ........................................ 137
        4.2.1.2 Conversion of Epoxides into Dihydrodiols by Aryl Epoxide Hydrolase ................................. 139
        4.2.1.3 Phenols ....................................... 142
        4.2.1.4 Reaction with Sulfuric Acid and Glucuronic Acid.... 143
        4.2.1.5 Quinones ..................................... 143
        4.2.1.6 $C_1$ Transfer ................................... 143
        4.2.1.7 Dihydromonols and Catechols ................... 143
        4.2.1.8 Side-Chain Oxidation .......................... 144
        4.2.1.9 Reactions with Glutathione Catalyzed by Glutathione-S-Epoxide Transferase (E.C.2.5.1.18) ............... 145
        4.2.1.10 Reaction with Proteins ......................... 146
        4.2.1.11 Reaction with Nucleic Acids .................... 146
        4.2.1.12 Metabolic Pathways of Benzo[a]pyrene ........... 148
    4.2.2 Methods of Detection ...................................... 148
        4.2.2.1 High-Pressure Liquid Chromatography (HPLC) .... 148
        4.2.2.2 Gas Chromatography (GLC) .................... 148
        4.2.2.3 Analytics of DNA Compounds .................. 151

References ................................................. 152

## 4.1 INTAKE AND DISTRIBUTION OF PAH

### F. Pott and G. Oberdörster

The intake of PAH and their distribution in the organism are dependent on numerous physiological, physical, and chemical factors which have still not been sufficiently elucidated. The various steps involved in the intake and distribution of PAH in the body are described so as to lead to a basic understanding of these processes.

### 4.1.1 Deposition of Particles in the Respiratory Tract

PAH with four and more rings, which are particularly suspected of being carcinogenic, are almost exclusively adsorbed onto solid particles when they occur in the atmosphere. The question is where do solid particles inhaled with the air remain? There are several possibilities. A considerable percentage of the particles of less than 5 μm diameter are immediately exhaled again into the atmosphere. This is seen clearly in the smoke stream exhaled by smokers after inhalation. Of obvious importance is the region in the respiratory tract where the particles are deposited. Regarding its functions, the respiratory tract can be divided into three compartments (for localization of organs see Figure 41):

1. "Upper" respiratory tract = nasopharyngeal area (extending to the epiglottis)
2. "Middle" respiratory tract = tracheobronchial area (extending from the larynx downward as far as the bronchi are coated with ciliated epithelium)
3. "Lower" respiratory tract = respiratory area [comprising the last branches of the respiratory system and the vesicular ends (pulmonary alveoli, Figure 42), total surface about 30 to 80 m$^2$].[91]

Based on their physical, physiological, and anatomical properties, the particles can be deposited according to three physical principles:

1. Impaction (collision) — The impact occurs due to mass inertia when the particles — in case of a change of flow direction — do not follow the air stream but continue to move straight ahead and collide with a wall. This applies predominantly to particles with a large aerodynamic diameter (>5 μm). The impaction mostly takes place at sharp bends where the air speed is high (nasopharyngeal space, main bronchi).
2. Sedimentation (settling) — Sedimentation of particles in the respiratory tract becomes more and more predominant the more narrow the airways grow, i.e., in the course of bronchial branching. It applies most of all to particles with an aerodynamic diameter of >1 μm.
3. Diffusion [deposition due to a change of direction following the collision of particles with gas molecules (Brownian movement)] — This type of deposition applies predominantly to very small particles (<0.5 μm), in the narrow branches and with longer retention times. It is largely independent of density (see later).

The percentage of inhaled particles which are deposited in these three regions of the respiratory tract depends on several physiological, physical, and anatomical parameters:

1. Physical properties of the particles:
- Aerodynamic diameter (derived from the rate of vertical descent of a particle)

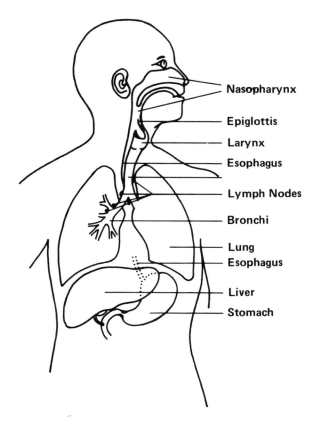

FIGURE 41. Topography of selected organs.

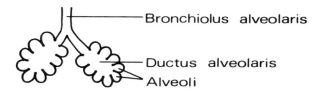

FIGURE 42. Diagrammatic view of the alveoli (pulmonary alveoli).[29]

- Geometrical diameter
- Shape (borderline case: spheres; the aerodynamic diameter of a sphere depends on the geometrical diameter and the material density only)
- Electric charge
- Material density (the density is irrelevant in the case of particles with an aerodynamic diameter below 0.5 μm; for these very small particles the geometrical diameter is crucial)
2. Air velocity in the respiratory tract (dependent on the respiratory rate, tidal volume, diameter of airways)
3. Branching of the respiratory tract

The Task Group on Lung Dynamics[107] presented a diagram established by measuring

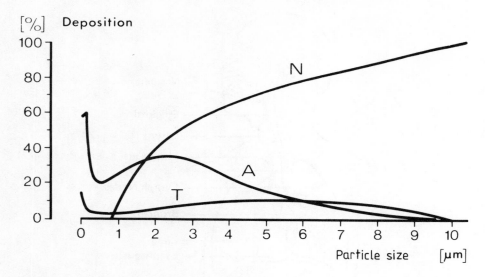

FIGURE 43. Deposition of particles in the different sections of the respiratory tract in relation to particle size; tidal volume: 750 m$\ell$ in 4 sec (= 15 inspirations per minute; minute volume: 11.2 $\ell$/min) spheric particle shape, material density = 1; N = nasopharyngeal compartment, T = tracheobronchial compartment, A = alveolar compartment.[107]

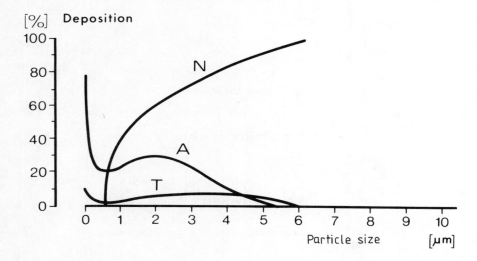

FIGURE 44. Diagram as in Figure 43, high tidal volume: 2150 m$\ell$/4 sec (= 15 inspirations per minute; minute volume: 32.3 $\ell$/min).[107]

and calculating the percentage of inhaled particles of different sizes which were precipitated in the different sections of the respiratory tract, depending on the tidal volume and respiratory rate. Figures 43 and 44 show two examples. It can be seen that particles with a diameter exceeding 10 μm do not penetrate into the middle and lower sections of the respiratory tract. Thus the danger that individual particles might block the respiratory tract does not arise since bronchioles and alveoli have a diameter of at least 200 to 300 μm. The so-called nose filter accomplishes its function by impaction of the bigger particles. Quite often the misleading statement has been made that particles >10 μm are not respirable because they do not reach the alveoli. They may, however, pen-

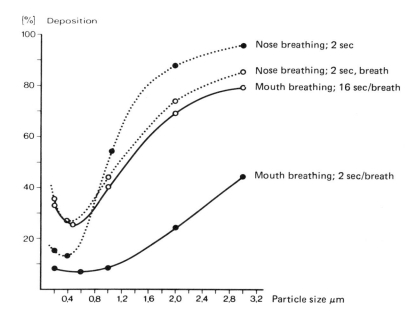

FIGURE 45. Total deposition of particles in the respiratory tract in relation to particle size and respiratory rate; tidal volume: 1000 mℓ.[42]

etrate into the nasopharyngeal space, i.e., depending on their properties they may also exert a biological activity. Particles >20 μ are rarely found in the atmosphere under light-wind conditions because they rapidly settle down after emission. If, however, dust is whirled up, particles up to about 100 μm may be inhaled into the nasal space where they are deposited.

Due to the high humidity in the upper respiratory tract small hygroscopic particles may increase in size and correspondingly alter their pattern of deposition.

Relatively few particles are deposited in the tracheobronchial region. The maximum and minimum values of deposition in the alveoli refer to particles of about 2 μm and about 0.5 μm, respectively. The minimum is due to the fact that deposition by sedimentation has already decreased considerably in this area whereas deposition by diffusion does not yet play a role.

The curves of total deposition show a minimum for particles of 0.4 to 0.6 μm. This was demonstrated by Heyder et al.[42] in well-defined investigations with volunteers, taking the breathing rate, depth of respiration, and oral or nasal respiration into account (see example in Figure 45). These results defined more closely the data on particle sizes between 0.2 and 3 μm previously presented by the Task Group on Lung Dynamics.[107]

### 4.1.2 Retention and Clearance of Deposited Substances

The retention of particulate matter after deposition depends largely on the section of the respiratory tract in which the particles were deposited:

1. After deposition in the upper respiratory tract:
   - Excretion with sputum or nasal secretion (within some minutes or a few hours)
   - Swallowing and transport into the gastrointestinal tract (largest amount) with subsequent excretion via feces or — after absorption of soluble components into blood and lymph — transport to the liver and excretion via bile ducts into the

bowels or distribution within the body via blood circulation and excretion through the kidneys
- Dissolution of the particle, or some of its constituents, and penetration of these substances into cells, lymph, blood (this step being of hardly any significance in the upper respiratory tract)
2. After deposition in the middle respiratory tract:
- Inclusion of the particle in macrophages
- Transport of the particle by ciliary movement of the ciliated epithelium, followed by coughing out with sputum or mostly by transport via the esophagus into the gastrointestinal tract (bronchial elimination of particles from the lung; duration less than 24 hr)
- Dissolution of the particle or of its constituents during transport by the ciliated epithelium and translocation of these substances into cells, lymph, blood
3. After deposition in the lower respiratory tract:
- Dissolution of the particle or of its constituents and translocation of these substances into cells, lymph, blood
- Movement of the undissolved particle included in alveolar macrophage to the "conveyor belt" of the ciliated epithelium with subsequent elimination via the gastrointestinal tract (within several days or months or even years); thus elimination of the majority of solid particles deposited in the lung
- Translocation of the particle through alveolar epithelial cells or junctions into the interstitial space of the lung, the parenchyma cells or the lymph and thus into the lymph nodes and perhaps into the blood or other organs

Insoluble dust is more rapidly cleared from the lung if its particle size is larger, because larger particles are mainly deposited in the upper and middle section of the respiratory tract. If, however, the inhaled dust consists predominantly of small particles, which are deposited to a larger extent into the alveolar region, then their mean retention time is usually much longer.

### 4.1.3 Retention and Clearance of Deposited PAH

The processes of elimination of particles deposited in the lung which have been described earlier in general will now be investigated specifically with respect to PAH and their action in the respiratory tract. The following questions arise:

- To what extent are those PAH which are of importance in carcinogenicity eliminated from the respiratory tract by the earlier described mechanisms, without being modified?
- In what parts of the respiratory tract does metabolism of PAH take place?
- What physiological processes in the respiratory tract, caused by deposited PAH, are of essential importance in tumor induction?
- What changes in processes occurring normally in the respiratory tract can enhance the induction of lung cancer initiated by inhaled PAH, and which influences can reduce the incidence of PAH-induced lung cancer?

The answers to these questions are still very fragmentary, but at least some aspects should be discussed here.

Of primary interest is the finding that in the lungs of deceased humans only very small amounts (0.3 to 15 μg), or no benzo[a]pyrene, was found.[28,60,114] On the other hand, PAH could be detected in the tissue of human lung carcinomas. This finding has not yet been explained satisfactorily.[109] According to the authors, at least benzo-

[a]pyrene is rapidly eliminated from the lung under normal conditions. This seems plausible, taking the action of PAH in the individual sections of the respiratory tract into consideration.

*4.1.3.1 Upper Respiratory Tract Compartment*

The mostly large particles deposited in this region contain relatively few PAH (see Chapter 3.5). They are soon eliminated. It is unknown to what extent PAH are separated from particles deposited on the mucosa. In any case, these particles or some of their constituents do not reach the middle region of the respiratory tract. At the most, they may have an effect on the gastrointestinal tract. However, according to present knowledge, the carcinogenic effect of inhaled PAH on the stomach is negligible. Since tumors occur relatively seldom in the nasopharyngeal space and in the oral cavity (except in the buccal cavity of smokers), this region is of minor interest to the discussion of the present problem.

*4.1.3.2 Tracheobronchial Compartment*

Due to the high tumor frequency observed, special attention is directed to this region of the respiratory tract. About 70% of all bronchial carcinomas develop near the hilus, i.e., in the upper region of the bronchi. Histologically, about 50% of these carcinomas are of the so-called parvicellular type, while 35 to 40% turn out to be squamous cell carcinomas with and without keratinization. The adenocarcinoma accounts for a relatively small percentage of 7%, the so-called alveolar cell carcinoma being considered a special type of adenocarcinoma. Finally the undifferentiated (anaplastic) bronchial carcinomas should be mentioned, these forming a collective group with different structures.[94]

There are several arguments in support of the assumption that despite the low percentage of the overall amount of particles deposited (see Figure 43), the concentration of carcinogens is high locally in the upper bronchial tree. The particles are precipitated predominantly by impaction on partitions and bends of the bronchi, i.e., unlike the alveolar region the "load" is not evenly distributed. Without giving exact data on the dimension of the surface area of the upper tracheobronchial tract, it can be said that this region accounts for only a small portion of the nearly 50-$m^2$ alveolar area, and therefore a relatively high deposition of particles takes place. Consequently, the "return flows" of the numerous particles deposited in the peripheral ciliated branches of the respiratory tract meet, and concentrate, on the relatively small surface area of the upper bronchi. Finally, local disorders of the ciliated epithelium may cause a congestion of particles and thus a more intensive impact of possible carcinogenic constituents. So far no results have been presented which might answer three important questions:

1. To what extent are deposited PAH separated from the particles under physiological conditions?
2. Where does the separation take place?
3. Do PAH really exert a more concentrated activity on the epithelial cells of certain parts of the upper bronchial region?

A positive correlation may be established, though, between the different frequencies of carcinomas in the five lobar bronchi and the amount of dust deposited in the corresponding areas of the respiratory tract. This correlation is supported by pathological anatomical findings from results obtained in corresponding model experiments (Figure 46).[89] These data suggest at least that certain particle-related carcinogens might act towards tumor induction.

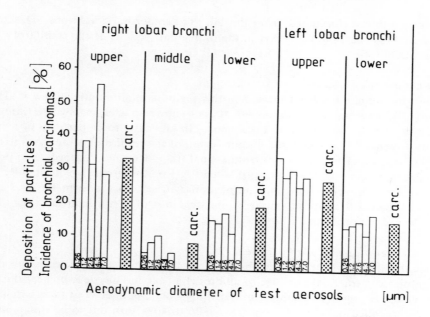

FIGURE 46. Deposition of particles of different size in the individual lobar bronchi in percentage of the amount of particles of a certain size deposited in all five lobar bronchi in a model experiment (air flow rate 30 ℓ/min) in relation to the incidence of bronchial carcinomas observed in the five lobar bronchi.[89]

### 4.1.3.3 Lower Respiratory Tract Compartment

The majority of particles deposited in the alveoli remains there for a relatively long period (days to months and years) until they are removed to the ciliated epithelium by fluid movement or, after inclusion into macrophages, by active movement of these cells, and then they are transported to the esophagus within 24 hr. In the alveoli not only are the conditions for a gas exchange between blood and respiratory tract and for the passage of liquid substances into the blood particularly favorable, but some substances with a low solubility in water are also readily dissolved in the fluid film (phospholipid film, "surfactant") coating the alveoli, and are then easily absorbed. This may apply, for instance, to lead and zinc compounds as well as PAH. Furthermore, PAH can be metabolized in alveolar macrophages, i.e., in the respiratory tract.[14,23,24,30,68,69] It is not known to what extent PAH are separated from deposited particulate matter in the lower respiratory tract and then metabolized or absorbed without being modified. Presumably, the percentage of separated PAH is so high that hardly any PAH deposited in the lower respiratory tract are eliminated via the bronchial route and thus cannot have any considerable quantitative effect on the bronchial epithelium.

### 4.1.4 Experimental Results on the Activity of PAH in the Respiratory Tract

Detailed data on the retention time of PAH — in particular of BaP — in the respiratory tract of experimental animals, their modification by different factors and the rate of PAH-induced tumors are reported in Chapter 5.2.2.1.

## 4.2 METABOLISM OF POLYCYCLIC AROMATIC HYDROCARBONS

### J. Jacob and G. Grimmer

#### 4.2.1 Metabolism of PAH
*4.2.1.1 Formation of Epoxides Catalyzed by the Monooxygenase System*

In mammalia, PAH are oxidized to epoxides by mixed-function oxidases such as cytochrome P 450, with NADPH being used as a cosubstrate. During this process the two atoms each of an oxygen molecule are concurrently transferred to the PAH and NADPH:

$$\text{benzene} + NADPH + O_2 \xrightarrow[\text{(Cytochrom P 450/448)}]{\text{Monooxygenase}} \text{arene oxide} + NADP^+ + OH^-$$

The monooxygenase system is located at the endoplasmic reticulum and seems to be linked to the membrane. Oxidations of aromatic hydrocarbons by microsomes can be carried out in vitro after cell fractionation. Activity of monooxygenase has been detected in the tissue (liver, lung, kidney, intestinal tract, skin) of different mammalia (rat, mouse, guinea pig, hamster, rabbit, dog, monkey, man). The complex enzyme system was isolated from rabbit liver by Coon et al.[20] According to these findings its function does not depend on a link to membranes, although this has been repeatedly claimed. Beside the hemoprotein cytochrome P 450, which has a molecular weight of about 50,000, the regenerating flavoprotein system (NADPH-cytochrome P 450 reductase), phospholipid, and a so-far unidentified factor (factor C), have been detected as functional units (for review see Ullrich[111]). The content in microsomal protein of cytochrome P 450, about 10%, is very high so that it has to be considered one of the most important enzymes of the endoplasmic reticulum. It is also possible to differentiate between a number of isoenzymes which obviously have different specificities for individual regions of aromatic hydrocarbons and can be induced by different substances, for instance, in the liver.[21,63,64,73] Their absorption peaks (CO-binding spectrum) clearly differ from each other (448 and 450 nm). Coon et al.[20] detected at least four enzymes in rabbit liver. Other authors described at least six different isocytochromes P 450, each possessing different but overlapping substrate specificities.[44] Tsuji et al.[110] have recently reported on the separation of different forms of cytochrome P 450 from pig liver microsomes. The ratio of the different monooxygenases varies from species to species as well as between strains and even sexes. Within a distinct species their ratio may vary from one tissue to another, and different subcellular localizations have been reported.[2,38,112]

According to our present knowledge, the P 450 system seems to preferentially oxidize olefinic double bonds or positions of increased electron localization in aromatic hydrocarbons whereas the P 448 system shows clear affinity to aromatic bond systems.

X-ray structural analysis has revealed that in aromatic systems the distances between C-C bonds vary and, therefore, individual bonds have a more or less olefinic character. Thus, for example, the following bond lengths were measured in anthracene.[16]

This means that not all possible mesomeric structures participate equally in the formation of the actual structure.

With benz[a]anthracene as substrate, incubations with liver microsomes of normal rats have produced 5,6-, 8,9-, and 10,11-dihydrodiols.[46] This supports the assumption that Structure I accounts to a greater degree than Structure II for the total status of linkage to the enzyme surface:

structure I    structure II

The chemical attack with $OsO_4$ really takes place at the 5.6 position. Liver microsomes of rats in which the cytochrome P 450-system is induced also show a preferred reaction at this bond site during incubation with benz[a]anthracene. This position is of a more olefinic character, similar to the 9.10 position of phenanthrene and was formerly designated as the K-region. Cytochrome P 448-induced animals, however, produce metabolites with different profiles in which other bond sites are preferentially attacked.[44]

On the other hand, there is no evidence that the positions C-2/C-3 or C-9/C-10 are attacked. Schmidt[90] and Clar[22] repeatedly hinted at a more or less marked, fixed localization of double bonds in aromatic hydrocarbons.

Monooxygenases can be induced by various chemicals as, for instance, phenobarbital, aminopyrine, chlordane, 5,6-benzoflavone ($\beta$-naphthoflavone), PCB and various PAH such as benzo[a]pyrene, 3-methylcholanthrene, 7,12-dimethylbenz[a]anthracene, dibenz[a,h]anthracene, benzo[k]fluoranthene.[9,10,18,19,33,45,70,92,117] Phenobarbital specifically induces the P 450 system whereas $\beta$-naphthoflavone and at least some PAH such as benz[a]anthracene, benzo[a]pyrene, and 3-methylcholanthrene specifically induce the P 448 system. PCB seem to be able to stimulate both systems.[6] A dosage-related induction of P 448 monooxygenase has been found with benzo[k]fluoranthene.[93] Argus et al.[1] demonstrated that PAH such as benz[a]anthracene or benzo[a]pyrene lose their inductive properties after being oxidized in the K- or bay-region, respectively. The inducibility of monooxygenases by benz[a]anthracene decreases to about ¼ after oxidation to 5,6-epoxide whereas 7,8-dihydrodiol and 7,8-dihydrodiol-9,10-epoxide of benzo[a]pyrene totally lack inducing properties. Depierre and Ernster,[25] and more recently Wiebel[116] presented an overview of monooxygenase induction. As expected, the extent of induction depends largely on the type of inducer.

Induction of monooxygenase by cigarette and marijuana smoke probably also has to be seen in connection with the monooxygenase-inducing capacity of PAH.[83,115]

Epoxidation of aromatic systems is not restricted to unsubstituted PAH. Monooxygenases can also convert previously oxidized aromatic hydrocarbons into epoxides. Thus, for instance, aromatic hydrocarbons which were already subjected to oxidation with subsequent hydration to trans-dihydrodiols by epoxide hydrolases (see Chapter 4.2.1.2) can again be used as monooxygenase substrate. This was proven in the case of benz[a]anthracene[7,8] and benzo[a]pyrene[119] (cf. Figure 47 for benzo[a]pyrene).

It is still unknown whether, in general, cytochrome P 448 or P 450 is the system responsible for the renewed reaction. Pezzuto et al.[86] proved that cytochrome P 448 exclusively participates in the epoxidation of trans-7,8-dihydroxy-7,8-dihydrobenzo[a]pyrene.

Contrary to observations of an increased reaction (induced, e.g., by 5,6-benzoflavone (β-naphthoflavone) there are several reports on a specific inhibition of the 3-methylcholanthrene-induced monooxygenase by 7,8-benzoflavone (α-naphthoflavone) in mouse skin,[54,55] in liver microsomes,[117] and in cell cultures.[26]

Aryl epoxides are very reactive compounds which readily react with different cell constituents, such as amino acids, proteins, and nucleic acids. In an acid medium, phenols are formed by spontaneous isomerization. However, it is still an unanswered question whether the conversion of epoxides into phenols, which can always be observed in vivo, takes place enzymatically or entirely spontaneously. A main metabolic pathway is the hydration of epoxides into trans-dihydrodiols, which are catalyzed by the aryl epoxide hydrolase (or hydratase) system.

### 4.2.1.2 Conversion of Epoxides into Dihydrodiols by Aryl Epoxide Hydrolase

Like the monooxygenase system, the epoxide hydrolase system is located at the endoplasmic reticulum of cells. During cell fractionation, it remains in the microsomal fraction which is therefore suitable for in vitro hydration of epoxides. Purified epoxide hydrolases have been isolated on several occasions.[3,57,65] The enzyme obtained from liver microsomes of phenobarbital-induced rats has a molecular weight of 50,000 in the presence of sodium dodecyl sulfate, while in the absence of this detergent, it is present as an aggregate of 600,000 to 650,000 daltons. Activity of epoxide hydrolase seems to exist in almost all tissues and in all mammalia. However, its titer varies considerably in the different tissues of a species.[74] However, Oesch et al.[78] could not detect any activity of epoxide hydrolase in the brain, heart, lung, spleen, nor the muscles of rats and guinea pigs.[82]

There is no evidence of a participation by NADPH and other cosubstrates in the enzymatic conversion of epoxides into trans-dihydrodiols.[76] Although at least two epoxide hydrolases (epoxide hydrolase EC 4.2.1.63 and arene oxide hydrolase EC 4.2.1.64) have been described, it is still an open question whether only one hydrolase with a widespread substrate specificity or several hydrolases, each with a high substrate specificity, catalyze the conversion of different epoxides.[79] The rate of in vitro hydration of different benzo[a]pyrene epoxides (4,5-, 7,8-, 9,10-, 11,12-oxide) in the presence of hydrolase obtained from rat liver microsomes can indeed be differently stimulated or inhibited by adding metyrapone (1,2-di-(3-pyridyl)-2-methyl-1-propanone) or cyclohexene oxide, respectively. So far, however, an electrophoretic resolution into different hydrolases or enzyme subunits has not been achieved. It is therefore not possible to decide whether a range of isoenzymes is involved or whether a hydrolase has different catalytic regions for individual substrates.[66] On the other hand, it is a fact that in different animal species epoxide hydrolases are present which are distinguishable by their structure.[66]

Pandov and Sims[85] observed, for instance, different rates of reaction for phenanthrene 9,10-oxide (rapid) and dibenz[a,h]anthracene 5,6-oxide (slow) catalyzed by hydrolase from rat liver microsomes. Incubation experiments carried out by Swaisland et al.[105] showed that a simple relationship between the carcinogenicity of a PAH and the rate of conversion of its epoxides cannot be established.

According to Oesch and Daly[77] the monooxygenase and hydrolase system detected in the liver of guinea pigs has to be considered as a multienzyme complex, the activities of which have to be assigned to distinguishable and separable individual structures.[81]

The hydrolase system can be inhibited in vitro by various substances (e.g., cyclohexene oxide, 3,4-dihydronaphthalene 1,2-oxide, 1,1,1-trichloropropene oxide).[80]

Selkirk et al.[95] completely blocked the hydration of benzo[a]pyrene epoxides by adding 1,1,1-trichloropropene oxide and thus studied the reactions effected in benzo[a]pyrene by monooxygenase alone. By adding purified hydrolase, dihydrodiols could again be detected.[43] A low inducibility by PAH (3-methylcholanthrene) and a distinct induction of epoxide hydrolase by phenobarbital was observed in the liver of guinea pigs.[77] In other experiments an inducibility of epoxide hydrolase by 3-methylcholanthrene could not be detected in the liver of guinea pigs[62] and mice.[72]

For a long time the hydration of PAH epoxides was considered a "detoxification" of PAH in the organism. However, the absolute correctness of this thesis has been questioned since it became known that dihydrodiols, on their part, may serve as substrates for monooxygenases, some of which have to be regarded as the actual effective carcinogens. The so-called "bay-region" dihydrodiol epoxides such as 7,8-dihydro-7,8-dihydroxybenzo[a]pyrene 9,10-oxide as well as other dihydrodiols with "bay-regions" are of particular interest as reactants ("ultimate carcinogens") with essential cell constituents (DNA)[61a] so that at least in some PAH the hydration of their epoxides has to be considered as a sort of toxification into carcinogenic precursors.

Some 'Bay-Region'-containing PAH and related Dihydrodiolepoxides being ultimate Carcinogens.

Chrysene → trans-1,2-Dihydroxy-3,4-epoxy-1,2-dihydrochrysene

Benz(a)anthracene → trans-3,4-Dihydroxy-1,2-epoxy-3,4-dihydrobenz(a)anthracene

3-Methylcholanthrene → trans-9,10-Dihydroxy-7,8-epoxy-3-methyl-9,10-dihydrocholanthrene

Benzo(a)pyrene → trans-7,8-Dihydroxy-9,10-epoxy-7,8-dihydrobenzo(a)pyrene

Dibenz(a,h)anthracene → trans-3,4-Dihydroxy-1,2-epoxy-3,4-dihydrodibenz(a,h)anthracene

The stereoselectivity of monooxygenases and epoxide hydrolases has been proved, with benzo[a]pyrene as an example resulting in exclusively (−)-enantiomers of trans-dihydrodiols. The second attack on the 9,10-position by monooxygenase results in two stereoisomeric forms of the 7,8-dihydroxy-9,10-epoxy-7,8,9,10-tetrahydrobenzo[a]-pyrene, the ratio of which depends on the balance of monooxygenase isoenzymes.

(−)trans-7,8-dihydrodiol →

(+)7β,8α-dihydroxy-9α,10α-epoxy-7,8,9,10-tetrahydrobenzo(a)pyrene
(+)anti-Diolepoxide

(−)7β,8α-dihydroxy-9β,10β-epoxy-7,8,9,10-tetrahydrobenzo(a)pyrene
(−)syn-Diolepoxide

The reaction velocities of the above-mentioned diolepoxides are different. Although it could be demonstrated that both isomers are bound to DNA and RNA, it seems that the antiform is the more relevant ultimate carcinogen.

There are some exceptions to the "bay-region" theory. Not all PAH with such regions are potent carcinogens, e.g., benzo[e]pyrene and chrysene with two bay-regions are not, or are only weakly carcinogenic, respectively. On the other hand, anthanthrene exhibits considerable carcinogenic activity although lacking a bay-region.

Unlike mammalia, microorganisms *(Beijerinckia, Pseudomonas)* have hydrolases which convert the intermediary epoxides into *cis*-dihydrodiols.[15,37,50,52]

**Further metabolic pathways** — Apart from epoxidation, hydration of the epoxides to trans-dihydrodiols, and reaction with proteins and nucleic acids as described in sections 4.2.1.10 and 4.2.1.11, a number of further partly tissue-specific metabolic pathways of PAH has been observed.

*4.2.1.3 Phenols*

It is generally believed that phenols, which regularly occur as main metabolites, are formed in vivo by spontaneous isomerization of epoxides although experimental evidence is still lacking. It is remarkable that, in general, only one of the two possible products of isomerization is detectable as a product of metabolism, e.g., 3-, 7-, and 9-hydroxybenzo[*a*]pyrene as derivatives of 2,3-, 7,8-, and 9,10-epoxide, respectively, but not the corresponding 2-, 8-, and 10-hydroxybenzo[*a*]pyrene. It was suggested that trans-dihydrodiols might be dehydrated in vivo to form phenols. However, according to investigations carried out with $^{18}$O-labeled dihydrodiols[119] this seems unlikely. Similarly, acid treatment of epoxides, as well as of transdihydrodiols, reveals a clear preference for the isomers detected in vivo even under in vitro conditions:

Trans-dihydrodiols often react with sulfuric acid in vivo and are excreted as monosulfates. For instance, Boyland and Sims[11] detected 9,10-dihydro-9-hydroxy-10-phenanthrylsulfate in urine of rats. Unlike this stable compound the sulfuric acid monoester of trans-1,2-dihydro-1,2-dihydroxynaphthalene is unstable and disintegrates into phenol and phenolsulfate. In this case, too, one isomer seems to be formed preferentially whereas only a small amount of the other one is present:

According to this mechanism, it may be assumed that due to the high stability of sulfuric acid monoesters of K-region dihydrodiols, the formation of K-region phenols is inhibited against the formation of other phenols. In the case of benzo[a]pyrene, also, considerable amounts of K-region phenols have not been found as metabolites (4- or 5-hydroxybenzo[a]pyrene).

### 4.2.1.4 Reaction with Sulfuric Acid and Glucuronic Acid

The excretion of phenols and dihydrodiols was observed repeatedly after reaction with sulfuric acid or glucuronic acid catalyzed by sulfotransferases or UDP glucuronyltransferases, respectively.[4,97] Recently benzo[a]pyrene-3-yl-hydrogen sulfate was also identified in cell cultures of human bronchial epithelium as a metabolite of benzo[a]pyrene.[17]

### 4.2.1.5 Quinones

Quinones in small amounts, are regularly found as metabolites of PAH, and benz[a]anthracene-7,12-quinone and benzo[a]pyrene-1,6-, 3,6-, and 6,12-quinone have been identified. With benzo[a]pyrene, following incubation experiments with specimens of lung microsomes of Tupaia, the above-mentioned three quinones were found at a comparatively high percentage (approximately 36% of all metabolites)[32] The metabolic pathways inducing the formation of quinones have not yet been elucidated, and it remains to be investigated to what extent mixed-function oxidases initiate reaction at the meso region (L-region) of PAH.

### 4.2.1.6 $C_1$ Transfer

Flesher and Sydnor,[3] Sloane,[100] and Sloane and Davis[102] reported on the reaction of benzo[a]pyrene in rat liver microsomes producing 6-hydroxymethylbenzo[a]pyrene. The $C_1$ transfer to benzo[a]pyrene in rat liver and lung microsomes is catalyzed by 6-hydroxymethylbenzo[a]pyrene synthetase which is different from cytochrome P 450, can be inhibited by 6-aminochrysene or stimulated by α-naphthoflavone, and needs NADPH as a cosubstrate.[101] It consists of an apoenzyme and a so-far undefined lipid-soluble factor, the function of which can apparently be substituted by vitamin K. Folic acid, as well as S-adenosylmethionine, could be excluded as $C_1$ fragments.

### 4.2.1.7 Dihydromonols and Catechols

In vivo reduction of epoxides of PAH possibly results in dihydromonols:

Dehydration of these compounds, which can be easily effected, would result in the original PAH. If this metabolic pathway were really of importance, this would mean a mechanism of PAH transport in the cell as well as a prolonged retention of the aromatic hydrocarbon due to recycling.

Dehydration of dihydrodiols by dehydrogenases in the liver has repeatedly been described and plays a role in the metabolism of at least some PAH.[75]

### 4.2.1.8 Side-Chain Oxidation

Irrespective of the ring oxidation catalyzed by rat liver microsomes, the alkyl chains of substituted PAH are oxidized to primary hydroxyl groups, which then might be oxidized to carboxyl groups. Thus, for instance, 7,12-dimethylbenz[a]anthracene is oxidized in rat liver first to 7-hydroxymethyl-12-methylbenz[a]anthracene[11a] and then to 12-methylbenz[a]anthracene 7-carboxylic acid, which is found as an excretion product in rat urine.

Possible side-chain oxidations of 7,12-dimethylbenz[a]-anthracene

There do not exist any reports of investigations of side-chain oxidation in the numerous methyl- and dimethyl-substituted PAH occurring in different matrices.

Among other reactions, oxidation of the methyl group of 3-methylcholanthrene to 3-hydroxymethylcholanthrene[98] and to cholanthrene-3-carboxylic acid[40] was observed in rat and mouse liver, respectively. However, the two methylene groups in the hydrated 5-membered-ring react preferentially to form 1- and 2-hydroxy-3-methylcholanthrene, cis- and trans-1,2-dihydroxy-3-methylcholanthrene and 1- and 2-keto compounds. It is not yet known whether this applies in general to partially hydrated PAH systems, which occur quite frequently.

### 4.2.1.9 Reactions with Glutathione Catalyzed by Glutathione-S-Epoxide Transferase (EC.2.5.1.18)

The reaction of PAH epoxides with glutathione is catalyzed by the glutathione-S-epoxide transferase system present in the cytoplasm. This system includes a yet unknown number of isoenzymes with very different substrate specificity.[39] Gelboin et al.[34] identified five different transferases in rat liver and two in human liver, all of which exhibit quite different rates of reaction with the K-region epoxide of benzo[a]pyrene (benzo[a]pyrene 4,5-oxide). This enzyme was detected in the livers and kidneys of rats[12] and later on in numerous other tissues. Pabst et al.[84] isolated glutathione S-epoxide transferase from rat liver. Its titer seems to vary considerably among different species, as well as among the different types of tissue of one animal species, e.g., the activity detected in lungs of rats and hamsters was significantly higher than in rat liver. This seems to be of vital importance for the reaction of PAH with glutathione, i.e., for the detoxification of PAH. In comparison with other epoxides, K-region epoxides are obviously preferentially transferred to glutathione by this transferase in rat liver.[8]

**Reaction of arene epoxides with glutathione:**

In rat kidneys the glutathione compound of naphthalene is enzymatically decomposed by elimination of glutamic acid to form the cysteinyl-glycine compound which is then metabolized to the cysteine compound by cleavage of the glycine. After acetylation this cysteine compound is excreted as mercapturic acid.

*4.2.1.10 Reaction with Proteins*

Reaction of a large number of PAH with proteins was proved both in vivo and in vitro. When comparing the individual types of metabolites, epoxides — as expected — are the ones incorporated most readily into proteins. The rate of incorporation of phenols is also substantial and always higher than that of dihydrodiols. Extensive comparative data[99] have shown that the incorporation of PAH into proteins quantitatively exceeds the reaction of PAH with nucleic acids. Kuroki and Heidelberger[59] reported on a specific conjugation of the K-region epoxides of dibenz[a,h]anthracene. In incubation experiments with benzo[a]pyrene and liver homogenates in which the PAH-protein compounds formed were chemically decomposed, chrysene was detected, i.e., indicating that benzo[a]pyrene 4,5-oxide (K-region) participated predominantly in the basic reaction.[87]

The preference given to K-region epoxides when binding to proteins parallels the reaction of PAH with glutathione. Since sufficient analytical material on stronger hydrophilic metabolites is still lacking, it cannot be decided at present whether this is a generally applicable principle.

*4.2.1.11 Reaction with Nucleic Acids*

There is a quantity of data supporting the assumption that reaction of PAH with deoxyribonucleic acid (DNA), which codifies the cell information, is the ultimate step to malignant cell transformation and induces the conversion of a normal cell into a cancerous one. The binding of PAH to cellular DNA has repeatedly been proved.[9,10,49] It is linked to the monooxygenase activity, is dependent on NADPH, can be stimulated by inducers of monooxygenase (e.g., benzo[a]pyrene, 3-methylcholanthrene), and inhibited by the inhibitors of this enzyme (7,8-benzoflavone). These findings clearly demonstrate that PAH have to be converted into their epoxides prior to reaction with DNA. Furthermore, an increased binding of benzo[a]pyrene and 3-methylcholanthrene to DNA was observed in vitro when inhibitors of epoxide hydrolase (1,1,1-trichloropropene oxide) were added simultaneously.[13] This is difficult to understand in view of the fact that dihydrodiols obviously have to be oxidized again to dihydrodiol epoxides to effect carcinogenesis. Vitamin A deficiency also seems to foster this conjugation.[35]

In vitro experiments have shown that epoxides of PAH readily react with polynucleotides, RNA and DNA, to a greater extent than any of the other metabolites of PAH (phenols, dihydrodiols). The only exception to this is 5-hydroxy-dibenz[a,h]anthracene which in V-79 cell cultures of Chinese hamsters showed higher rates of reaction with DNA than the corresponding epoxide. In general, purine nucleotides have greater rates of reaction than pyrimidine nucleotides. Thus conjugation carried out with model nucleotides in vitro decreases in the following order: polyguanylic acid > polyadenylic acid > polyxanthylic acid > polyinosinic acid > polyuridylic acid > polycytidylic acid; i.e., purines carrying amino groups react preferentially with activated PAH. The rate of reaction with double-stranded DNA is greater than with single-stranded DNA and with RNA. This is in accordance with the assumption that insertion of PAH into the double strand of DNA (intercalation) has to be regarded as a primary process of the covalent binding of PAH to nucleic acids (or purine bases).

In the case of benzo[a]pyrene, it was proven in vitro, as well as in vivo, that the "ultimate carcinogen" 7β,8α-dihydroxy-7,8,9,10-tetrahydrobenzo[a]pyrene-9α,10α-epoxide reacts with the guanine of RNA or DNA, the linkage taking place between the C-10 atom of benzo[a]pyrene and the C-2 amino group of guanine.[51,58,71,113]

The distribution of PAH between proteins and nucleic acids within the cell varies considerably from PAH to PAH. Furthermore, not all epoxides of PAH react with nucleic acids in a like manner. The K-region epoxide of benzo[a]pyrene does not play an obvious role in the reaction with cellular DNA. In the case of a number of PAH (e.g., benz[a]anthracene, 7-methylbenz[a]anthracene, 7,12-dimethylbenz[a]anthracene, 3-methylcholanthrene), "bay-region" or dihydrodiol epoxides seem to react predominantly, if not exclusively, with DNA. (review by Marquardt[67]). Different factors can be responsible for the initiation of this crucial reaction with cellular DNA. However, it is not yet possible to finally assess the significance of these factors.

1. The lifetime of epoxides of PAH has to be sufficiently long to permit penetration to the nucleus without prior reaction with other cell components.
2. Metabolized PAH (e.g., dihydrodiols) reach the nucleus where microsomal monooxygenases convert them into the actually effective "ultimate carcinogens" (e.g., bay-region epoxides).
3. In contrast to reaction in the endoplasmic reticulum (ER) oxidation near the nucleus produces different epoxides which are characterized by a closer affinity to DNA than are epoxides formed in the ER.

The critical cell event depends on the balance of the various monooxygenase isoenzymes and hydrolases, which, moreover, can be influenced by inducers or inhibitors.

In any case, PAH activated by epoxidases have to get to the cellular DNA prior to conjugation. Rogan et al.[88] have recently detected different enzyme activity in the nucleus and the endoplasmic reticulum. Thus, for instance, 3-methylcholanthrene induces monooxygenase and cytochrome P 450 in the nucleus at a significantly greater rate than in the endoplasmic reticulum and acts as a stronger inducer than benzo[a]pyrene or phenobarbital. Benzo[a]pyrene reacts at the 6-position rather than at any of the other positions.[88] At present the mechanism of oxidation in the nucleus, and reaction of PAH with DNA, is still unknown. New impulses are expected to come from analytical data on conjugated PAH-DNA hydrolysates, which can be separated from non-conjugated nucleotides by chromatography on Sephadex® LH 20.

*4.2.1.12 Metabolic Pathways of Benzo[a]pyrene*

A schematic overview of metabolites of benzo[a]pyrene identified by various authors follows, and a rough assessment of metabolic pathways relevant for malignant transformation is attempted (Scheme 1 and 2). Scheme 2 includes references to those steps at which a modification may take place. Both schemes reflect only part of our present state of knowledge, the majority of which is well supported by evidence. They are not claimed to be complete and final. This has to be stressed all the more so since there is increasing evidence that metabolic pathways near the nucleus are possibly very different from those occurring in the endoplasmic reticulum and the cytosol. The stereochemistry of epoxidation of 7,8-dihydroxy-7,8-dihydrobenzo[a]pyrene and hydration of the resulting epoxide has not been taken into consideration here although it has been intensively investigated recently.[119]

**4.2.2 Methods of Detection**

Whereas the profile analysis of metabolites obtained from animal tissue requires a highly specific analytical method in which the lipophility and aromaticity of PAH and their metabolites are used for enrichment, the incubation of microsome samples and composition of cell cultures involve relatively few disturbing factors so that isolation of the metabolites is simpler. So far two methods have proved useful for the separation and detection of metabolites of PAH: (1) high-pressure liquid chromatography and (2) gas chromatography.

*4.2.2.1 High-Pressure Liquid Chromatography (HPLC)*

Phenols, quinones, and dihydrodiols could be separated from unconverted PAH, in 25- to 30-cm columns, by gradient elution with nonpolar solvents on mineral carriers, or even better, by "reverse phase" techniques with aqueous phases, e.g., on Bondapak® C-18. Due to the limited number of theoretical plates it is difficult in some cases to separate isomers from one another. However, the 3- and 9-phenols, as well as the three isomers 4,5-, 7,8-, and 9,10-dihydrodiol of benzo[a]pyrene could be resolved and isolated without difficulty.[61,96,103] The advantage of HPLC is that it permits the collection of individual fractions without difficulty, and their determination by UV photometry, whereas mass-spectrometric identification is possible only after formation of derivatives. This method also permits the detection of highly polar metabolites.[120] To permit a quantitative evaluation, however, the detection by UV spectrometry requires response factors determined by standardization using pure substances. The latter, however, are usually not available in weighable amounts.

*4.2.2.2 Gas Chromatography (GLC)*

Gas-chromatographic separation of PAH metabolites have been described repeatedly.[5,44,104] The enrichment of metabolites and unconverted PAH can be achieved by

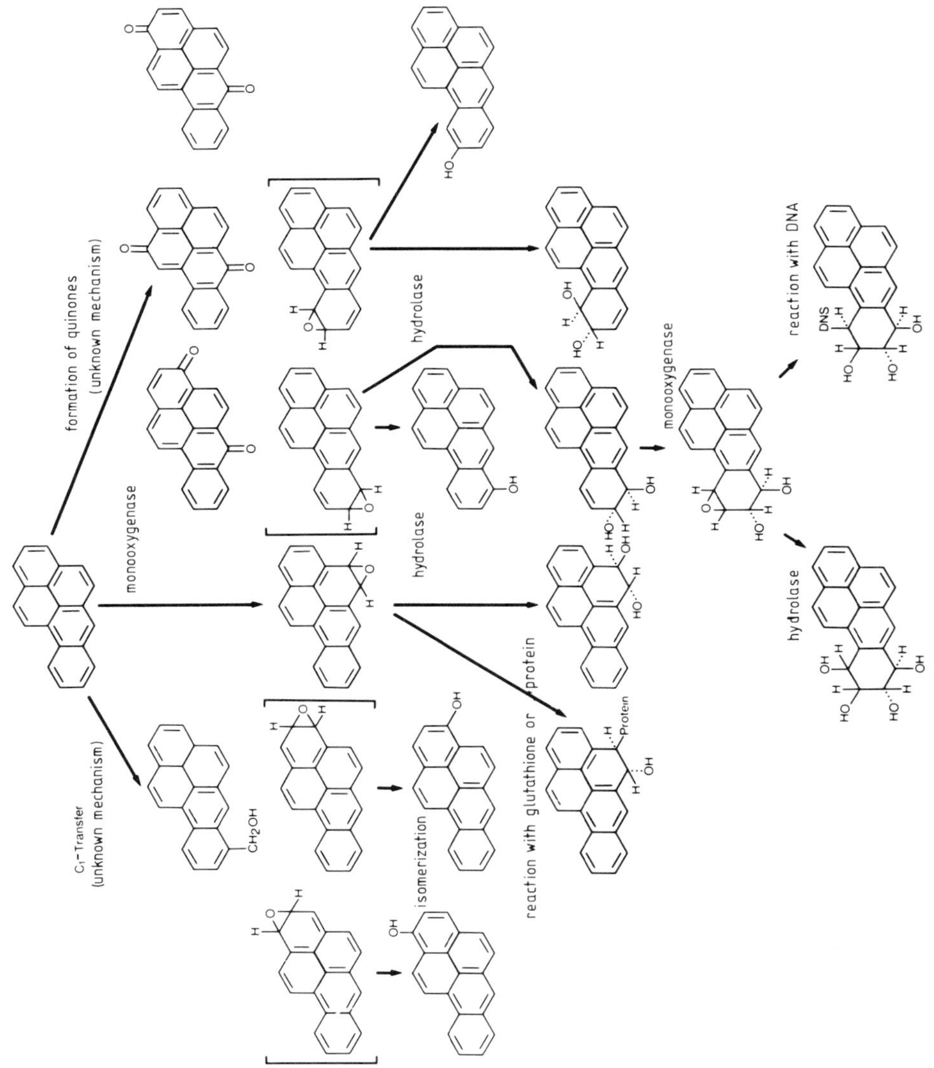
formation of quinones
(unknown mechanism)
monooxygenase
C1-Transfer
(unknown mechanism)
CH2OH
hydrolase
isomerization
reaction with glutathione or protein
Protein
monooxygenase
hydrolase
reaction with DNA
DNS

SCHEME 1.

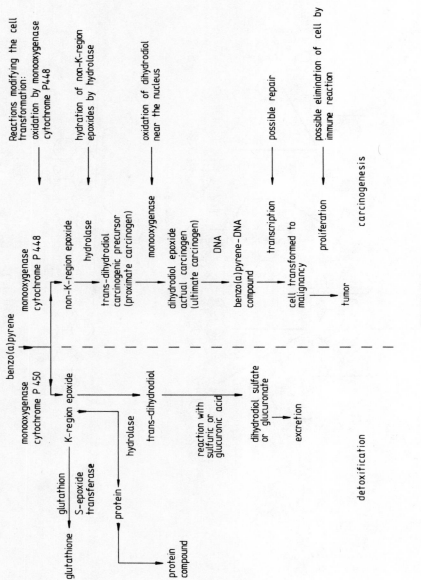

SCHEME 2.

extraction with ethyl acetate and subsequent chromatography on Sephadex® LH 20. Silylation results in trimethylsilyl ethers which can readily be separated by gas-liquid chromatography on silicone phases.[44-47] A separation of the TMS ethers of benz[a]anthracene phenols and trans-dihydrodiols is shown in Figure 47. Even higher oxidized PAH metabolites, e.g., triols (dihydrodiolphenols) and tetrahydrotetrols can be determined after silylation.

On the basis of the calculated number of theoretical plates, the resolution capacity of packed high-performance columns or of glass capillaries considerably exceeds that of the HPLC method. Furthermore, GLC can easily be combined with mass spectrometry, so that valuable data on the structure of individual metabolites can be obtained simultaneously. However, isolation of individual metabolites is more complicated. Although HPLC may be recommended for the detection of highly polar metabolites, the GLC detection of the triols and tetrols of benz[a]anthracene as their TMS-ether derivatives, from rat liver microsome incubation has been reported recently.[44-47]

*4.2.2.3 Analytics of DNA Compounds*

A method for separating PAH-containing and PAH-free nucleosides formed after enzymatic cleavage of nucleic acids has been reported.[106] Column chromatography on Sephadex® LH 20 with methanol-water gradients permits the separation of these two nucleoside fractions. The reaction of benzo[a]pyrene with DNA in epithelial cell culture systems[118] as well as the reactions in vivo of 7,12-dimethylbenz[a]anthracene and benzo[a]pyrene with DNA of fetal and maternal rat tissue have been studied.[27,56]

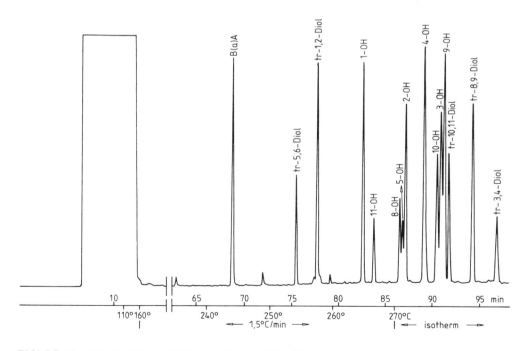

FIGURE 47. Gas chromatographic separation of TMS ethers of benz[a]anthracene phenols and trans-dihydrodiols (50 m capillary column [glass] CP [TM] sil 5).

## REFERENCES

1. Argus, M. F., Myers, S. C., and Arcos, J. C., Apparent absence of requirement of hydrocarbon metabolism for induction and repression of mixed-function oxidases, *Chem. Biol. Interact.*, 29, 247, 1980.
2. Bend, J. R. and Hook, G. E. R., Hepatic and extrahepatic mixed-function oxidases, in *Handbook of Physiology. Reactions to Environmental Agents,* Lee, D. H. K., Ed., Williams & Wilkins, Baltimore, 1977, 419.
3. Bentley, P. and Oesch, F., Purification of rat liver epoxide hydratase to apparent homogeneity, *FEBS Lett.*, 59, 291, 1975.
4. Berenbom, M. and Young, L., Biochemical studies of toxic agents. III. The isolation of 1- and 2-naphthylsulphonic acid and 1- and 2-naphthylglucuronide from the urine of rats dosed with 1- and 2-naphthol, *Biochem. J.,* 49, 165, 1951.
5. Bettencourt, A., Lhoest, G., Roberfroid, M., and Mercier, M., Gaschromatographic and mass fragmentographic assays of carcinogenic polycyclic hydrocarbon epoxide hydrase activity, *J. Chromatogr.*, 134, 323, 1977.
6. Bickers, D. R., Kappas, A., and Alvares, A. P., Differences in inducibility of cutaneous and hepatic drug metabolizing enzymes and cytochrome P450 by polychlorinated biphenyls and 1,1,1-trichloro-2,2-bis(*p*-chlorophenyl)-ethane (DDT), *J. Pharmacol. Exp. Ther.*, 188, 300, 1974.
7. Booth, J. and Sims, P., 8,9-Dihydro-8,9-dihydroxybenz[a]anthracene-10,11-oxide: a new type of polycyclic aromatic hydrocarbon metabolite, *FEBS Lett.*, 47, 30, 1974.
8. Booth, J. and Sims, P., Metabolism of benz[a]anthracene epoxides by rat liver, *Biochem. Pharmacol.*, 23, 2547, 1974.
9. Bowden, G. T., Shapas, B. G., and Boutwell, R. K., Binding of 7,12-dimethylbenz[a]anthracene to replicating and nonreplicating DNA in mouse skin, *Chem. Biol. Interact.*, 8, 379, 1974.
10. Bowden, G. T., Slaga, T. J., Shapas, B. G., and Boutwell, R. K., The role of aryl hydrocarbon hydroxylase in skin tumor initiation by 7,12-dimethylbenz[a]anthracene and 1,2,5,6-dibenzanthracene using DNA binding and thymidine-$^3$H incorporation into DNA as criteria, *Cancer Res.*, 34, 2634, 1974.
11. Boyland, E. and Sims, P., Metabolism of polycyclic compounds. XXII. The metabolism of (±)-trans-9,10-dihydro-9,10-dihydroxyphenanthrene in rats, *Biochem. J.*, 84, 583, 1962.
11a. Boyland, E. and Sims, P., The effect of pretreatment with adrenal protecting compounds on the metabolism of 7,12-dimethylbenz[a]anthracene and related compounds by rat liver homogenates, *Biochem. J.*, 104, 394, 1967.
12. Boyland, E. and Williams, K., An enzyme catalyzing the conjugation of epoxides with glutathion, *Biochem. J.*, 94, 190, 1965.
13. Buerki, K., Stoming, T. A., and Bresnick, E., Effects of an epoxide hydrase inhibitor on in vitro binding of polycyclic hydrocarbons to DNA and on skin carcinogenesis, *J. Natl. Cancer Inst.*, 52, 785, 1974.
14. Cantrell, E., Busbee, D., Warr, G., and Martin, R., Induction of aryl hydrocarbon hydroxylase in human lymphocytes and pulmonary alveolar macrophages — a comparison, *Life Sci.*, 13, 1649, 1973.
15. Catterall, F. A., Murray, K., and Williams, P. A., The configuration of the 1,2-dihydroxy-1,2-dihydronaphthalene formed in the bacterial metabolism of naphthalene, *Biochim. Biophys. Acta*, 237, 361, 1971.
16. Cruickshank, D. W. J., A detailed refinement of the crystal and molecular structure of anthracene, *Acta Crystallogr.*, 9, 915, 1956.
17. Cohen, G. M., Haws, S. M., Moore, B. P., and Bridges, J. W., Benzo[a]pyrene-3-yl hydrogen sulphate, a major ethyl acetate-extractable metabolite of benz[a]pyrene in human, hamster, and rat lung cultures, *Biochem. Pharmacol.*, 25, 2561, 1976.
18. Conney, A. H., Miller, E. C., and Miller, J. A., Substrate-induced synthesis and other properties of benzopyrene hydroxylase in rat liver, *J. Biol. Chem.*, 228, 753, 1957.
19. Conney, A. H. and Burns, J. J., Biochemical pharmacological considerations of zoxolamine and chlorzoxazone metabolism, *Ann. N.Y. Acad. Sci.*, 86, 167, 1960.
20. Coon, M. J., Haugen, D. A., Guengerich, F. P., Vermillion, J. L., and Dean, W. L., Liver microsomal membranes, in *The Structural Basis of Membrane Function,* Hatefi, Y. et al., Eds., Academic Press, New York, 1976, 409.
21. Cooper, D. Y., Schleyer, H., Rosenthal, O., Levin, W., Lu, A. Y. H., Kuntzman, R., and Conney, A. H., Inhibition by CO of hepatic benzo[a]pyrene hydroxylation and its reversal by monochromatic light, *Eur. J. Biochem.*, 74, 69, 1977.
22. Clar, E., *Polycyclic Hydrocarbons,* Vol. 1, Academic Press, New York, 1964, 130.
23. Dehnen, W., Untersuchungen über den Metabolismus von Benzo[a]pyren in Alveolarmakrophagen. I. Aufnahme von Benzo[a]pyren und Induktion der Benzo[a]pyren-Hydroxylase, *Zentralbl. Bakteriol. Parasitenkd. Infektionskr. Hyg., Abt. 1, Orig. Reihe B,* 160, 191, 1975.

24. Dehnen, W., Untersuchungen über den Metabolismus von Benzo[a]pyren in Alveolarmakrophagen. II. Kinetik des Abbaus und Charakterisierung der Metabolite, *Zentralbl. Bakteriol. Parisitenkd. Infektionskr. Hyg., Abt. 1, Orig. Reihe B*, 161, 1, 1975.
25. Depierre, J. W. and Ernster, L., The metabolism of polycyclic hydrocarbons and its relationship to cancer, *Biochim. Biophys. Acta*, 473, 149, 1978.
26. Diamond, L., McFall, R., Miller, J., and Gelboin, H. V., The effect of two isomeric benzoflavones on aryl hydrocarbon hydroxylase and the toxicity and carcinogenicity of polycyclic hydrocarbons, *Cancer Res.*, 32, 731, 1972.
27. Doerjer, G. H., Diessner, H., Bücheler, J., and Kleihues, P., Reaction of 7,12-dimethylbenz[a]anthracene with DNA of fetal and maternal rat tissues in vivo, *Int. J. Cancer*, 22, 288, 1978.
28. Falk, H. L., Kotin, P., and Markul, I., The disappearance of carcinogens from soot in human lungs, *Cancer (Philadelphia)*, 11, 482, 1958.
29. Ferlinz, R., *Lungen- und Bronchialerkrankungen. Ein Lehrbuch der Pneumologie*, Georg Thieme Verlag, Stuttgart, 1974.
30. Fisher, A. B., Huber, G. A., and Furia, L., Cytochrome P-450 content and mixed-function oxidation by microsomes from rabbit alveolar macrophages, *J. Lab. Clin. Med.*, 90, 101, 1977.
31. Flesher, J. W. and Sydnor, K. L., Possible role of 6-hydroxy-methylbenzo[a]pyrene as a proximate carcinogen of benzo[a]pyrene and 6-methylbenzo[a]pyrene, *Int. J. Cancer*, 11, 433, 1973.
32. Freudenthal, R. I., Hundley, S. G., and Cattaneo, S. M., A comparison of the metabolites of benzopyrene by lung mixed function oxides from rat, rhesus, and humans, in *Polynuclear Aromatic Hydrocarbons*, Jones, P. W. and Freudenthal, R. I., Eds., Raven Press, New York, 1978.
33. Gelboin, H. V., Kinoshita, N., and Wiebel, F. J., Microsomal hydroxylases: induction and role in polycyclic hydrocarbon carcinogenesis and toxicity, *Fed. Proc. Fed. Am. Soc. Exp. Biol.*, 31, 1298, 1972.
34. Gelboin, H. V., Okuda T., Selkirk, J., Nemoto, N., Yang, S. K., Wiebel, F. J., Whitlock, J. P., Rapp, H. J., and Bast, R. C., Benzo[a]pyrene metabolism: enzymatic and liquid chromatographic analysis and application to human liver, lymphocytes and monocytes, in Screening Tests in Chemical Carcinogenesis, *IARC Sci. Publ.*, No. 12, 225, 1976.
35. Genta, V. M., Kaufman, D. G., Harris, C. C., Smith, J. M., Sporn, M. B., and Saffiotti, U., Vitamin A deficiency enhances binding of benzo[a]pyrene to tracheal epithelial DNA, *Nature (London)*, 247, 48, 1974.
36. Gibson, D. T., Roberts, R. L., Wells, M. C., and Kobal, V. M., Oxidation of diphenyl by a Beijerinckia species, *Biochem. Biophys. Res. Commun.*, 50, 211, 1973.
37. Gibson, D. T., Mahadevan, V., Jerina, D. M., Yagi, H., and Yeh, H. J. C., Oxidation of the carcinogenic benzo[a]pyrene and benz[a]anthracene to dihydrodiols by a bacterium, *Science*, 189, 295, 1975.
38. Gram, T. E., Comparative aspects of mixed function oxidation by lung and liver of rabbits, *Drug Metab. Rev.*, 2, 1, 1973.
39. Habig, W. H., Pabst, M. J., and Jacoby, W. B., Glutathion-S-transferases, *J. Biol. Chem.*, 249, 7130, 1974.
40. Harper, K. H., The intermediary metabolism of polycyclic hydrocarbons, *Br. J. Cancer*, 13, 718, 1959.
41. Haugen, D. A. and Coon, M. J., Properties of electrophoretically homogenous phenobarbital-inducible and β-naphthoflavone-inducible forms of liver microsomal cytochrom P 450, *J. Biol. Chem.*, 251, 7929, 1976.
42. Heyder, J., Armbruster, L., Gebhart, J., Grein, E., and Stahlhofen, W., Total deposition of aerosol particles in the human respiratory tract for nose and mouth breathing, *J. Aerosol Sci.*, 6, 311, 1975.
43. Holder, G., Yagi, H., Dansette, P., Jerina, D. M., Levin, W., Lu, A. Y. H., and Conney, A. H., Effects of inducers and epoxide hydrase on the metabolism of benzo[a]pyrene by liver microsomes and a reconstituted system: analysis by high pressure liquid chromatography, *Proc. Natl. Acad. Sci. U.S.A.*, 71, 4356, 1974.
44. Jacob, J., Grimmer, G., and Schmoldt, A., Metabolitenprofile von polycyclischen aromatischen Kohlenwasserstoffen (PAH) nach Vorbehandlung mit verschiedenen Induktoren mikrosomaler Monooxygenasen der Rattenleber, *Hoppe Seylers Z. Physiol. Chem.*, 360, 1525, 1979.
45. Jacob, J., Grimmer, G., Richter-Reichhelm, H. B., and Emura M., Gas-chromatographische Profilanalyse von PAH-Metaboliten aus Mikrosomen-Präparartionen und Gewebekulturen, *VDI Ber. (Ver. Dtsch. Ing.)*, 358, 273, 1980.
46. Jacob, J., Schmoldt, A., and Grimmer, G., Time course of oxidative benz[a]anthracene metabolism by liver microsomes of normal and PCB-treated rats, *Carcinogenesis*, 2, 395, 1981.
47. Jacob, J., Schmoldt, A., and Grimmer, G., Glass-capillary-gas-chromatography/mass spectrometry data of mono- and polyhydroxylated benz[a]anthracene. Comparison with benz[a]anthracene metabolites from rat liver microsomes, *Hoppe Seylers Z. Physiol. Chem.*, 362, 1021, 1981.

48. James, M. O., Foureman, G. L., Law, F. C., and Bend, J. R., The perinatal development of epoxide metabolising enzyme activities in liver and extrahepatic organs of guinea pig and rabbit, *Drug Metab. Dispos.*, 5, 19, 1977.
49. Janss, D. H., Moon, R. C., and Irving, C.C., Binding of 7,12-dimethylbenz[a]anthracene to mammary parenchyma DNA and protein in vivo, *Cancer Res.*, 32, 254, 1972.
50. Jeffrey, A. M., Yeh, H. J. C., Jerina, D. M., Patel, T. R., Davey, J. F., and Gibson, D. T., Initial reactions in the oxidation of naphthalene by *Pseudomonas patida, Biochemistry*, 14, 575, 1975.
51. Jeffrey, A. M., Jennette, K. W., Blobstein, S. H., Weinstein, I. B., Beland, F. A., Harvey, R. G., Kasai, H., Miura, I., and Nakanishi, K., Benzo[a]pyrene-nucleic acid derivative found in vivo: structure of benzo[a]pyrenetetrahydrodiol epoxide-guanosine adduct, *J. Am. Chem. Soc.*, 98, 5714, 1976.
52. Jerina, D. M., Daly, J. W., Jeffrey, A. M., and Gibson, D. T., Cis-1,2-dihydroxy-1,2-dihydronaphthalene: a bacterial metabolite from naphthalene, *Arch. Biochem. Biophys.*, 142, 394, 1971.
53. Jerina, D. M., Selander, H., Yagi, H., Wells, M. C., Davey, J. F., Mahadevan, V., and Gibson, D. T., Dihydrodiols from anthracene and phenanthrene, *J. Am. Chem. Soc.*, 98, 5988, 1976.
54. Kinoshita, N. and Gelboin, H. V., Aryl hydrocarbon hydroxylase and polycyclic hydrocarbon tumorgenesis. Effect of the enzyme inhibitor 7,8-benzoflavone on tumorgenesis and macromolecule binding, *Proc. Natl. Acad. Sci. U.S.A.*, 69, 824, 1972.
55. Kinoshita, N. and Gelboin, H. V., Role of aryl hydrocarbon hydroxylase in 7,12-dimethylbenz[a]anthracene skin tumorgenesis. Mechanism of 7,8-benzoflavone inhibition of tumorgenesis, *Cancer Res.*, 32, 1329, 1972.
56. Kleihues, P., Doerjer, G., Ehret, M., and Guzman, J., Reaction of benzo[a]pyrene and 7,12-dimethylbenz[a]anthracene with DNA of various rat tissues in vivo. Quantitative aspects of risk assessment in chemical carcinogenesis, *Arch. Toxicol.*, 3 (suppl.), 237, 1980.
57. Knowles, R. G. and Burchell, B., A simple method for purification of epoxide hydratase from rat liver, *Biochem. J.*, 163, 381, 1977.
58. Koreeda, M., Moore, P. D., Wislocki, P. G., Levin, W., Conney, A. H., Yagi, H., and Jerina, D. M., Binding of benzo[a]pyrene-7,8-diol-9,10-epoxides to DNA, RNA, and protein of mouse skin occurs with high stereoselectivity, *Science*, 199, 778, 1978.
59. Kuroki, T. and Heidelberger, C., Determination of the h-protein in transformable and transformed cells in culture, *Biochemistry*, 11, 2116, 1972.
60. Kutscher, W., Tomingas, R., and Weisfeld, H. P. Untersuchung von Humanlungen auf ihren Staub- und Benzpyrengehalt im Raum Heidelberg, *Arch. Hyg.*, 151, 666, 1967.
61. Leber, P., Kerchner, G., and Freudenthal, R. I., A comparison of benzo[a]pyrene metabolism by primates, rats, and miniature swine, in *Polynuclear Aromatic Hydrocarbons*, Jones, P. W. and Freudenthal, R. I., Eds., Raven Press, New York, 1976, 33.
61a. Lehr, R. E., Yagi, H., Thakker, D. R., Levin, W., Wood, A. W., Conney, A. H., and Jerina, D. M., The bay region theory of polycyclic aromatic hydrocarbon-induced carcinogenicity, in *Carcinogenesis*, Vol. 3, Jones, P. W. and Freudenthal, R. I., Eds., Raven Press, New York, 1978, 231.
62. Leutz, J. C. and Gelboin, H. V., Benzo[a]pyrene-4,5-oxide hydratase: assay, properties and induction, *Arch. Biochem. Biophys.*, 168, 722, 1975.
63. Lu, A. Y. H. and Levin, W., Partial purification of cytochrome P 450 and P 448 from rat liver microsomes, *Biochem. Biophys. Res. Commun.*, 46, 1334, 1972.
64. Lu, A. Y. H., Kuntzman, R., West, S., Jacobson, M., and Conney, A. H., Reconstituted liver microsomal enzyme system that hydroxylates drugs, other foreign compounds and endogenous substrates, *J. Biol. Chem.*, 247, 1727, 1972.
65. Lu, A. Y. H., Ryan, D., Jerina, D. M., Daly, J. W., and Levin, W., Liver microsomal epoxide hydrase: solubilization, purification and characterization, *J. Biol. Chem.*, 250, 8283, 1975.
66. Lu, A. Y. H., Levin, W., Thomas, P. E., Jerina, D. M., and Conney, A. H., Enzymological properties of purified liver microsomal cytochrome P-450 system and epoxide hydrase, in *Polynuclear Aromatic Hydrocarbons*, Jones, P. W. and Freudenthal, R. I., Eds., Raven Press, New York, 1978, 243.
67. Marquardt, H., Malignant transformation in vitro: a model system to study mechanisms of action of chemical carcinogens and to evaluate the oncogenic potential of environmental chemicals, in *Screening Tests in Chemical Carcinogens*, *IARC Sci. Publ.*, No. 12, 389, 1976.
68. McLemore, T. L. and Martin, R. R., In vitro induction of aryl hydrocarbon hydroxylase in human pulmonary alveolar macrophages by benz[a]anthracene, *Cancer Lett.*, 2, 327, 1977.
69. McLemore, T. L., Warr, G. A., and Martin, R. R., Induction of aryl hydrocarbon hydroxylase in human pulmonary alveolar macrophages and peripheral lymphocytes by cigarette tars, *Cancer Lett.*, 2, 161, 1977.
70. Mullen, J. O., Juchau, M. R., and Fouts, J. R., Interaction of 3,4-benzpyrene, 3-methylcholanthrene, chlordane, and methyltestosterone on stimulators of hepatic microsomal enzyme systems in the rat, *Biochem. Pharmacol.*, 15, 137, 1966.

71. Nakanishi, K., Kasai, H., Cho, H., Harvey, R. G., Jeffrey, A. M., Jennette, K. W., and Weinstein, I. B., Absolute configuration of a ribonucleic acid adduct formed in vivo by metabolism of benzo[a]pyrene, *J. Am. Chem. Soc.*, 99, 258, 1977.
72. Nebert, D. W., Benedict, W. F., Gielen, J. E., Oesch, F., and Daly, J. W., Aryl hydrocarbon hydroxylase, epoxide hydrase and 7,12-dimethylbenz[a]anthracene-produced skin tumorigenesis in the mouse, *Mol. Pharmacol.*, 8, 374, 1972.
73. Nebert, D. W., Heidema, J. K., Strobel, H. W., and Coon, M. J., Genetic expression of aryl hydrocarbon hydroxylase induction. Genetic specificity resides in the fraction containing cytochromes P 448 and P 450, *J. Biol. Chem.*, 248, 7631, 1973.
74. Oesch, F., Mammalian epoxide hydrases: inducible enzyme catalyzing the inactivation of carcinogenic and cytotoxic metabolites derived from aromatic and olefinic compounds, *Xenobiotica*, 3, 305, 1973.
75. Oesch, F., Biochemistry of polycyclic aromatic hydrocarbons, *VDI Ber. Ver. Dtsch. Ing.*, 358, 251, 1980.
76. Oesch, F. and Daly, J., Solubilization, purification and properties of hepatic epoxide hydrase, *Biochim. Biophys. Acta*, 227, 692, 1971.
77. Oesch, F. and Daly, J., Conversion of naphthalene to trans-naphthalene-dihydrodiol. Evidence for the presence of a coupled aryl monooxygenase-epoxide hydrase system in hepatic microsomes, *Biochem. Biophys. Res. Commun.*, 46, 1713, 1972.
78. Oesch, F., Jerina, D. M., and Daly, J., A radiometric assay for hepatic epoxide hydrase activity with 7-$^3$H styrene oxide, *Biochim. Biophys. Acta*, 227, 685, 1971.
79. Oesch, F., Jerina, D. M., and Daly, J., Substrate specificity of hepatic epoxide hydrase in microsomes and in purified preparation: evidence of homologous enzymes, *Arch. Biochem. Biophys.*, 144, 253, 1971.
80. Oesch, F., Kaubisch, N., Jerina, D. M., and Daly, J., Hepatic epoxide hydrase. Structure-activity relationship for substrates and inhibitors, *Biochemistry*, 10, 4858, 1971.
81. Oesch, F., Jerina, D. M., Daly, J. W., Lu, A. Y. H., Kuntzman, R., and Conney, A. H., A reconstituted microsomal enzyme system that converts naphthalene to trans-1,2-dihydroxy-1,2-dihydronaphthalene via naphthalene-1,2-oxide: presence of epoxide hydrase in cytochrome P 450 and P 448 fractions, *Arch. Biochem. Biophys.*, 153, 62, 1972.
82. Oesch, F., Jerina, D. M., Daly, J. W., and Rice, J. M., Induction, activation, and inhibition of epoxide-hydrase. Anomalous prevention of chlorobenzene-induced hepatotoxicity by an inhibitor of epoxide hydrase, *Chem. Biol. Interact.*, 6, 189, 1973.
83. Okamoto, T., Chan, P. C., and So, B. T., Effect of tobacco, marijuana, and benzo[a]pyrene on aryl hydrocarbon hydroxylase in hamster lung, *Life Sci.*, 11, 733, 1972.
84. Pabst, M. J., Habig, W. H., and Jacoby, W. B., Mercapturic acid formation. Several glutathion transferases of rat liver, *Biochem. Biophys. Res. Commun.*, 52, 1123, 1973.
85. Pandov, H. and Sims, P., Conversion of phenanthrene-9,10-oxide and dibenz[a,h]anthracene-5,6-oxide into dihydrodiols by rat liver microsomal enzymes, *Biochem. Pharmacol.*, 19, 299, 1970.
86. Pezzuto, J. M., Yang, C. S., Yang, S. K., McCourt, D. W., and Gelboin, H. V., Metabolism of benzo[a]pyrene and (−)-trans-7,8-dihydroxy-7,8-dihydrobenzo[a]pyrene by rat liver nucleic and microsomes, *Cancer Res.*, 38, 1241, 1978.
87. Raha, C. R., Gallagher, C. H., and Shubik, P., Chemical reactions producing chrysene as an artifact of some K-region metabolites of benzo[a]pyrene, *Proc. Soc. Exp. Biol. Med.*, 143, 531, 1973.
88. Rogan, E., Roth, R., and Cavalieri, E., Enzymology of polycyclic hydrocarbon binding to nucleic acids, in *Polynuclear Aromatic Hydrocarbons*, Jones, P. W. and Freudenthal, R. I., Eds., Raven Press, New York, 1978, 265.
89. Schlesinger, R. B. and Lippmann, M., Selective particle deposition and bronchogenic carcinoma, *Environ. Res.*, 15, 424, 1978.
90. Schmidt, O., Die Charakterisierung der einfachen und krebserzeugenden aromatischen Kohlenwasserstoffe durch die Dichteverteilung bestimmter Valenzelektronen (B-Elektronen), *Z. Phys. Chem. Abt. B.* 42, 83, 1939.
91. Schmidt, W., Versuch einer Berechnung der Alveolaroberfläche der menschlichen Lunge, *Z. Anat. Entwicklungsgesch.*, 125, 119, 1966.
92. Schmoldt, A., Benthe, H. F., and Frühling, R., Induction of rat liver enzymes by polychlorinated biphenyls (PCBs). Independent of the dose and chlorine content, *Arch. Toxicol.*, 32, 69, 1974.
93. Schmoldt, A., Jacob, J., and Grimmer, G., Dose-dependent induction of rat liver microsomal aryl hydrocarbon monooxygenase (AHH) by benzo[k]fluoranthene, *Cancer Lett.*, 13, 249, 1981.
94. Seifert, G., Atmungsorgane, in *Lehrbuch der allgemeinen Pathologie und der pathologischen Anatomie*, Eder, M. and Gedigk, P., Eds., Springer-Verlag, New York, 1977, 587.
95. Selkirk, J. K., Croy, R. G., Roller, P. P., and Gelboin, H. V., High-pressure-liquid-chromatography analysis of benzo[a]pyrene metabolism and covalent binding and the mechanism of action of 7,8-benzoflavone and 1,2-epoxy-3,3,3-trichloropropane, *Cancer Res.*, 34, 3474, 1974.

96. Selkirk, J. K., Croy, R. G., and Gelboin, H. V., Benzo[a]pyrene metabolites: efficient and rapid separation by high-pressure liquid chromatography, *Science,* 184, 169, 1974.
97. Sims, P. Metabolism of polycyclic compounds. XIV. The conversion of naphthalene into compounds related to trans-1,2-dihydro-1,2-dihydroxynaphthalene by rabbits, *Biochem. J.,* 73, 389, 1959.
98. Sims, P., The metabolism of 3-methylcholanthrene and some related compounds by rat liver homogenates, *Biochem. J.,* 98, 215, 1966.
99. Sims, P. and Grover, P. L., Epoxides in polycyclic aromatic hydrocarbon metabolism and carcinogenesis, *Adv. Cancer Res.,* 20, 265, 1974.
100. Sloane, N. H., Hydroxymethylation of the benzene ring. I. Microsomal formation of phenol via prior hydroxymethylation of benzene, *Biochim. Biophys. Acta,* 107, 599, 1965.
101. Sloane, N. H., Uzgiris, V. I., and Seeley, D., 6-Hydroxymethyl-benzo[a]pyrene synthetase: resolution of the enzyme system into apoenzyme and lipid C-1 donor cofactor, in *Polynuclear Aromatic Hydrocarbons,* Jones, P. W. and Freudenthal, R. I., Eds., Raven Press, New York, 1978, 303.
102. Sloane, N. H. and Davis, T. K., Hydroxymethylation of the benzene ring. Microsomal hydroxymethylation of benzo[a]pyrene to 6-hydroxymethylbenzo[a]pyrene, *Arch. Biochem. Biophys.,* 163, 46, 1974.
103. Soedigdo, S., Angus, W. W., and Flesher, J. W., High-pressure liquid chromatography of polycyclic aromatic hydrocarbons and some of their derivatives, *Anal. Biochem.,* 67, 664, 1975.
104. Stoming, T. A. and Bresnick, E., Gas chromatographic assay of epoxide hydrase activity with 3-methylcholanthrene-11,12-oxide, *Science,* 181, 951, 1973.
105. Swaisland, A. J. and Grover, P. L., Properties of K-region epoxides of polycyclic aromatic hydrocarbons, *Biochem. Pharmacol.,* 22, 1547, 1973.
106. Swaisland, A. J., Hewer, A., Pal, K., Keysell, G. R., Booth, J., Grover, P. L., and Sims, P., Polycyclic hydrocarbon epoxides: the involvement of 8,9-dihydro-8,9-dihydroxybenz[a]anthracene-10,11-oxide in reactions with the DNA of benz[a]anthracene-treated hamster embryo cells, *FEBS Lett.,* 47, 34, 1974.
107. International Radiological Protection Commission, Task Group on Lung Dynamics, Deposition of retention models for internal dosimetry of the human respiratory tract, *Health Phys.,* 12, 173, 1966.
108. Thakker, D. R., Levin, W., Stoming, T. A., Conney, A. H., and Jerina, D. M., Metabolism of 3-methylcholanthrene by rat liver microsomes and a highly purified monooxygenase system with and without epoxide hydrase, in *Polynuclear Aromatic Hydrocarbons,* Jones, P. W., and Freudenthal, R. I., Eds., Raven Press, New York, 1978, 253.
109. Tomingas, R., Pott, F., and Dehnen, W., Polycyclic aromatic hydrocarbons in human bronchial carcinoma, *Cancer Lett.,* 1, 189, 1976.
110. Tsuji, H., Muta, E., and Ullrich, V., Separation and purification of liver microsomal monooxygenases from induced and untreated pigs, *Hoppe Seylers Z. Physiol. Chem.,* 361, 681, 1980.
111. Ullrich, V., Cytochrome P 450 and biological hydroxylation reactions, *Top. Curr. Chem.,* 83, 67, 1979.
112. Ullrich, V., Roots, I., Hildebrandt, A., Estabrook, R. W., and Conney, A. H., Eds., *Microsomes and Drug Oxidations,* Pergamon Press, New York, 1977.
113. Weinstein, I. B., Jeffrey, A. M., Jennette, K. W., Blobstein, S. H., Harvey, R. G., Harris, C., Autrup, H., Kasai, H., and Nakanishi, K., Benzo[a]pyrene diol epoxides as intermediates in nucleic acid binding in vitro and in vivo, *Science,* 193, 592, 1976.
114. Weisz, H., Brockhaus, A., and Könn, G., Untersuchungen über den 3,4-Benzypyrengehalt von Menschenlungen, *Zentralbl. Bakteriol. Parisitenkd. Infektionskr. Hyg. Abt. I B,* 155, 142, 1971.
115. Welch, R. M., Cavallito, J., and Loh, A., Effect of exposure to cigarette smoke on the metabolism of benzo[a]pyrene and acetophenetidin by lung and intestine of rats, *Toxicol. Appl. Pharmacol.,* 23, 749, 1972.
116. Wiebel, F. J., Activation and inactivation of carcinogens by microsomal monooxygenases. Modification by benzoflavones and polycyclic aromatic hydrocarbons, in *Modifiers of Chemical Carcinogenesis,* Slaga, T. J., Ed., Raven Press, New York, 1980.
117. Wiebel, F. J., Leutz, J. C., and Gelboin, H. V., Aryl hydrocarbon (benzo[a]pyrene) hydroxylase. Inducible in extrahepatic tissues of mouse strains not inducible in liver, *Arch. Biochem. Biophys.,* 154, 292, 1973.
118. Wigley, C. B., Thompson, M. H., Brookes, P., The nature of benzo[a]pyrene binding to DNA in an epithelial cell culture system, *Eur. J. Cancer,* 12, 743, 1976.
119. Yang, S. K., Roller, P. P., and Gelboin, H. V., Benzo[a]pyrene metabolism: mechanism in the formation of epoxides, phenols, dihydrodiols, and the 7,8-diol-9,10-epoxides, in *Polynuclear Aromatic Hydrocarbons,* Jones, P. W. and Freudenthal, R. I., Eds., Raven Press, New York, 1978, 285.
120. Yang, S. K., Gelboin, H. V., Trump, B. F., Autrup, H., and Harris, C. C., Metabolic activation of benzo[a]pyrene and binding to DNA in cultured human bronchus, *Cancer Res.,* 37, 1210, 1977.

Chapter 5

# BIOLOGICAL ACTIVITY

## TABLE OF CONTENTS

| | | | |
|---|---|---|---|
| 5.1 | Toxicity — D. Schmähl | | 158 |
| 5.2 | Carcinogenicity — D. Schmähl | | 160 |
| | 5.2.1 | Short-Term Tests — R. P. Deutsch-Wenzel and H. Brune | 161 |
| | | 5.2.1.1 Ames Test | 163 |
| | |     5.2.1.1.1 Methods | 163 |
| | |     5.2.1.1.2 Results and Discussion | 164 |
| | | 5.2.1.2 Cell Transformation Test | 166 |
| | |     5.2.1.2.1 Methods | 170 |
| | |     5.2.1.2.2 Results and Discussion | 170 |
| | | 5.2.1.3 Sebaceous Gland Suppression Test | 170 |
| | |     5.2.1.3.1 Methods | 171 |
| | |     5.2.1.3.2 Results and Discussion | 171 |
| | | 5.2.1.4. General Discussion | 174 |
| | | 5.2.1.5. Summary | 177 |
| | 5.2.2 | Long-Term Experiments | 178 |
| | | 5.2.2.1 Site of Application: Respiratory Tract — P. Schneider and U. Mohr | 178 |
| | |     5.2.2.1.1 Implantation | 178 |
| | |     5.2.2.1.2 Intratracheal Installation | 179 |
| | |     5.2.2.1.3 Inhalation | 183 |
| | |     5.2.2.1.4 Conclusions | 185 |
| | | 5.2.2.2 Site of Application: Skin — M. Habs and D. Schmähl | 186 |
| | | 5.2.2.3 Site of Application: Subcutaneous Tissue — F. Pott | 190 |
| | |     5.2.2.3.1 Introduction | 190 |
| | |     5.2.2.3.2 Importance of the Methodology | 191 |
| | |     5.2.2.3.3 Carcinogenicity of Individual PAH | 192 |
| | |     5.2.2.3.4 Carcinogenicity of Mixtures of PAH | 193 |
| | |     5.2.2.3.5 Carcinogenicity of PAH-Containing Mixtures | 194 |
| | |     5.2.2.3.6 Summary | 201 |
| | | 5.2.2.4 Oral Application — D. Steinhoff | 203 |
| References | | | 206 |

## 5.1 TOXICITY

### D. Schmähl

Toxicology is the science of toxins, their occurrence, and their mode of action.[2] Formerly, toxicology was integrated into the field of pharmacology. Only recently, because toxicology has acquired increasing importance in the last few years, have some universities in West Germany begun to establish professorial posts in toxicology.

As to the pharmacodynamic mechanisms of action, the activities of toxins can be divided into those which rapidly manifest themselves (after seconds, minutes, hours, or days) and others which become evident only after many weeks, months, or even years.[3] Thus we have to differentiate between "acute" toxic effects and "late" toxic effects. A typical representative of an acute toxic effect is inhaled hydrocyanic acid, which can lead to the death of the poisoned individual within seconds, or at the most within minutes, whereas carcinogenic substances are considered to be typical representatives of long-term toxic effects which often exhibit their activity only after decades (e.g., bladder cancer induced in employees of the dyestuff industry by contact with $\beta$-naphthylamine, average induction time: 22 years).

When discussing toxic effects, we distinguish between "concentration toxins", whose effects "depend on their immediate concentration in the system, and "cumulative toxins", whose effects are due to the accumulation of small, sub-potent doses of the toxin. With "concentration toxins", the effect is related directly to the concentration of the toxin in the blood (Figure 48). A characteristic feature of "concentration toxins" is that, in principle, their effect is rapidly reversible. Typical representatives of this type of action are inhalation narcotics.

With "cumulative toxins" (cumulus = the pile), their effect is characterized in that the abatement phase of the toxic effect occurs slowly. If the toxic activity of the second, third, and further doses is added to the slowness of abatement of the toxicity, this will bring about a "cumulative effect". Thus, depending on the velocity at which the effect abates, the risk of an accumulation of toxic effects increases if the intervals between the individual additions are not long enough. Typical representatives of this type of toxic effects are digitalis and strychnine. Finally, carcinogenic substances, because of their mode of action, were designated "summation toxins" by Druckrey and Küpfmüller.[1] There is practically no reversibility of their effect so that every new application of a carcinogen, even of low doses, is added to the remaining effect of the previous dose. Thus a "summation" is effected. Although we know today that repair mechanisms may have an impact on toxic effects, most pharmacological and toxicological investigations on the pharmacodynamic mechanism of action of carcinogenic substances indicate that the effect of these compounds is to a large extent irreversible. Consequently, the mechanism of summation toxins still has to be considered an imminent hazard.

In order to exhibit its toxic activity, the toxin has to penetrate into the body and to its site of action. The toxin can be absorbed through the gastrointestinal tract, the respiratory tract, or the skin. In the course of medical treatment, a substance may also reach the body by intravenous, intramuscular, or subcutaneous injection. Having reached the bloodstream, the toxin invades the different tissues of the organism where it is metabolized and finally discharged. A distinction has to be made between the excretion of the substance and the onset or disappearance of its action. Both time parameters may vary. As a rule, the toxin is excreted via the urogenital system or the intestinal tract, but it may also be secreted through the respiratory tract or by perspiration. Secretion in breast milk should not be neglected, since, via this mechanism, neonates may be confronted with the toxins. In every case it is important for the toxi-

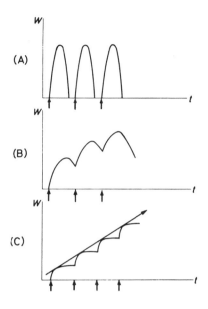

FIGURE 48. Abscissa: time, ↑ toxin intake. Ordinate: potency of the effect. (A) Rapid onset and rapid abatement of the toxic effect of a "concentration-type toxin". Rapid and complete reversibility. (B) Slow abatement of the toxic effect and, thus, the risk of its "cumulation". The effect is principally reversible. (C) Irreversibility of the toxic effect of "summation toxins". No reversibility possible.

cologist to know the essential mechanisms of absorption, excretion, and the metabolization of the substance being investigated.

These reflections belong to the field of "general toxicology". Apart from knowing these general toxicological aspects it is of course important that the toxicologist has specialized knowledge which, due to the enormous expansion of this field in the last years, is quite often restricted to a limited group of substances.

In the case of PAH the acute toxicity, for example, after intravenous injection of the substances, is limited predominantly to necrosis of the cortex of the suprarenal gland and, to a lesser extent, to the marrow of the suprarenal gland. This entails the corresponding consequences for the organism which — depending on the level of the dose applied — may bring about the death of the experimental animals. Besides these effects upon the suprarenal gland, damage to antigens have been described following application of high doses so that a passing immunosuppression with increased susceptibility to infections of the experimental animals was observed. Finally damages to the leukopoiesis and a retardation in growth have been described. All these effects appear, however, only after application of comparatively high doses which are by far higher than the dose needed for the induction of cancer (local carcinogenesis).

## 5.2 CARCINOGENICITY

### D. Schmähl

When describing substances which induce malignant tumors, the term "carcinogenic" has been used throughout this documentation, meaning "cancer-inducing". In the literature, events leading to the development of tumors are also described by the terms "tumorigenic", "oncogenic", "blastomogenic", and "cancerogenic" as well as the term employed by us. They are used almost synonymously. There is no uniformity as to whether only the induction of malignant tumors is meant or also the development of benign tumors. The exclusive use of the term "carcinogenic" in the manner explained, i.e., malignant tumors, should help to avoid misunderstandings. Useful proposals concerning the nomenclature in the field of chemical carcinogenesis are to be found in the monographs of the International Agency for Research on Cancer, Lyon, France.

In animal experiments we speak of carcinogenic activity when, after treatment of a certain population with a test substance, the frequency of malignant tumors in this treated population is statistically significantly higher than in the control groups. These malignant tumors can be localized either in a certain organ or affect different localizations (multipotent carcinogens). In this respect the definition of the term "carcinogenic" is clear and undisputed.

It has also been suggested that we speak of carcinogenic action when the tumors induced are not malignant, but of a benign nature, i.e., lacking symptoms of malignant growth. Here no carcinomas or sarcomas (malignant tumors originating from the epithelium or connective tissue of the body) are induced, but benign tumors are. Such tumors can often become malignant in the course of their growth or behave like malignant tumors in their clinical development despite their histological benignity. Therefore, demands have been voiced to also classify compounds which induce benign tumors "only" as carcinogenic substances.

Finally, it has also been proposed to classify as carcinogenic compounds those substances which can induce earlier occurrence of normally arising "spontaneous" tumors that appear with a more or less high frequency and after certain time intervals in experimental animals, depending on the type of species. In this case it is not the transformation of a normal cell that is the decisive criterion, but the acceleration of the tumor growth manifesting itself in shorter induction times of certain "spontaneously" occurring types of tumors.

The last two examples broaden the term "carcinogenic" to cover every carcinogenic activity that, in the cellular transformation of a cell, might induce benign tumor cells or an acceleration of their growth. Thus no half-way suspicious substance is neglected. It remains to be seen whether this broad definition will prove to be acceptable in the future.

The meaning of "carcinogenic", as used in this documentation, covers only the definition given initially, i.e., it characterizes activities that can transform normal cells into malignant cells.

Such an effect is normally dependent on alterations in the genetic apparatus of a cell. It is probably caused, for example, by an irreversible alteration of nucleic acids as a consequence of a mutative incident, and this irreversible alteration can also be transferred to daughter cells. This alteration might be effected by a "one-impulse incident", but also by a "multiple-impulse incident", i.e., one impulse alone might suffice to bring about such a cancerous transformation or many such impulses might be necessary. At least as far as the carcinogenesis in man is concerned, it is not clear and has not been investigated to what extent "repair mechanisms" can influence such mutative processes, i.e., can again normalize cell processes once their cancerization has started.

Besides this "specific" activity of carcinogenic substances which attack — as stated earlier — the genetic structure of cells, there seem to be "unspecific" carcinogenic effects. This is proved primarily by observations in man. Thus, for example, chronic fistulas, which for years remain unchanged, often develop into fistular carcinomas when the fistula breaks through the skin. Furthermore, malignant alterations of the respiratory tract, caused by dusts, are attributed to such "unspecific" irritative mechanisms. This is the more likely since we know that not only asbestos fibers, but also other fibers,[5] e.g., fibers of glass, may induce "mesotheliomas" if they have the same physical structure (certain length, certain diameter, certain form and size of these fibers). In this case the carcinogenic activity manifests itself because, in the course of constantly repeated regeneration on the basis of a permanent irritation, the structures that are responsible for regeneration are finally so exhausted that a "defective regeneration"[4] takes place. This can lead to malignant growth. Here the cancer develops on the basis of a permanent chronic, unspecific irritation. Such cases seem to occur only seldom, but they should be mentioned.

Accordingly, in general, a carcinogenic action is effected because a specific carcinogenesis irreversibly alters cell structures and codifies them towards malignant growth. On the other hand, an unspecific cancer induction is conceivable, in which such specific alterations in the genome do not have to exist.

Finally, it should be mentioned that in classifying carcinogenic compounds relative to their toxicological type of activity, at least two groups have to be distinguished. One type induces tumors at the site of their direct topical application, e.g., in the skin. Such compounds are designated as substances with *local* carcinogenic activity. The hydrocarbons described later in detail are almost exclusively of this type of activity. Secondly, there are other substances that induce cancer in certain parenchymatous organs after application to the skin or absorption into the gastrointestinal tract. These compounds are called *systemically* active carcinogens. Frequently it is not possible to clearly separate these types of activity from each other. Nevertheless, for the sake of systematic order, the division into locally and systemically active carcinogens is expedient.

### 5.2.1 Short-Term Tests

#### R. P. Deutsch-Wenzel and H. Brune

Various short-term test models have been developed in the last decade, since it was deemed desirable to replace the time- and cost-consuming long-term animal experiments for the detection of carcinogenic activities of environmental chemicals by short-term tests. So far, however, none of these tests has been able to prove the carcinogenicity or noncarcinogenicity of a substance or mixture of substances with absolute certainty.

The short-term tests carried out to assess the biological activity of polycyclic aromatic hydrocarbons (PAH) have yielded insufficient data which do not permit the classification of a compound as carcinogenic. Thus short-term tests cannot replace long-term experiments, but may provide valuable additional information for the pre-screening of existing and newly developed environmental chemicals. They are to be included in the proposed "three-tier protocol for the biological testing of compounds":[18,20]

- Tier 1 consists of testing as many substances as possible in different short-term tests, each test having a high predictive value for selection for long-term carcinogenesis tests.
- Tier 2 comprises both short-term and long-term tests on mammals. Priority for long-term investigations is given to substances which were positive in Tier 1.
- Tier 3 is designed to evaluate as quantitatively as possible the hazards to man from environmental chemicals tested particularly in long-term but also short-term tests, including epidemiological studies.

### Metabolic Activation of Known Carcinogenic PAH

Most carcinogenic and mutagenic compounds do not exert their activity as such, but only after metabolic activation into their ultimate active structure ("ultimate carcinogen") in the mammalian organism.[64,65] When using short-term tests, it is endeavored to simulate in vivo conditions as closely as possible. This is achieved by adding a metabolizing component to the test system. Suited for this purpose are, for example, isolated liver cells with a high enzymatic activity, cell fractions such as liver microsomes containing the enzymes necessary for activation, and the S-9 fraction which contains the microsome fraction and is obtained by centrifuging liver homogenates. In order to enhance various enzymatic activities, the animals are pretreated with different inducers such as phenobarbitone, 3-methylcholanthrene, or polychlorinated biphenyls (PCB).[9,13]

As described in detail in Chapter 4.2.1, PAH are oxidized to epoxides, their reactive metabolites, in the mammalian organism by mixed function oxidases such as cytochrome P 450 and P 448 with NADP as a cosubstrate (Ashby et al., 1977). The effects of PAH observed in short-term tests are elucidated, using benzo[a]pyrene, for example. The carcinogen benzo[a]pyrene, has to be activated by metabolic transformation prior to yielding a carcinogenic or mutagenic effect.

Recent investigations have shown that the 7,8-dihydroxy-9,10-epoxy-7,8-dihydro derivative is probably the cancer-inducing compound.[13,60]

The 7,8-dihydrodiol is formed as an intermediate by reaction of epoxide hydratases with the 7,8-epoxide. Consequently the metabolizing component used for the activation of benzo[a]pyrene in in vitro test systems has to contain these oxidative enzymes, on the one hand for the epoxide formation (cytochrome P 450 and P 448) and on the other hand for the hydration of the epoxide formed (epoxide hydratase). Beside glucuronidases, the S-9 fraction of mammalian livers also contains various glutathione-S-epoxide transferases which can react with epoxide groups to inactivate benzo[a]pyrene.

It has to be mentioned that the activation of benzo[a]pyrene into a mutagen varies from species to species. Thus, when investigating benzo[a]pyrene in the Ames test, no mutagenic activity is observed when, for example, the S-9 fraction of livers of uninduced guinea pigs is used instead of uninduced mouse or rat liver.[13] This demonstrates that it is possible to modify the mutagenic effect of benzo[a]pyrene in in vitro tests by varying the relative concentrations or activities of enzymes, by using different inducers, and by adding the S-9 fraction of livers of different animal species. Consequently, it is very difficult to make quantitative predictions about the effect of benzo[a]pyrene in vivo.

The description of three selected short-term tests and their predictive value for the carcinogenic activity of PAH follow.

*5.2.1.1 Ames Test*

A large number of in vitro test systems developed in the last few years are based on the assumption that there is a correlation between carcinogenicity and mutagenicity.[15] These test systems can show damages to the deoxyribonucleic acid (DNA) which induce inheritable modifications (mutations).[37]

Over a period of about 10 years, a system of testing mutagenic substances has been developed, using different strains of *Salmonella typhimurium*.[6,8-10] Mutagenic test substances effect a reverse mutation of histidine-deficient mutants into histidine-phototrophic bacteria. The strains TA 1535 and TA 1538 are reverted into their wild type, which can grow in the absence of histidine, by substances inducing base-pair substitutions or frameshift mutations, respectively. These strains contain two additional types of mutation:

1. The first one is characterized by a reduced content of lipopolysaccharides in the bacterial cell wall, which facilitates the passage of substances or their metabolites.
2. The second mutation is based on the damage to the DNA repair system.

These two properties increase the sensitivity of the tester strains when investigating mutagenic compounds.

It was possible to achieve resistance to antibiotics in two further strains (TA 98 and TA 100) by inserting a resistance factor (R factor), in this case a plasmid.[63] As a consequence, the sensitivity to various mutagenic substances, e.g., polycyclic aromatic hydrocarbons, was increased.

By adding the S-9 fraction of mammalian liver to the test system, it is possible to detect those chemical compounds which are potential carcinogens or mutagens and are activated only through metabolism in mammals. Thus, the mutagenicity of several carcinogenic substances which need metabolic activation by microsomal enzymes could be proved.

*5.2.1.1.1 Methods*

In the Ames test, the bacteria, the test substance, and the S-9 mixture needed for activation are added to a semiliquid top agar. This mixture is poured onto plates with a histidine-free agar medium and incubated at 37°C for 2 to 3 days to allow growth of mutants. Then the colonies are counted and compared with the number of colonies in untreated controls. Colonies growing on the agar plates are bacteria reverted to their wild type, since only these can grow on this agar.

*5.2.1.1.2 Results and Discussion*

An investigation of more than 300 compounds with different chemical structures has proved a correlation between carcinogenicity and mutagenicity for up to 90%.[62] This study included a number of polycyclic aromatic compounds. *Salmonella typhimurium* TA 100 served as a tester strain.

Coombs et al.[30] tested the mutagenicity of 54 polycyclic hydrocarbons, for which carcinogenicity data was already available, partly from skin painting experiments with mice and partly from injection experiments with rats. The 37 carcinogens tested, and 1-methylchrysene which is known to be an initiator, exhibited mutagenic activity. Even so-called weak carcinogens proved to be mutagenic. However, a correlation between carcinogenic and mutagenic potency of a compound could not be detected. In analogy to Ames's investigations, it was demonstrated that strong carcinogens were often less mutagenic than weaker ones.

In a study describing the investigation of 120 compounds (58 carcinogens and 62 noncarcinogens) in different test systems, Purchase et al.[80,81] could show that in the Ames test up to 91% of the carcinogenic compounds displayed a correlation between carcinogenic and mutagenic activity. Up to 94% of the noncarcinogenic compounds were negative in the Ames test (see Table 40). Testing of polycyclic aromatic hydrocarbons also resulted in a high rate of correlation between carcinogenic and mutagenic activity or noncarcinogenic and nonmutagenic activity, respectively (Table 36).

In his review article, Bridges[19] reports on the correlation of carcinogenic and mutagenic activity of polycyclic aromatic hydrocarbons with and without metabolic activation.[62]

Table 37 shows that mutagenic activity was observed in 26 of a total of 27 carcinogenic polycyclic aromatic hydrocarbons. Of the nine noncarcinogenic polycyclic aromatics tested, seven did not show mutagenic activity.

Andrews et al.[11] found a considerably lower correlation between mutagenicity and carcinogenicity when testing the mutagenic activity of 25 polycyclic hydrocarbons in comparison with results from carcinogenicity experiments. The correlation was 58% for positive and 41% for negative results.

Complex mixtures, such as smoke condensates from cigarettes, cigars, and pipe tobacco, proved to be mutagenic after activation with rat liver microsomes, using the strains *S. typhimurium* TA 100 and TA 98 in the Ames test.[86]

Among other compounds, various PAH could be isolated from cigarette smoke condensates.[56,90] The mutagenicity of these condensates is, however, not induced solely by the PAH contained, particularly benzo[*a*]pyrene; N-PAH[36] and nitrosamines and pyrolysates of proteins[86] have to be considered as further substances with mutagenic activity.

Recent investigations of automobile exhaust, using the Ames test, have shown that this mixture contains direct acting mutagens, i.e., mutagens which do not need metabolic activation. These active constituents seem to be N-substituted PAH.[99]

Dehnen et al.[32] investigated air filter extracts using the Ames test. The total extract had a mutagenic effect. This extract was fractionated to find out whether the contained polycyclic aromatic hydrocarbons alone accounted for this mutagenic activity. However, only a small portion of the total mutagenic activity could be attributed to the fraction which contained the polycyclic compounds. Other compounds, not all of which have yet been identified, must be of decisive importance.

Tables 38 and 39 comprise results obtained in the Ames test with PAH which occur frequently, only in low concentrations, or not at all, in the environment. These compounds are chosen in accordance with Tables 2 and 3 of Chapter 2.

Table 36
CORRELATION BETWEEN CARCINOGENIC ACTIVITY OF
COMPOUNDS AND THEIR ACTIVITY IN THREE DIFFERENT
SHORT-TERM TESTS IN PER CENT[80]

|  | Polycyclic aromatic hydrocarbons | Arylamines | Alkylating | Others | Total agents |
|---|---|---|---|---|---|
| Ames test | 95 | 94 | 83 | 94 | 93 |
| Cell transformation test | 95 | 97 | 94 | 92 | 94 |
| Sebaceous gland suppression test | 90 | 62 | 67 | 57 | 65 |

Table 37
CORRELATION OF ANIMAL CARCINOGENICITY AND BACTERIAL
MUTAGENICITY WITH AND WITHOUT METABOLIC ACTIVATION[62]

| Group of compounds | Carcinogens detected as bacterial mutagens | Noncarcinogens not mutagenic to bacteria | Compounds of uncertain carcinogenicity detected as mutagens |
|---|---|---|---|
| Aromatic amines | 23/25 | 10/12 | 5/7 |
| Alkyl halides | 17/20 | 1/3 | 1/1 |
| Polycyclic aromatics | 26/27 | 7/9 | 1/1 |
| Esters, epoxides, carbamates | 13/18 | 5/9 | 0/1 |
| Nitro aromatics and heterocycles | 28/38 | 1/4 | 0/2 |
| Miscellaneous organics | 1/6 | 13/13 | 0/1 |
| Nitrosamines | 20/21 | 2/2 | 1/1 |
| Fungal toxins and antibiotics | 8/9 | 5/5 | — |
| Mixtures (cigarette smoke condensate) | 1/1 | — | — |
| Miscellaneous heterocycles | 1/4 | 7/7 | — |
| Miscellaneous nitrogen compounds | 7/9 | 2/4 | — |
| Azo dyes and diazo compounds | 11/11 | 2/3 | 3/3 |
| Common laboratory biochemicals | — | 46/46 | — |
| Total | 157/178 | 101/117 | 11/17 |

Since some of the substances in Tables 38 and 39 have not yet been subjected to investigation in the Ames test, it would seem premature to assess the rate of correlation between their carcinogenic activity and bacterial mutagenicity. The relatively high number of false-positive results, in comparison with animal tests, should give rise to repeated critical examination of these results, particularly because different laboratories using identical test substances obtained completely different results. Although basically the tests were carried out in keeping with Ames's experimental design, minor modifications were often undertaken. Furthermore, the following inaccuracies can be frequently observed.

- Insufficient concentrations of buffers and cofactors are used.
- Concentrations of enzymes and substrate in the initial mixture are not optimal.
- The amount of enzymes used and their activity are inconsistent.
- No attention is paid to the fact that for testing different substances, different amounts of protein have to be added.[41]

Appropriate regulative measures for standardization, quality control, and statistical evaluation of results are urgently needed to permit predictions on the carcinogenic properties of environmental substances from such a routine test. The strict observation of the following details is to be recommended: constant implementation of control procedures, permanent control of indicator bacteria used, maintainence of constant activities of mammalian oxygenases used, consideration of the different properties of oxygenases obtained from different species, and, after addition of different inducers, consideration of solvent activities and the efficiency of different variants of the enterobacteriaceae oxygenase test.[71]

The present short-term test model may be used as a rapid and economical method for screening pure substances only if the above details, particularly the observation of consistent test conditions, are met so that the test results are reproducible. Still ongoing investigations with multicomponent mixtures have so far shown that results obtained with mixtures in the Ames test have to be interpreted with caution.[82] Interactions were detected which strongly promoted or neutralized the mutagenic effect of individual components.

*5.2.1.2 Cell Transformation Test*

Berwald and Sachs[16] were the first to report on the transformation of cell cultures by chemical carcinogens. They observed this effect in embryo cells of the Syrian hamster following treatment with 3-methylcholanthrene and benzo[a]pyrene.

The tumor take of a graft of transformed cells on juvenile syngeneic experimental animals is considered to be proven evidence of malignant transformation.[87] The ability of transformed cells to grow as cell clones in "soft" agar is regarded to be a clear indication of transformation.[48,67,101]

Transformed cells differ from untreated control cells in many respects:[87]

- Their ability to form clones (colony growth in "soft" agar)
- Reduction or complete lack of contact inhibition of dividing cells growing as "monolayers", in combination with formation of multiple layer sheets of cells
- Morphological changes of the cells as such and of their growth pattern (crisscross and piling-up growth)
- Ability to grow in vitro in unlimited sequence of generations
- Ability to form local tumors after grafting onto compatible (syngeneic) or conditioned receiver animals

In order to induce a corresponding effect in the cell transformation test, the test compounds often have to be converted into their active form. This may be achieved by

1. Addition of S-9 microsomal fraction as metabolizing component[93]
2. Isolation of hamster embryo cells following treatment of the pregnant mother.[20]
3. Cultivation of cell cultures together with other metabolizing cells[57]

These feeder cells are used for activation of the compounds, whereas the effect induced by the active metabolites is indicated by the other cells used in the test system.

Furthermore, a test system has been developed which on the one hand comprises the enzymes needed for activation of the test substance and on the other hand can show the effect of the reactive intermediate products.[83]

*5.2.1.2.1 Methods*

The following cells are used, among others, in the cell transformation test: embryo

Table 38
CORRELATION OF ANIMAL CARCINOGENICITY (SUBCUTANEOUS AND TOPICAL APPLICATION) AND BACTERIAL MUTAGENICITY OF POLYCYCLIC AROMATIC HYDROCARBONS FREQUENTLY OCCURRING IN THE ENVIRONMENT

| Compound | Carcinogenic activity | | | | Mutagenic activity | |
|---|---|---|---|---|---|---|
| | Subcutaneous | Ref. | Topical | Ref. | Ames test | Ref. |
| 01 Fluorene | − | 12 | − | 12 | − | 53,62,63,94 |
| 02 Phenanthrene | − | 84 | − | 85 | − | 23,53,62,63,70 |
| 03 Anthracene | − | 93a | − | 14a | − | 53,62,63,70,81 |
| 04 Fluoranthene | − | 24 | − | 46,47 | + | 98 |
| | | | | | − | 68 |
| 05 Pyrene | − | 12 | − | 39 | − | 62,63,70,94 |
| 06 Benzo[ghi]fluoranthene | | − | − | 103 | | — |
| 07 Cyclopentadieno-[cd]pyrene | + | 79 | − | 21,44 | | — |
| 08 Benz[a]anthracene | + | 79,93a,b | + | 51,88,92 | + | 30,62,63,70,81 |
| 09 Triphenylene | | — | − | 12 | + | 68,98 |
| | | | (+) | 47 | | |
| 10 Chrysene | − | 12 | + | 38a,44a,45a,88 | + | 62,63,70,98 |
| | + | 79 | − | | | |
| 11 Benzo[b]fluoranthene | + | 24 | + | 103 | + | 52b,72 |
| | | | − | 66 | | |
| | | | + | 44 | | |
| 12 Benzo[k]fluoranthene | + | 24,79 | − | 44,66,103 | − | 68 |
| 13 Benzo[j]fluoranthene | | — | + | 103 | + | 52b,72 |
| | | | − | 66 | | |
| | | | + | 44 | | |
| 14 Benzo[e]pyrene | − | 79 | − | 39 | + | 62,63 |
| | | | (+) | 88 | − | 97 |
| | | | | | + | 11,70,98 |

Table 38 (continued)
CORRELATION OF ANIMAL CARCINOGENICITY (SUBCUTANEOUS AND TOPICAL APPLICATION) AND BACTERIAL MUTAGENICITY OF POLYCYCLIC AROMATIC HYDROCARBONS FREQUENTLY OCCURRING IN THE ENVIRONMENT

| Compound | Carcinogenic activity | | | | Mutagenic activity | |
|---|---|---|---|---|---|---|
| | Subcutaneous | Ref. | Topical | Ref. | Ames test | Ref. |
| 15 Benzo[a]pyrene | + | See Chapter 5.3.2.3. | + | See Chapter 5.2.2.2. | + | 8,11,62,63,81,94,97, 98 |
| 16 Perylene | | — | — | 38,47 | + | 81 |
| 17 Indeno[1,2,3-cd]-fluoranthene | | — | + | 45 | | — |
| 18 Indeno[1,2,3-cd]-pyrene | + | 12,79 | + | 45 | | |
| | | | — | 44 | | |
| 19 Benzo[ghi]perylene | — | 69 | — | 54,69 | + | 11,68,98 |
| 20 Anthanthrene | — | 12 | + | 28 | — | 98 |
| | | | — | 45,55 | + | 11 |
| 21 Coronene | — | 12 | — | 44 | + | 68 |

Table 39
CORRELATION OF ANIMAL CARCINOGENICITY (SUBCUTANEOUS AND TOPICAL APPLICATION) AND BACTERIAL MUTAGENICITY OF POLYCYCLIC AROMATIC HYDROCARBONS OCCURRING IN THE ENVIRONMENT ONLY IN LOW CONCENTRATIONS OR NOT AT ALL

| Compound | Carcinogenic activity | | | | Mutagenic activity | |
|---|---|---|---|---|---|---|
| | Subcutaneous | Ref. | Topical | Ref. | Ames test | Ref. |
| 22 Benzo[a]fluorene | − | 12 | − | 12 | + | 98 |
| 23 Benzo[b]fluorene | − | — | − | — | − | — |
| 24 Benzo[c]fluorene | − | 12 | − | 12 | − | — |
| 25 7,12-Dimethylbenz[a]anthracene | + | 12 | + | 12,83a,92 | + | 8,11,30,62,63,70,81 |
| 26 3-Methyl-1,2-dihydrobenz[j]aceanthrylene | + | 12 | + | 12 | + | 8,11,62,63,70,81,97 |
| 27 Dibenz[a,c]anthracene | − | 12 | − | 12 | + | 11,62,63,70,81,98 |
| 28 Dibenz[a,h]anthracene | + | 22,74,78,93a,93b | + | 45,49 | + | 62,63 |
| | | | | | − | 97 |
| 29 Dibenz[a,j]anthracene | | — | − | 12 | + | 11,70 |
| 30 Dibenzo[a,l]pyrene | + | 24,52a,100 | + | 45 | + | 11 |
| 31 Dibenzo[a,i]pyrene | + | 24 | + | 45 | + | 62,63,70,81 |
| 32 Dibenzo[a,h]pyrene | + | 24 | + | 28,45 | − | 97 |
| | | | | | + | 52b |
| 33 Dibenzo[a,e]pyrene | + | 24 | + | 45 | − | 97 |
| 34 Dibenzo[e,l]pyrene | − | 24 | | — | | — |

169

cells of the Syrian golden hamster,[16,26,33,81,83] fibroblastic cultures of mouse cell lines,[34,50,58] and human fibroblasts of liver cells.[31,81,102]

In the test model described[80] long-term cultivated kidney fibroblasts of neonatal Syrian golden hamsters, long-term cultivated diploid human lung fibroblasts, or long-term cultivated human liver cells are mixed with the test substance in vitro in a liquid culture medium. The test substance is metabolically activated by S-9 mix, as it is added to the culture medium in the Ames test. A small number of cells is allowed to grow in a liquid medium for 6 to 8 days to determine the survival rate. The cells used for transformation are poured onto semisolid agar in which only transformed cells can grow. A 2.5-fold increase in growth of transformed cells, compared with controls, is considered a positive result.

This method is but one of a large variety of cell transformation tests which offer, for example, the possibilities of using different cell cultures, feeder cells, or S-9 mix as activating components, or adding viruses to the test system.[25,89]

Several polycyclic aromatic hydrocarbons, such as benzo[a]pyrene, dimethylbenz[a]anthracene, and 3-methylcholanthrene, may convert fibroblastic cultures even without additional metabolic activation systems.[26,34,50] This is due to the fact that some cell lines contain aryl epoxidases[7] which permit oxidation of polycyclic hydrocarbons to active metabolites.

### 5.2.1.2.2 Results and Discussion

The cell transformation test has permitted the correct classification of the biological activity of 94% of 120 compounds tested, including polycyclic aromatic hydrocarbons, arylamines, and alkylating compounds (see Table 36).[81]

As can be seen from Table 40, 91% of the carcinogenic compounds tested yielded positive results in the cell transformation test. Of the noncarcinogenic compounds, 97% did not induce cell transformation. The correlation between cell-transforming and carcinogenic effects, or non-cell-transforming and noncarcinogenic activities, respectively, of the polycyclic aromatic hydrocarbons tested was 95% (cf. Table 36).

When investigating the biological activity of non-K-region dihydrodiols of several PAH, such as 7,8-dihydroxy-7,8-dihydrobenzo[a]pyrene, these proved to be more active in inducing malignant transformation in vitro than the parent compounds.[59,61] In vivo investigations carried out simultaneously demonstrated that the non-K-region dihydrodiols were also more active than their parent compounds in initiation-promotion experiments on mouse skin.

Of the diol epoxides of benzo[a]pyrene tested, 7,8-dihydroxy-9,10-epoxy-7,8-dihydrobenzo[a]pyrene exhibited the strongest cell-transforming activity.[60] This dihydrodiol epoxide is known to be carcinogenic.[91]

Tables 41 and 42 summarize the results obtained in the cell transformation test with PAH which occur frequently in the environment (Table 41), or only at low concentrations or not at all (Table 42).

In this case, too, it would be premature to draw conclusions from the correlation of carcinogenic and cell-transforming activity of the compounds listed in Tables 41 and 42, since the majority of PAH has not yet been investigated in this test system. It may, however, be anticipated that there is a good correlation between the carcinogenic potential, particularly of PAH and their in vitro transforming activity.

The cell transformation test is distinguished from other test models in that it offers the possibility of studying effects of test substances on human cells.

### 5.2.1.3 Sebaceous Gland Suppression Test

When carcinogenic PAH had been isolated from coal tar for the first time and synthesized,[29] it became possible to start morphological studies on early reactions of tissue

Table 40
CORRELATION OF THE CARCINOGENIC OR
NONCARCINOGENIC ACTIVITY OF
COMPOUNDS AND THEIR EFFECT IN THREE
DIFFERENT SHORT-TERM TESTS[81]

| Test system | Results (%) | |
|---|---|---|
| | Carcinogens | Noncarcinogens |
| Ames test | 91 | 94 |
| Cell transformation test | 91 | 97 |
| Sebaceous gland suppression test | 67 | 66 |

to topical application of these pure substances. In the course of these investigations, the so-called sebaceous gland phenomenon aroused more and more interest and finally led to the development of the sebaceous gland suppression (SGS) test.[17,43,52,95,104]

When investigating mouse skin to which BaP had been applied topically by means of fluorescent microscopy, it was found that even only a few minutes after the first application, a selective concentration of this PAH took place in the sebaceous glands.[40] Other authors confirmed this observation, expanding it to further PAH and other animal species.[52]

The PAH are selectively concentrated in sebaceous glands because of the lipoid content of the latter and the lipoid solubility of PAH. The suppression itself starts when the number of pycnotic nuclei increases beyond the apical region of the glandular coat in which the sebaceous secretion takes place. The plasma becomes increasingly eosinophilic, the glandular acini lose their turgor tension and adhere to the hair root sheath. A coarsely fragmental cell degeneration sets in, progressing to the glandular floor. The fibrous tissue coat of the gland starts to shrink and becomes invisible or regresses into the horizontal tissue of the corium.

Regeneration of the sebaceous gland sets in about 7 to 10 days after the last application of PAH at the outer root sheath, i.e., in reverse order to the suppression process, and is completed after a few days with the formation of intact sebaceous glands.

*5.2.1.3.1 Methods*

To groups of 10 mice, each dose of test substance is applied to two sites per mouse. It is expedient to start the experiment in the second resting phase of the hair growth cycle of mice.

The animals are kept separately under conventional conditions. On days 1, 3, and 5, 0.05 m$\ell$ of the test solution is dropped onto shaved skin areas of 1 × 1.5 cm in the dorso-lumbar region. The mice are killed by cervical dislocation 3 days after the last application. Treated skin areas are excised, fixed in formalin, and after embedding in paraffin are stained with hematoxylin/eosin. For evaluation, 100 follicles and the corresponding sebaceous glands are counted per specimen. The rate of suppression is given in per cent and the median values are calculated for each experimental group.

*5.2.1.3.2 Results and Discussion*

The fraction of automobile exhaust condensate which contains PAH with four and more rings is separated into seven subfractions. In the sebaceous gland suppression test only fraction 2, 3, and 7 show activities which can be clearly ascribed to the contained carcinogenic PAH (Figure 49).

## Table 41
### CORRELATION OF ANIMAL CARCINOGENICITY (SUBCUTANEOUS AND TOPICAL) AND CELL-TRANSFORMING ACTIVITY OF POLYCYCLIC AROMATIC HYDROCARBONS OCCURRING FREQUENTLY IN THE ENVIRONMENT

| Compound | Carcinogenic activity | | | | Activity in cell transformation test | Ref. |
|---|---|---|---|---|---|---|
| | Subcutaneous | Ref. | Topical | Ref. | | |
| Phenanthrene | − | 84 | − | 85 | − | 27,35,50,75 |
| Anthracene | − | 93a | − | 14a | − | 33—35,81 |
| Pyrene | − | 12 | − | 39 | − | 27,33—35,50 |
| Benz[a]anthracene | + | 79,93a,93b | + | 51,88,92 | + | 81 |
| Benzo[a]pyrene | + | See Chapter 5.3.2.3 | + | See Chapter 5.2.2.2 | + | 27,33—35,50,81,83,89 |
| Perylene | | — | − | 38,47 | − | 81 |

## Table 42
### CORRELATION OF ANIMAL CARCINOGENICITY (SUBCUTANEOUS AND TOPICAL) AND CELL-TRANSFORMING ACTIVITY OF POLYCYCLIC AROMATIC HYDROCARBONS OCCURRING ONLY AT LOW CONCENTRATIONS OR NOT AT ALL IN THE ENVIRONMENT

| Compound | Carcinogenic activity | | | | Activity in cell transformation test | Ref. |
|---|---|---|---|---|---|---|
| | Subcutaneous | Ref. | Topical | Ref. | | |
| 7,12-Dimethylbenz[a]anthracene | + | 12 | + | 12,83a,92 | + | 27,33—35,81,89 |
| 3-Methyl-1,2-dihydrobenz[j]aceanthrylene | + | 12 | + | 12 | + | 27,33—35,50,75,76,81 |
| Dibenz[a,c]anthracene | − | 12 | − | 12 | + | 27,81 |
| Dibenz[a,h]anthracene | + | 22,74,78 93a,93b | + | 45,49 | + | 27,89 |
| Dibenzo[a,l]pyrene | + | 24 | + | 45 | + | 81 |

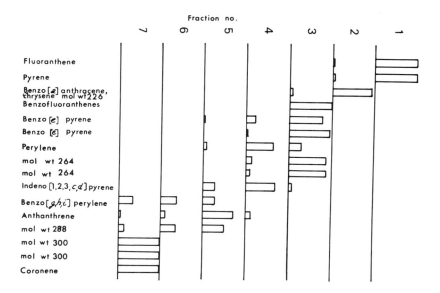

FIGURE 49. Separation on Sephadex® LH20 of PAH with four and more rings contained in the nitromethane fraction.[42]

According to Purchase et al.[80] there is a 90% congruence between sebaceous gland suppression and carcinogenic activity in experimental animals for PAH and PAH-containing mixtures (cf. Table 36).

As was to be expected, this rate of congruence is lower for other chemical classes. Taking all the 120 compounds of different organic structure tested by Purchase into account, the percentage of accurate predictions of carcinogenicity is only 65% in the sebaceous gland suppression test (Table 36).

Within the scope of environmental toxicology and carcinogenesis, tests have to be carried out not only with defined pure substances but, increasingly, also with PAH-containing multicomponent systems, such as mineral oil products, auxiliary and waste products of the plastic industry, tobacco smoke condensates, and other classes of emissions contaminating air, water, and soil, such as automobile exhaust condensate. The results of corresponding investigations are listed in Table 43.[21] The results of the sebaceous gland test are given in per cent. A so-called basic suppression of about 20% has to be taken into account, i.e., even in histological specimens of controls, only 80 of 100 sebaceous glands can be detected.

Thus BaP induces a suppression of about 50% only when applied in total doses above 75 $\mu$g, i.e., 3 × 25 $\mu$g BaP in 0.05 m$\ell$ of acetone solution. A total of 800 $\mu$g per year corresponds to only 15.4 $\mu$g per application; this would be too little to prove any activity in the sebaceous gland suppression test.

Tables 44 and 45 present a review of PAH occurring more or less frequently in the environment, which have so far been investigated in the SGS test.

For 22 of the 23 PAH tested in the sebaceous gland suppression test, the carcinogenicity correlates with their activity in the SGS test. Dibenz[a,c]anthracene, which is known to be noncarcinogenic in long-term experiments, produced a false-positive result in the sebaceous gland suppression test, as well as in the Ames test and the cell transformation assay when compared with animal tests. The results described earlier recommend this test system to be used for detecting carcinogenic properties of PAH. A final evaluation should not be made, however, until the other relevant environmental PAH are investigated in the SGS test.

## Table 43
## INVESTIGATION ON THE CARCINOGENIC BURDEN BY AIR POLLUTION ON MAN

A. Comparison of investigations on automobile exhaust condensate and its fractions in animal experiments with long-term topical application and in the SGS test

| Test substance and solvent | Annual dose BaP-equivalent/animal | Local carcinomas | Median SGS (%) |
|---|---|---|---|
| Total exhaust con- | 15.6 µg | 3 | 19.6 |
| densate | 46.8 µg | 22 | 16.8 |
| (75% DMSO/25% Ac) | 140.4 µg | 57 | 45.6 |
| Methanol phase | 62.4 µg | 1 | 16.3 |
| (75% DMSO/25% Ac) | 187.2 µg | 2 | 13.8 |
| Cyclohexane phase I | 31.2 µg | 10 | 18.1 |
| (10% DMSO/90% Ac) | 93.6 µg | 49 | 45.8 |
| Cyclohexane phase II | 31.2 µg | 3 | 18.3 |
| (10% DMSO/90% Ac) | 93.6 µg | 0 | 14.7 |
| Nitromethane phase | 31.2 µg | 13 | 15.5 |
| (75% DMSO/25% Ac) | 93.6 µg | 49 | 29.4 |
| Recombined condensate | 31.2 µg | 6 | 14.1 |
| (75% DMSO/25% Ac) | 93.6 µg | 43 | 39.4 |
| Benzo[a]pyrene | 200.0 µg | 6 | 14.8 |
| controls | 400.0 µg | 8 | 16.9 |
| (75% DMSO/25% Ac) | 800.0 µg | 26 | 16.8 |
| Solvent (75% DMSO/25% Ac) | 10.5 mℓ | — | 14.8 |

B. Dose-response relationship in the SGS test of the PAH fraction of automobile exhaust condensate in comparison with the synthetic mixture of PAH (15 pure substances)

| | | |
|---|---|---|
| PAH fraction (≥4 r)[a] of | 1 : 500 | 89.6 |
| automobile exhaust | 1 : 1000 | 34.1 |
| gas condensate in | 1 : 2000 | 25.9 |
| acetone | | |
| Synthetic mixture | 1 : 500 | 91.0 |
| of PAH as above | 1 : 1000 | 37.2 |
| in acetone | 1 : 2000 | 27.3 |

[a] ≥4r means equal to or greater than 4 aromatic rings.

The sebaceous gland suppression test is considered a screening system for PAH "which should permit the selection of compounds to be tested in long-term carcinogenicity experiments".[96] It merely has to be added that useful predictive values can be expected when PAH, N-heterocyclics, and multicomponent mixtures containing potentially carcinogenic PAH and N-heterocyclics are investigated. False-positive and false-negative results are obtained in the sebaceous gland suppression test as well as in the other test models described when compared with long-term animal studies.

*5.2.1.4 General Discussion*

A discussion of the extent to which carcinogenic properties of polycyclic aromatic hydrocarbons can be predicted from the biological activities seen in the three short-term tests just described follows.

## Table 44
### CORRELATION OF ANIMAL CARCINOGENICITY (SUBCUTANEOUS AND TOPICAL) AND SEBACEOUS GLAND SUPPRESSION ACTIVITY OF POLYCYCLIC AROMATIC HYDROCARBONS OCCURRING FREQUENTLY IN THE ENVIRONMENT

| Compound | Carcinogenic activity | | | | | Activity in sebaceous gland suppression test | Ref. |
|---|---|---|---|---|---|---|---|
| | Subcutaneous | Ref. | Topical | Ref. | | | |
| Phenanthrene | − | 84 | − | 85 | | − | 20b |
| Anthracene | − | 93a | − | 14a | | − | 20b,81 |
| Fluoranthene | − | 24 | − | 46,47 | | − | 20b |
| Pyrene | − | 12 | − | 39 | | − | 20b |
| Cyclopentadieno[cd]pyrene | + | 79 | − | 21,44 | | + | 20b |
| Benz[a]anthracene | + | 79,93a,93b | + | 51,88,92 | | + | 17,80 |
| | | | | | | (+) | 14,20b |
| Chrysene | − | 12 | + | 38a,44a,45a,88 | | (+) | 14,17,20b |
| | + | 79 | | | | | |
| Benzo[b]fluoranthene | + | 24 | + | 44,103 | | + | 20a |
| | | | − | 66 | | | |
| Benzo[k]fluoranthene | + | 24,79 | − | 44,66,103 | | + | 20a |
| Benzo[j]fluoranthene | | — | + | 44,103 | | (+) | 20a |
| | | | − | 66 | | | |
| Benzo[e]pyrene | − | 79 | − | 39 | | − | 20b |
| | | | (+) | 88 | | | |
| Benzo[a]pyrene | + | See Chapter 5.3.2.3 | + | See Chapter 5.2.2.2 | | + | 14,17,20b,81,96 |
| Perylene | | — | − | 38,47 | | − | 81 |
| Indeno[1,2,3-cd]pyrene | + | 12,79 | + | 45 | | + | 20a |
| | | | − | 44 | | | |
| Anthanthrene | − | 12 | + | 28 | | − | 20a |
| | | | − | 45,55 | | | |

Table 45
CORRELATION OF ANIMAL CARCINOGENICITY (SUBCUTANEOUS AND TOPICAL) AND SEBACEOUS GLAND SUPPRESSION ACTIVITY OF POLYCYCLIC AROMATIC HYDROCARBONS OCCURRING IN THE ENVIRONMENT ONLY AT LOW CONCENTRATIONS OR NOT AT ALL

| Compound | Carcinogenic activity | | | | Activity in sebaceous gland suppression test | Ref. |
|---|---|---|---|---|---|---|
| | Subcutaneous | Ref. | Topical | Ref. | | |
| Benzo[a]fluorene | − | 12 | − | 12 | − | 20b |
| Benzo[b]fluorene | − | — | − | — | − | 20b |
| Benzo[c]fluorene | − | 12 | − | 12 | − | 20b |
| 7,12-Dimethylbenz[a]anthracene | + | 12 | + | 12,83a,92 | + | 17,81,96 |
| 3-Methyl-1,2-dihydrobenz[j]aceanthrylene | + | 12 | − | 12 | + | 14,17,81,96 |
| Dibenz[a,c]anthracene | − | 12 | − | 12 | + | 20b,81 |
| Dibenz[a,h]anthracene | + | 22,74,78 93a,93b | + | 45, 49 | + | 17 |
| Dibenzo[a,l]pyrene | + | 24 | + | 45 | + | 81 |

Three test systems were selected from the large number of known models; discussed were the Ames test, cell transformation assay, and sebaceous gland suppression test. These are preferentially used for investigating PAH because the number of congruent correlations between the findings of long-term carcinogenesis experiments and the results obtained in these tests are higher than with results of other short-term tests.

In all three test systems a certain number of noncarcinogens are ranked as mutagenic, cell-transforming, or positive in the SGS test, and, conversely, carcinogens are shown to be nonmutagenic, non-cell-transforming, or negative in the SGS test, i.e., these compounds are classified wrongly. The percentage wrongly classified is the crucial factor when using these models for prescreening environmental chemicals. A false-positive result found in these test systems may cause a substance to be wrongly classified as hazardous, and a false-negative result may be responsible for ranking a carcinogen as harmless.[77]

When interpreting results of short-term tests, the following criteria have to be taken into consideration:

- Results obtained in short-term tests do not yet permit extrapolation to man.
- Short-term tests do not supply any information on the organotropism of tumor incidences.
- Results of short-term tests do not give any indication of the potency of a carcinogen.

In this connection the question arises as to the usefulness and relevance of short-term tests for detecting carcinogenic compounds. Today they are used as prescreening tests in the preliminary assessment of the potential carcinogenic activity of a compound, but they do not replace animal experiments in determining the degree of carcinogenicity of the compound. Furthermore, the Ames test and the cell transformation assay may give valuable information on the metabolism of substances, i.e., whether they act directly or have to be activated first via mammalian metabolism.

In general, one short-term test alone is not sufficient to investigate and evaluate possible carcinogenic substances. At present scientists recommend carrying out a combination of short-term tests in which several biological effects are recorded. Thus the number of false-negative and false-positive results which the individual test systems produce can be reduced to a minimum. The results obtained from such a battery of tests may then play an important role in the determination of priorities for selecting compounds to be investigated in long-term carcinogenesis experiments.

To sum up it can be said that the three short-term tests described represent useful models for detecting the biological activities of relevant environmental PAH. Their efficiency in predicting carcinogenic properties of relevant environmental PAH cannot yet be finally assessed, because many of these substances have still not been tested.

*5.2.1.5 Summary*

Three different short-term tests and their predictive value on the carcinogenic activity of polycyclic aromatic hydrocarbons (PAH) are described:

1. The Ames test, which is carried out to detect mutagenic activities of carcinogens using bacteria as an indicator
2. The cell transformation test, by which the carcinogen-induced transformation of normal cells to tumor cells is determined in cell cultures from humans and experimental animals
3. The sebaceous gland suppression test, in which the short-term exposure of mouse skin to carcinogens causes a measurable damage to the sebaceous glands of hair follicles

Many working groups have carried out investigations on polycyclic aromatic hydrocarbons which show that there are associations between biological effects, seen in these test systems, and carcinogenic activities of these compounds in long-term animal experiments. At present, however, it is not possible to assess the final value of these short-term tests as predictors of carcinogenicity.

## 5.2.2 Long-Term Experiments
### 5.2.2.1 Site of Application: Respiratory Tract

**P. Schneider and U. Mohr**

The rate of lung cancer mortality has increased considerably in the last few years.[106,123] It is essential that research on the causes of this increase be carried out, but investigations of the mechanisms of carcinogenesis in the respiratory tract of man are impossible on ethical grounds. Therefore, animal models have been developed to elucidate etiologic factors as well as problems in the histo- and pathogenesis of benign and malignant neoplasms of this region. Roe[157] and Nettesheim[151] outlined the requirements to be met by such models.

The following problems can be examined using animal models:

1. Identification of carcinogenic substances active in the respiratory tract
2. Definition of the conditions under which substances become effective carcinogens (e.g., mode of application, dose, distribution in the respiratory tract, physical and chemical properties)
3. Existence of "cofactors" which modify the carcinogenic effect
4. Determination of species-specific influences on activity

In comparison with the rat, the Syrian golden hamster is the preferred model in experiments on the carcinogenic effects of substances in the respiratory tract because

1. It rarely develops spontaneous lung tumors.[156]
2. It is not prone to infections of the respiratory tract.
3. The morphology of induced neoplasms seems to be similar to tumors detected in the respiratory tract of man.[116,161]

A successful carcinogenesis study with coal tar was first described by Yamagiwa and Ichiwaka.[177] Since then, and particularly in the last few years, a number of different methods for the experimental induction of tumors in the respiratory tract by PAHs have been developed.

The most common techniques are intratracheal instillation of compounds in solutions or suspensions, implantation of beeswax pellets including the substances, and finally inhalation of chemicals as gases or aerosols. Unlike investigations of systemic effects, the test substances are brought directly into the respiratory tract by these methods.

#### 5.2.2.1.1 Implantation
Implantation methods proved to be useful for testing the carcinogenicity of locally active substances. The carcinogen is applied to the target organ together with a carrier, mostly beeswax and tricaprylin (a mixture of equal portions of beeswax and tricaprylin

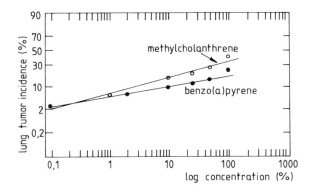

FIGURE 50. Dose-response relationship after intrapulmonary implantation of different doses of benzo[a]pyrene and 3-methylcholanthrene in rats.[144]

heated up to 70°C). Following anesthesia and intercostal thoracotomy, 0.05 to 0.1 m$\ell$ of this mixture are slowly injected into a lobe of the lung under visual control until a wax pellet is formed. After dressing of the surgical wound, no further postoperative measures are necessary. Whereas it is expected that the carcinogen diffuses into the lung tissues in the course of time, pellet remainders can be detected even after 2 years.

The first successful implantation experiment was carried out by Aervont.[105] After intrapulmonary implantation of silk threads impregnated with dibenz[a,h] anthracene, adenomas, adenocarcinomas, and some squamous cell carcinomas of the lung were induced in mice. After repeated injections of a suspension of dibenz[a,h]anthracene in olive oil via the trachea into the lung of rats, Niskanen[152] observed squamous cell carcinomas in 6 out of 25 animals. The direct intrapulmonary injection of benzo[a]pyrene suspended in olive oil after thoracotomy induced a large number of sarcomas of the lung and some squamous cell carcinomas in rats.[107] Furthermore it was possible to induce bronchiogenic carcinomas in up to 50% of all animals after intrapulmonary implantation of dimethylbenz[a]anthracene (DMBA) and benzo[a]pyrene (BaP) in mice and rats.[141-143] A good dose-response relationship could be established between the implanted amount of benzo[a]pyrene and the number of lung tumors found in rats[142,144] (Figure 50). Stanton et al.[173] reported work in which the effect of tobacco smoke condensate on the respiratory tract of Osborne-Mendel rats was investigated.

Thus, it could be demonstrated that tumors are induced locally after intrapulmonary implantation of substances. Although "high concentrations" of a compound can be introduced into the target organ, a number of method-dependent disturbing factors have to be taken into account which may distort the interpretation of the effects of a substance. These include chemical and traumatic (thermal) damages and secondary tissue reactions caused by these damages (cicatrization, granulomas, etc.).

*5.2.2.1.2 Intratracheal Instillation*

Intratracheal instillation is a relatively simple technique, and therefore is frequently used in the testing of chemicals which are potentially carcinogenic to the respiratory tract. As with implantation, instillation has the advantage of permitting exact determination of the administered dose per animal. On the other hand the distribution of substances in the pulmonary region is more homogenous in comparison with implantation. Instillation permits the application of relatively high doses in comparatively short time periods while the contamination of the skin is reduced to a minimum. It is even possible to apply a substance into one lobe of the lung only. Furthermore, the effect of particles and substances which are normally inhaled can be investigated, thus avoiding the inevitable nasal breathing of rodents. Consequently, the size and shape of particles are of minor importance in this type of application. When assessing the carcinogenic response, it seems to be important to examine whether the effect of a substance is increased or decreased by other less or noncarcinogenic substances. It is also necessary to investigate whether substances which are not carcinogenic as such can produce a carcinogenic effect after concurrent application with unspecific substances. Here the simultaneous application of carriers is of particular importance.

The method of intratracheal instillation of locally active carcinogens, such as, for instance, most PAH, in Syrian golden hamsters was tested for the first time by Della Porta et al.,[116] and this method was later modified by several investigators.

To carry out the instillation, the animal, anesthetized with either barbiturates or ether, is fixed to an appropriate holding device and a bulb-headed cannula or a thin plastic catheter is passed through the larynx into the trachea, so that the test substance can be applied directly below the larynx. It is also possible to push the cannula or catheter up to the bifurcation of the trachea (approximately 5.5 cm) so that the test substance can be dispersed in the main bronchi.

Obviously, influences of solvents, suspending agents, and particle-shaped carriers are of importance in the carcinogenesis induced by intratracheally applied substances in the respiratory tract.

The following media were investigated: physiological saline,[121,122,127,135,140] a suspension of sodium dodecyl phosphate, $Na_2$ ethylenediaminetetraacetate (EDTA) and TRIS-HCl buffer,[130] water,[174] carboxymethyl cellulose, polysorbite 80 and benzyl alcohol in physiological saline,[165] infusine,[113] as well as physiological saline and TRIS-HCl buffer or bovine serum albumin.[137,138]

Further experiments were carried out using gelatin and acetone in water,[116] gelatin in 0.9% saline,[128] Tween 60 (polyoxyethylene sorbite monostearate),[129,145,154] tricaprylin,[154] as well as olive oil[129] and mineral oil.[124]

Further investigations examined particulate carriers, such as ferric oxide in 0.9% saline,[126,148,161-163] whereas other investigators used carbon black in combination with infusine (a mixture of sodium, calcium and magnesium chloride, sodium hydrogen carbonate, casein, and water) and plasma-expanding agents,[170] furfural,[121] atmospheric particulate matter,[154] carbon black alone,[113,126,168] magnesium oxide,[175] and charcoal powder.[178]

It seems that suspending agents and soluble additives can directly increase the effect of substances or indirectly influence the reactivity or condition of experimental animals.[135,137,162] Furthermore the carcinogenicity of certain substances seems to depend on the size, composition, shape, and density of the applied particles.[128,164] Particles with an aerodynamic diameter of 10 to 20 μm are precipitated predominantly in the upper respiratory tract (nose, pharynx, larynx), whereas particles with diameters of 2 to 10 μm or 0.1 to 2 μm are precipitated in the tracheobronchial tree or the lung (as far as the bronchioles or to the alveoli), respectively.[164]

Detailed investigations were carried out to assess the effect of particulate carriers and BaP in hamsters. Long-term application of a certain dose yielded positive dose-response relationships, taking time and number of tumors into account.[121,161] Long-term intratracheal instillation of different doses (0.0625 to 20 mg) also resulted in dose-related tumor development depending on the suspending agent or carrier used (Table 46).[122,130,136,137,163,170]

Different treatment times and intervals were also investigated.[116,122,154,163] In these experiments the tumor frequency varied between 0 and 100%.[163,170] After single intratracheal application of different doses of BaP (4 to 37.5 mg), dose-related effects could not be observed. The tumor rates observed in golden hamsters ranged from 3 to 52%.[135,148]

## Table 46
## TUMOR INDUCTION IN SYRIAN GOLDEN HAMSTERS AFTER INTRATRACHEAL INSTILLATION OF BaP AND $Fe_2O_3$ AT DIFFERENT DOSE LEVELS[163]

| Group | Treatment BaP:$Fe_2O_3$ (mg) | Effective number of hamsters | Respiratory tract | | | | Forestomach | |
|---|---|---|---|---|---|---|---|---|
| | | | Tumor-bearing animals | Total no. of tumors | Lesion-bearing animals | Total no. of lesions | Papilloma-bearing animals | Total no. of papillomas |
| 1 | 2.0:2.0 | 28 ♂ | 17 | 28 | 9 | 12 | 13 | 17 |
| | | 29 ♀ | 17 | 30 | 10 | 16 | 8 | 12 |
| 2 | 1.0:1.0 | 33 ♂ | 22 | 31 | 6 | 7 | 8 | 11 |
| | | 34 ♀ | 20 | 28 | 4 | 5 | 9 | 12 |
| 3 | 0.5:0.5 | 33 ♂ | 10 | 11 | 7 | 7 | 5 | 7 |
| | | 30 ♀ | 9 | 11 | 5 | 9 | 2 | 2 |
| 4 | 0.25:0.25 | 47 ♂ | 6 | 6 | 6 | 9 | 2 | 2 |
| | | 41 ♀ | 4 | 4 | 3 | 3 | 4 | 4 |
| 5 | 0:2.0 | 47 ♂ | 0 | 0 | 1 | 1 | 2 | 2 |
| | | 45 ♀ | 0 | 0 | 0 | 0 | 0 | 0 |
| 6 | Untreated | 97 ♂ | 0 | 0 | 0 | 0 | 5 | 5 |
| | | 96 ♀ | 0 | 0 | 0 | 0 | 2 | 2 |

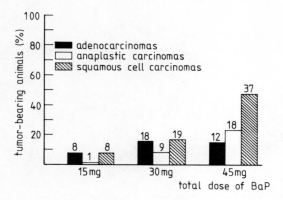

FIGURE 51. Distribution of tumors of the respiratory tract according to their histological type after treatment of Syrian golden hamsters with BaP and $Fe_2O_3$.[160]

It is striking that the concurrent application of ferric oxide (hematite) increased the carcinogenic effect of BaP.[161] Syrian golden hamsters showed tumors of the respiratory tract as early as 15 weeks after commencing weekly intratracheal instillations of a mixture of BaP and ferric oxide, when a cumulative dose of 45 mg BaP and 45 mg of ferric oxide was reached. Dose-related bronchiogenic squamous cell, anaplastic, and adenocarcinomas were induced[160] (Figure 51). The tumor induction times showed a negative dose correlation.

Pott et al.[155] investigated the influence of an almost permanent inhalation of $SO_2$ and $NO_2$ on the frequency of tumors induced by intratracheal instillation of benzo[a]pyrene and dibenz[,h]anthracene. The results are presented in Table 47.

Treatment with other PAH also produced neoplasms of the respiratory tract. Repeated intratracheal instillations of 7,12-dimethylbenz[a]anthracene in 1% gelatin induced squamous cell and adenocarcinomas of the lung in Syrian golden hamsters.[116,124] Furthermore, 7H-dibenzo[c,g]carbazole (7H-DBc) which is contained in tobacco tar[150,167] and dibenzo[a,i]pyrene (DBP)[174] were administered to hamsters. Chronic treatment of Syrian golden hamsters with 7H-DBc at two different dose levels (3 mg 7H-DBc + 3 mg ferric oxide for 15 weeks and 0.5 mg 7H-DBc + 0.5 mg ferric oxide for 30 weeks) resulted in approximately the same incidence of bronchiogenic tumors (30/35 at the high dose and 40/45 at the low dose). After treatment with DBP (1 mg for 12 weeks and 0.5 mg for 17 weeks), predominantly squamous cell carcinomas were induced in the trachea and bronchi (65% tumors at 0.5 mg and 75% tumors at 1 mg).

Chronic instillation of 7,12-dimethylbenz[a]anthracene in rats produced squamous cell carcinomas of the lung in 30% of the animals.[170,171] Of the treated animals, 67% showed neoplasms in the lung after instillation of benzo[a]pyrene. Schreiber et al.[166] carried out comparative investigations on rats and hamsters. Both animal species reacted to a corresponding treatment by developing tumors in the respiratory tract.

Mixtures of PAH were also tested for their carcinogenic activity after intratracheal instillation. Multiple adenomas of the lung were observed in 100% of the golden hamsters treated by instillation of 2.5 and 5 mg of automobile exhaust condensate at 2-week intervals.[146]

With respect to the induction of neoplasms of the respiratory tract after intratracheal application of PAH, it was suggested[116,129,169] that inert dust particles might play a role in the transport of carcinogens. After phagocytosis these particles should give off the adhering carcinogen so that it can act on the epithelium of the respiratory tract.[159] The particle size of the carrier dust and the carcinogen modify the penetration, diffusion,

Table 47
TUMOR INDUCTION IN LARYNX, TRACHEA, AND LUNG OF SYRIAN GOLDEN HAMSTERS AFTER INTRATRACHEAL INSTILLATION OF BaP AND DBahA, AS WELL AS INHALATION OF 29 mg/m³ $SO_2$ + 19 mg/m³ $NO_2$[155]

| Instillation | Animals with tumors (%) | |
|---|---|---|
| | With purified air | After inhalation of $SO_2$ + $NO_2$ |
| NaCl solution | 3 | 3 |
| 6 mg BaP | 17 | 20 |
| 18 mg BaP | 68 | 58 |
| 6 mg DBahA | 55 | 68 |
| 18 mg DBahA | 65 | 71 |

and biological response. Whereas a correlation between the size of coal particles and the amount of adsorbed carcinogen was found, the particle size of ferric oxide seems to be of minor importance.[126] It was demonstrated that although larger particles of ferric oxide hardly phagocytize with BaP, the carcinogenesis is accelerated and the tumor frequency increases.[128] Evidence was also given that BaP, which is only loosely adsorbed to carriers, can be eliminated from the lungs by the mucociliary clearance mechanism.[134] The differences in binding to different carriers possibly accounts for the varying tumor frequencies observed after administration of equal doses of carcinogens. Further experiments proved however that dust particles are of minor importance in carcinogenesis because tumors are also induced by PAH administered at comparative dose levels in suspensions without carriers.[121,130,154] In this case the latency periods were longer and the tumor rate comparatively low. Tumors developing in the upper region of the respiratory tract (larynx, trachea) can be autotransplanted into the lung by repeated intratracheal manipulation.[168]

*5.2.2.1.3 Inhalation*

The carcinogenic effect of a substance is possibly also dependent on the concentration of receptors in the target organ. The distribution kinetics of a substance after implantation and intratracheal instillation, and consequently its concentration at certain sites of the respiratory tract varies when compared with continuous inhalation, which is the usual type of human exposure to atmospheric contaminants. Experiments with labeled substances proved that test substances are more evenly distributed in the respiratory tract by inhalation than by intratracheal instillation.[110]

Under the normal conditions of human exposure the nasal cavity, larynx, trachea, and lungs are probably not contaminated in the same manner. Similarly, the substances are absorbed and metabolized differently at different sites of the respiratory tract. For the investigation of the carcinogenic effect of airborne contaminants, the conditions of such exposure have to be simulated as close to reality as possible. Inhalation experiments with animals have therefore acquired special importance.[112]

Utilizing the latest technological developments, it was possible to construct suitable inhalation devices.[119,131,132,158] These installations have so far found only a limited use in animal experiments because of the high costs incurred by the precautionary measures necessary to protect the personnel.

In long-term inhalation studies the experimental animals are either exposed for a certain time (e.g., 6 to 7 hr daily) on 5 days of the week (intermittent exposure) or they inhale for 22 to 24 hr every day throughout the week (continuous exposure).

Kimmerle[139] developed an inhalation chamber which consists essentially of an aerosol generator (Stöber), a room for keeping animals, and an adjacent exposure room. This installation basically corresponds to the construction of an isolator for keeping germfree animals. It comprises a room for animal keeping and the inhlalation chamber and can house 24 Syrian hamsters in isolation. During the experiment only the respiratory tract of animals is exposed to the test substances. Contamination of the fur is largely avoided by using a tube system. After inhalation of BaP for 10 weeks (total dose: 67 mg/animal) and for 26 weeks (total dose: 23 mg/animal) no treatment-related neoplastic changes could be observed in the respiratory tract within a time period of 94 weeks.

A 2-year exposure to benzo[a]pyrene (2, 10, and 50 mg/m³ air) by inhalation, 3 hr/day for 5 days of the week, did not induce any macroscopically detectable tissue changes. Histologically, only circumscribed hyperplasia was observed in the bronchi. Tumors (papillomas and squamous cell carcinomas) were detected predominantly in the pharynx, epiglottis, and larynx. Individual doses of 2 mg/m³ air did not exert a carcinogenic effect in these organs.

When evaluating these experiments the fact that BaP was tested as a pure substance has to be taken into account. Further findings obtained in other animal species have to be considered, too, in order to permit an assessment of the effect of this substance on man. The effect of BaP when administered in combination with other substances to experimental animals should also be investigated because the effect of inhaled BaP, alone or as part of a mixture, insufficiently known.

When investigating the activity of polycyclic aromatic hydrocarbons in rats and hamsters, the combined inhalation of $SO_2$ and benzo[a]pyrene induced bronchiogenic squamous cell carcinomas in rats only.[142,144] $SO_2$ administered alone was inactive. During the combined application both animal species were exposed to a mixture of 3.5 ppm $SO_2$ and 10 mg BaP per cubic meter air for 1 hr a day (Table 48). In rats the tumor rate was 5/21 = 24% (Table 49). Since very high concentrations of the two test substances were used in these experiments, a human health hazard cannot be deduced merely from their occurrence in the atmosphere.

**Animal experiments with tobacco smoke and tobacco smoke condensate** — A large number of experiments have been carried out to investigate the effect of tobacco smoke and tobacco smoke condensate on the respiratory tract of experimental animals. All the essential experimental results have been summarized by Wynder and Hoffmann[176] and Mohr and Reznik.[147] Malignant tumors were induced in 11% of the treated animals by implantation of tobacco smoke condensate (CSC) into the lung of rats after thoracotomy.[108] Preneoplastic lesions and epidermoid carcinomas were observed in 31 of 106 female Osborne-Mendel rats after implantation of CSC in beeswax.[173] However, implantation experiments with CSC in Syrian golden hamsters were negative.[138] Tobacco smoke condensate and certain fractions were also tested for activity in the lung of laboratory animals after intratracheal administration.[114,116,117,172] In spite of fairly complex studies,[115] tumors could be induced in the respiratory tract at only a very low rate after intratracheal instillation of CSC. Borisjuk[109] found bronchiogenic carcinomas in only 2 of 43 treated animals after instillation of 500 to 600 mg CSC per animal — corresponding to approximately 15 to 20 cigarettes. In Syrian golden hamsters the lifelong weekly intratracheal instillation of 0.07 CSC per week resulted in papillary

Table 48
INHALATION EXPERIMENT WITH $SO_2$ AND/OR BENZO[a]PYRENE[144]

Period of Exposure — 794 Calendar Days

| Type of exposure | Pattern of exposure (5 days/week) | | Exposure (days) | |
| --- | --- | --- | --- | --- |
| | Irritant ($SO_2$) 6 hr/day | Carcinogen (BaP) + $SO_2$ 1 hr/day | Irritant | Carcinogen |
| Fresh air | | | | |
| Fresh air + carcinogen | | 10 mg/m³ B.p. + 3.5 ppm $SO_2$ | | 494 |
| Irritant | 10 ppm $SO_2$ | | 534 | |
| Irritant + carcinogen | 10 ppm $SO_2$ | 10 mg/m³ B.p. + 3.5 ppm $SO_2$ | 534 | 494 |

Table 49
PATHOLOGICAL CHANGES IN THE LUNG OF RATS AFTER INHALATION OF SULFUR DIOXIDE AND/OR BENZO[a]PYRENE[144]

| Type of exposure | Number of animals | Advanced squamous cell metaplasia | Squamous cell carcinoma |
| --- | --- | --- | --- |
| Fresh air | 3 | 0/3 | 0/3 |
| Fresh air + carcinogen | 21 | 1/21 | 2/21 |
| Irritant | 3 | 0/3 | 0/3 |
| Irritant + carcinogen | 21 | 2/21 | 5/21 |

tumors of the trachea in 7 of the 37 treated animals.[117] Inhalation of cigarette smoke effected an increase in the lung tumor rate in treated mice in comparison with control animals.[111,120,133] Inhaled cigarette smoke also showed a weak carcinogenic effect in rats.[125,149] In Syrian golden hamsters the inhalation of cigarette smoke induced tumors of the larynx in up to 20% of the treated animals.[118,133]

*5.2.2.1.4 Conclusions*

In experiments in which the methods described earlier are applied, the biological activity of environmental contaminants in the respiratory tract is investigated. The results obtained are to serve as a basis for the assessment of human risk even though the findings of animal experiments cannot be simply extrapolated to man.

Experimental implantation of carcinogenic substances into the respiratory tract could establish whether a local effect is exerted and which tissues and cells react. Apart from chemically induced alterations, method-related mechanical and thermal damages may occur in the tissue so that secondary reactions can occur (inflammation, fibrosis, granulomas). This complicates an assessment of the substance-related effect. A further disadvantage of this technique is that the exposure of animals and the method-dependent reaction of the lung tissue cannot be compared with the conditions of human exposure. It is advantageous, however, that even substances of low solubility can be investigated locally in high concentrations.

During instillation of substances via the trachea into the lung the substances are partly distributed over the surface of the respiratory epithelium. However, this technique involves mechanically and chemically induced effects, also, such as lesions caused by the cannula or suspending agent. As a consequence the consistency and extent of the mucosal lining of the bronchi can be altered. The treatment is relatively uncomplicated and could possibly determine whether substances that are exogenously applied to the respiratory epithelium would prove to be biologically active, i.e., induce tumors.

Inhalation is the method which, to our knowledge, comes closest to normal human exposure. It permits investigation of the effect of particles and gases, excluding treatment-related traumatic factors. However inhalation experiments involve complicated technology which means that these investigations are very expensive. Consequently, the primary method for investigation of the biological activity of substances in the respiratory tract is intratracheal instillation.

If epidemiological studies, investigations at place of work, and animal experiments have raised questions about a substance or class of substances, the animal model should be used and the techniques described earlier applied to investigate the biological effects and thus possibly permit assessment of the human risk. A valuable contribution could be the development of novel inhalation devices which should be standardized and easy to handle.

The extrapolation of test results of animal experiments is further limited by species-specific patterns of reaction. In this respect the special anatomical circumstances of the respiratory tract of rodents and the time factor in tumor development have to be taken into consideration. According to our present knowledge the testing of noxious substances should include several different animal species so that parameters are obtained which permit an assessment of the human situation.

### 5.2.2.2. Site of Application: Skin

#### M. Habs and D. Schmähl

The skin as a biological system for testing local carcinogenic effects offers some advantages compared with other organs. Beside the low costs, the relatively low need of test substance has to be mentioned, as well as the possibility of direct, exact inspection and documentation (photographs), which can be repeated as often as necessary. The evident value of epicutaneous testing is, however, restricted in that only locally active carcinogens can definitely be detected. Penetration of the test substance into the skin is a prerequisite for the development of carcinogenic activities. For instance, asbestos is inactive after topical application because there is no bioavailability. Substances with a low vapor pressure may also not be detected or only be detected to a small extent.[184] Furthermore, inflammatory or necrotizing effects may influence the carcinogenesis and aggravate the evaluation of test results.[221] If a metabolization is necessary for the development of the active form of a carcinogen, then considerable quantitative differences may arise due to different enzyme patterns and enzyme inducibilities of the epidermis of different species and strains. The skin has a particularly selective activity in the case of carcinogens adsorbed to particles, as, for instance, in the testing of condensates. Only the activity of detached and resorbed constituents is investigated, whereas particles remaining on the epidermis do not have any effect at all. This may possibly explain the different results obtained in epicutaneous and subcutaneous investigations of condensates (cf. the next to last paragraph of this chapter). The solvent used, too, may have an extremely strong impact on the tumor yield. In order to make comparable statements possible, it is necessary to compile numerous parameters, each of which can considerably modify the experimental results. These are, for instance, animal strain used, sex, solvent, amount and concentration of the individual dose, frequency and duration of treatment and observation, method of hair clipping, technique of application (painting or dropping with a micropipette, calibrated syringe or dosimeter, the latter techniques being better suited for qualitative investigations), documentation of the treated area, data on the biological end point (time of occurrence of tumors, differentiation between papilloma and carcinoma at the site of application). As in all long-term experiments, further effects can arise from the mode of housing (e.g., individual housing, group housing), the diet, etc.

The tumor-inducing effect of PAH-containing mixtures on human skin was described for the first time by Percivall Pott in 1775. In the past 60 years the tumor-inducing activity of a large number of chemical substances was demonstrated in topical painting or dropping tests.[181] As when testing by subcutaneous administration, the mouse is the experimental animal most commonly used in the epicutaneous testing of PAH and PAH-containing mixtures or condensates when determining biological activities. The basic sensitivity of the skin of other species to carcinogens has already been demonstrated, e.g., in the case of man,[188,209] rat,[220] hamster,[234] guinea pig,[223] rabbit,[246] and rhesus monkey.[237] The experimental design of cutaneous carcinogenesis experiments is referred to in a number of reviews and books.[180,183,196,205,206,228,232,239,244,245,247]

The topical application of carcinogenic PAH predominantly causes tumors of epithelial origin at the site of application. Induced are benign tumors such as papillomas, as well as carcinomas which histologically prove to be mainly keratinizing squamous cell carcinomas. Sarcomas are observed only in rare cases. In mice of certain inbred strains the topical application of cancer-inducing PAH effects the development of systemic tumors. Most frequently, increased incidences of lung adenomas and hepatomas are observed.[196,222] Since mouse strains mostly have a high rate of "spontaneous" tumors,[199] untreated and solvent-treated controls are indispensible. There is still no generally accepted standardization of test methods for topical application of substances relative to possible carcinogenic effects. Due to the variety of modifications in the design of epicutaneous test models described earlier, it is difficult to compare the results obtained by different working groups.

The scientific findings obtained so far do not show how to extrapolate the results of epicutaneous tests to changes observed in the lung, the most important target organ of cancer-inducing PAH (cf. Chapter 8).

The experimental results obtained with PAH as pure substances, mixtures of PAH, or PAH-containing condensates after topical application are outlined in the following overview. The list presented is restricted to those compounds which occur in relevant concentrations in the environment and of which at least sufficient experimental data exist, which permit making reliable statements on the quality of the carcinogenic effect in epicutaneous tests. A classification according to the carcinogenic potency in this experimental animal model can be carried out only on the basis of experimental results obtained under identical conditions.

Table 50 lists the carcinogenicity of some PAH which have been investigated in epicutaneous tests. The references given are meant to be representative only; completeness is not aimed at. We have included long-term experiments in which the test substance was applied to the skin chronically as well as in so-called initiator-promoter systems. In the latter case, the alleged carcinogenic PAH is applied to the skin once or several times followed by a local treatment with a very weak or noncarcinogenic compound, such as a phorbol ester. (See Berenblum[182] and Schmähl[228] with regard to the design of these experiments.)

When comparing the dose-response relationship of benzo[*b*]fluoranthene, benzo[*j*]fluoranthene, and benzo[*a*]pyrene obtained under standardized conditions in an epi-

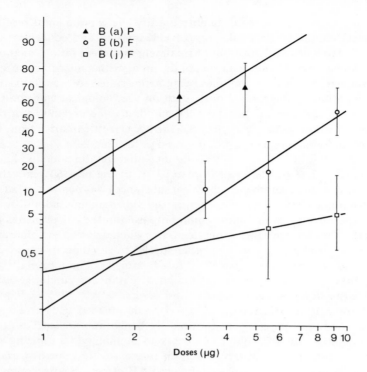

FIGURE 52. Dose-response relationship of benzo[a]pyrene, benzo[b]fluoranthene and benzo[j]fluoranthene after application to the skin of female NMRI mice.

cutaneous test system,[197] it becomes evident that their dose-response relationships are not parallel. Consequently it may be stated that the relation of the carcinogenic activity of these compounds to one another varies as a function of the doses applied. As in the subcutaneous test model, the differences in potency are more marked at higher doses than at lower doses (cf. Figure 52).

Several working groups have investigated PAH mixtures and condensates and their fractions in dropping tests on mouse skin relative to carcinogenic effects.[203,210] For instance, Wallcave et al.[238] described the cancer-inducing activity of petroleum pitch, coal tar, and other pitches as a function of the PAH content of the corresponding product. The condensate of automobile exhaust and its fractions were dealt with by Hoffmann et al.,[202] Kotin et al.,[211] and our own group.[185] Detailed investigations were carried out with cigarette smoke condensate and its fractions.[189-191,240,241] Dontenwill et al.[190] proved in their investigations that in the topical dropping test the carcinogenic activity of cigarette smoke condensate can be predominantly attributed to the PAH. A fraction, which amounts to 0.4% of the initial condensate and in which the PAH are 250 times concentrated, accounts for 50% of the carcinogenic activity of the total condensate, on the assumption of substitutive interactions of the individual fractions. Of the total carcinogen activity 7% were attributed to a PAH-free fraction. Schmidt et al.[231] investigated PAH mixtures obtained from smokehouse soot.

Noncarcinogenic fractions, mixtures of substances, or pure PAH have only a slight effect on the tumor-inducing activity of carcinogenic PAH. In many experiments, enhancing (so-called cocarcinogenic or promoting effects) and inhibiting effects did not play a relevant role in the dose ranges investigated in the present test model.[185,217,229,231] One exception was the investigations carried out by Hoffmann et al.[202] who used fractions of automobile exhaust condensates to which hydrocarbons had been added. Sig-

Table 50
## CARCINOGENICITY IN EPICUTANEOUS TESTS OF PAH OCCURRING IN AIR PARTICULATE MATTER

| Substance | Carcinogenicity | Ref. |
|---|---|---|
| Anthanthrene | + | 187 |
|  | − | 200,215 |
| Benzo[a]pyrene and metabolites | + | 187,200,212—214,230,242 |
| Dibenz[a,h]anthracene | + | 200,207 |
| Dibenzo[a,e]pyrene | + | 200 |
| Dibenzo[a,i]pyrene | + | 200 |
| Dibenzo[a,h]pyrene | + | 187,200 |
| Dibenzo[a,l]pyrene | + | 200 |
| Benzo[b]fluoranthene | + | 243,197 |
|  | − | 218 |
| Benzo[j]fluoranthene | + | 243,197 |
|  | − | 218 |
| Benzo[k]fluoranthene | − | 197,218,243 |
| Benz[a]anthracene | + | 208,233,235 |
| Chrysene | + | 192,198,202a,233 |
| Cyclopentadieno[c,d]pyrene | − | 185,197 |
| Anthracene | (+) | 233 |
|  | − | 181 |
| Phenanthrene | (+) | 227,233,241 |
|  | − | 226 |
| Fluoranthene | − | 201,204 |
| Benzo[e]pyrene | − | 186,195 |
|  | (+) | 233 |
| Pyrene | − | 195,226 |
|  | (+) | 204 |
| Perylene | − | 193,204 |
| Indeno[1,2,3-cd]pyrene | + | 200 |
|  | − | 197 |
| Coronene | − | 197 |
| Fluorene | − | 179 |
| Benzo[ghi]fluoranthene | − | 243 |
| Triphenylene | (+) | 204 |
| Indeno[1,2,3-cd]fluoranthene | + | 200 |
| Benzo[ghi]perylene | − | 219 |
| Benzo[a]fluorene | − | 179 |
| Benzo[c]fluorene | − | 179 |
| 7,12-Dimethylbenz[a]anthracene | + | 225,235 |

Note: + = Carcinogenicity proven.
(+) = Evidence of possible carcinogenicity.
− = Carcinogenicity not proven.

nificant inhibitory effects of weak or noncarcinogenic PAH on 7,12-dimethylbenz-[a]anthracene- and benzo[a]pyrene-induced tumors have recently been reported by Slaga et al.[236]

Summarizing and generalizing, the following may be stated:

1. All condensate samples investigated in epicutaneous tests show carcinogenic activity.
2. The observed carcinogenicity can predominantly be attributed to the PAH-containing fractions, the four- to six-ring structures showing the strongest effect.
3. On the assumption of additive effects, benzo[a]pyrene accounts only for a small proportion (in general 2 to 10%) of this biological activity.

### 5.2.2.3 Site of Application: Subcutaneous Tissue

#### F. Pott

#### 5.2.2.3.1 Introduction

The subcutaneous injection of chemical substances is the most common method for the investigation of their carcinogenicity.

Compared to all the other methods, this test has a number of essential advantages:

1. It is relatively simple and can be carried out with little effort.
2. Contamination of the laboratory with carcinogenic substances can easily be avoided.
3. The test is sensitive so that, for instance, one injection of PAH may suffice to induce tumors, if suitable liquid carriers are chosen (see later discussion). Consequently, only small amounts of test substances are needed; this is of importance if the isolation of these substances from emissions or from air pollutants or their synthesis meets with difficulties.
4. Local carcinogenic and systemic carcinogenic effects can easily be distinguished from one another.

All these properties make the subcutaneous test a suitable experimental method which can produce essential information on a substance which is only in its early phases of testing.

Large subcutaneous inoculations of chemically inert substances can induce local sarcomas in animal experiments.[252,281,282] However, the possibility of extrapolating these findings to man has not yet been confirmed. This fact has wrongly discredited the subcutaneous test in the opinion of some authors. Grasso particularly rejected the subcutaneous application as an irrelevant method;[262,264,266,267,270] In our opinion, a test which is sensitive to chemical carcinogens cannot lose its significance because it is not at the same time a good indicator of mechanical stimuli. It is, of course, understood that test results obtained after subcutaneous application have to be analyzed and interpreted with great care, as is the case with all the other carcinogenesis test methods.

Years ago Druckrey et al.[257] and van Duuren et al.[259] expressed a positive opinion on the subcutaneous test method. Tomatis[302] recently confirmed this assessment when he stated in a critical overview of test results obtained with 102 substances, that the subcutaneous test yielded the same percentage of false-positive and false-negative results as all the other relevant test systems. Arcos et al.[248] and Weisburger[306] summed up general instructions on the testing of carcinogens in animal experiments.

Finally, the acute and subacute toxicity of test substances which may limit the applicable dose — as in other tests too — must be mentioned: substances or mixtures of substances which — when applied at the maximum injectable dose — either kill the animal after absorption (e.g., nicotine in cigarette smoke condensate) or necrotize the subcutaneous tissue, and the adjacent skin can be investigated for carcinogenicity only by using lower concentrations and perhaps by repeated injections.

After subcutaneous injection of carcinogenic PAH, tumors are induced almost exclusively at the site of application unless neonatal animals are used. Histologically almost all these tumors are sarcomas (fibrosarcomas, rhabdomyosarcomas, hemangiosarcomas) and, very seldom, carcinomas. Only if very large amounts of the test substance are administered, tumor growth may also be observed in other organs (see Chapter 5.2.2.4). However, some substances which do not belong to the class of PAH induce tumors more frequently away from the site of application and must be considered accordingly.

Subcutaneous injection of PAH or PAH-containing mixtures into pregnant animals induced cancer in the newborn. The subcutaneous application of these substances to newborn mice produced lung adenomas, hepatomas, and occasionally tumors in other organs.[249,260,275]

Below, the subcutaneous test is reviewed in a more narrow sense concerning the induction of tumors at the site of application. Essentially, results of determining dose-response relationships are presented. These dose-response relationships indicate how many individuals (or the percentage of individuals) of a group developed a tumor in response to the applied dose. The dose-response relationship can be expressed as a curve only if at least three values have been determined. In case of a linear percent graduation, an S-shaped curve is normally obtained which becomes a straight line after probit transformation ("regression line"; for examples, see Figure 54); in this presentation the distances of the percent graduation are spread more and more with increasing distance to the 50% mark.

Dose-response relationships are a special case of dose-effect relationships. The term "dose-effect relationship" implies all possible dose-dependent possibilities of a response including changes of the mode, site, and degree of the effect, as well as changes of induction time, the incidence, and time of manifestation of an exactly defined effect. When testing the carcinogenicity of a substance in a certain experimental model, such as, for example, the subcutaneous test, usually a certain mode of action (induction of a tumor at the site of application) is investigated with regard to the frequency of incidences in relation to the dose applied. If possible, the tumor induction time is also included in the evaluation.

*5.2.2.3.2 Importance of the Methodology*

The dose-response relationship of a carcinogen in a test system depends to a large extent on numerous methodical factors which include in particular the solvent or suspending medium when testing PAH after subcutaneous administration. Some aspects should therefore be outlined here.

In mice, PAH suspended in 0.1 to 0.5 m$\ell$ of a liquid carrier are injected under the dorsal skin. If the PAH is suspended in NaCl solution, there is basically no risk that the substance will leak out of the cutaneous place of perforation. If, however, oil is used as a carrier, the place of perforation has to be sealed carefully by a tissue adhesive in order to guarantee exact doses and avoid loss of substance.[294]

The type of carrier liquid used for the injection of PAH is of great importance for the tumor incidence. The carrier liquid may perhaps exert a chemical influence on the PAH or accompanying substances (reference review by Berenblum[251]). At least it essentially determines the formation of a depot and the retention time of PAH under the skin.[261,278,284,292,305] Liquids, which readily dissolve BaP, on the other hand, are rapidly distributed within the body after subcutaneous injection (dimethylsulfoxide, Lutrol®), transport the dissolved BaP when they are absorbed, and largely prevent the formation of a depot at the site of application. Consequently, the tumor incidence is low at this place. NaCl solution is also rapidly absorbed, but leaves the suspended BaP which is of low water solubility at the place of injection. The tissue fluid then gradually dissolves and removes the substance. By using oil — mostly tricaprylin — it is easy to form a depot of PAH. Tricaprylin is absorbed only very slowly from the subcutaneous place of injection and at the same time retains the PAH due to its good lipoid solubility. Within a few days a connective-tissue sheath envelops the oil drop at the place of injection. In the course of some months the PAH continuously diffuse into these connective tissue cells. The cyst diminishes very slowly and usually can be observed more than 2 years after injection. The PAH however are eliminated after

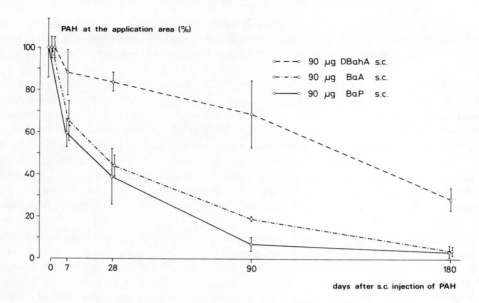

FIGURE 53. Retention times of benzo[a]pyrene (BaP), benz[a]anthracene (BaA) and dibenz[a,h]anthracene (DBahA) at the application area in mice after subcutaneous injection of 90 µg pure PAH in 0.5 ml of tricaprylin.[303]

some months (Figure 53). A high tumor incidence results in relation to the applied dose. After application of pure PAH, tumors develop predominantly from the 4th to the 12th month after injection.

The amount of solvent used has also a distinct influence on the tumor rate: 50 µg BaP in 0.1 ml of tricaprylin resulted in subcutaneous tumors in only 15% of the mice within 300 days compared with 60% after injection of 1 ml of tricaprylin. (Pott et al., 1973a). The tumor incidence was higher when the total dose of BaP was not administered once in 0.1 ml of tricaprylin, but was divided into 12 monthly doses each suspended in 0.1 ml of tricaprylin.[283]

Concerning the dose-response relationship, it seems to be important that the rates of elimination of some PAH from the site of application clearly differ from each other.[269,303] Dibenz[a,h]-anthracene, for instance, is absorbed particularly slowly or metabolized at the site of injection. (Figure 53). In the lung, the circumstances may be quite different from those in the subcutaneous tissue.

These remarks may suffice to explain the great variations in the findings of different authors using different experimental techniques. On the other hand they demonstrate that different PAH or PAH-containing mixtures can be ranked with respect to their carcinogenic potency only on the basis of test results which were obtained under identical conditions.

### 5.2.2.3.3 Carcinogenicity of Individual PAH

Without regard to dose-response relationships and details of methodology, in the following list of PAH are those which have been detected in airborne particulate matter and proved to be carcinogenic in the subcutaneous test.

In the thirties, first attempts were made to isolate individual PAH from extracts of soot and coal-tar and test them subcutaneously in order to detect tumorigenic substances (reference review by Badger[250]); these experiments had been preceded by skin-painting tests (see Chapter 5.2.2.2). Beside the known strong carcinogens,

Table 51
RELEVANT AIRBORNE PAH INVESTIGATED FOR CARCINOGENICITY
AFTER SUBCUTANEOUS ADMINISTRATION

| Substance | Molecular weight | Carcinogenicity | Ref. |
|---|---|---|---|
| Phenanthrene | 178 | − | 301 |
| Anthracene | 178 | − | 301 |
| Fluoranthene | 202 | − | 256 |
| Cyclopentadieno[cd]pyrene | 226 | + | 280,295 |
| Benz[a]anthracene | 228 | + | 295,300,301 |
| Chrysene | 228 | + | 295,300,301 |
| Benzo[b]fluoranthene | 252 | + | 256 |
| Benzo[k]fluoranthene | 252 | + | 256,295 |
| Benzo[e]pyrene | 252 | − | 295 |
| Benzo[a]pyrene | 252 | + | 254,263,272,277 |
| 11-H-Cyclopenta[qrs]-benzo[e]pyrene (= 1,12-Methylbenzo[e]pyrene) | 264 | + | 295 |
| 10-H-Cyclopenta[mno]benzo[a]pyrene (= 10,11-Methyl-benzo[a]pyrene) | 264 | + | 295 |
| Benzo[ghi]perylene | 276 | − | 279,295 |
| Indeno[1,2,3-cd]pyrene | 276 | + | 295 |
| Dibenz[a,h]anthracene | 278 | + | 254,255,285,291,300,301 |
| 1,12-Methylenebenzo[ghi]perylene | 288 | + | 298 |
| Dibenzo[def,p]chrysene (= Dibenzo[a,l]pyrene) | 302 | + | 256,304 |
| Benzo[rst]pentaphene (= 3,4,9,10-Dibenzpyrene = Dibenzo[a,i]pyrene) | 302 | + | 256 |
| Dibenzo[b,def]chrysene (= 3,4,8,9-Dibenzpyrene = Dibenzo[a,h]pyrene) | 302 | + | 256 |
| Naphtho[1,2,3,4-def]chrysene (= 1,2,4,5-Dibenzpyrene = Dibenzo[a,e]pyrene) | 302 | + | 256 |

benzo[a]pyrene and dibenz[a,h] anthracene, numerous other PAH were tested which had been synthesized to elucidate the correlation between chemical structure and carcinogenicity. Meanwhile the carcinogenicity of BaP has been proved following subcutaneous administration in mice, rats, golden hamsters, guinea pigs, monkeys, and newts (reference review in *IARC Monographs*[272]). In cattle and pigs the intramuscular injection of very high doses of BaP did not induce tumors.[274] As in skin-painting experiments, the mouse is the preferred animal species for the subcutaneous testing of PAH.

Table 51 gives a list of relevant airborne PAH investigated in the subcutaneous test.

The dose-response relationships of of PAH determined by the same method show different gradients of the regression line (Figure 54). In these cases the carcinogenic activity of different PAH can be correlated at a fixed numerical ratio only for the incidence of certain tumors. In many cases the differences in the frequency of carcinogenic action at high doses are greater than at low doses. The prevailing role of BaP in the medium and upper range of tumor incidence is undoubted. In the lower range (tumor rate < 10%, it is difficult to correlate the differences in tumor incidence due to possible statistical uncertainties. Only DBahA shows clearly a stronger activity at lower doses (< 30µg) than BaP, i.e., the dose-response relationship of BaP forms a steeper gradient than that of DBahA.[285]

*5.2.2.3.4 Carcinogenicity of Mixtures of PAH*

Several authors have investigated the question whether the subcutaneous injection

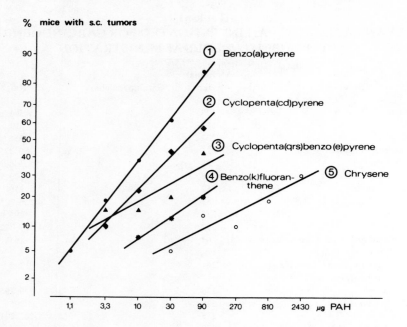

FIGURE 54. Dose-response relationships of benzo[a]pyrene, cyclopenta-[cd]pyrene, cyclopenta[qrs]benzo[e]pyrene, benzo[k]fluoranthene, and chrysene (Pott et al., 1978). The differences in the dose-response relationships cannot be correlated at a fixed numerical ratio in several cases. The gradients of the regression lines differ significantly as shown in the following comparisons:

regression lines 1:2 = significant (<1% probability of error)
regression lines 1:3 = significant (<1% probability of error)
regression lines 1:4 = significant (<0.1% probability of error)
regression lines 1:5 = significant (<0.1% probability of error)
regression lines 2:5 = significant (<5% probability of error)

of two or more PAH yields results different from that to be expected by adding the individual effects, and whether interactions of PAH can be observed.[261,288,295,300,301] Various possibilities were observed: the tumor rate was occasionally higher and occasionally lower than was expected from summation of the individual tumor incidences. No modifying influence could be observed, and the direction and strength of interactions were not always dependent on the dose. Figures 55 to 57 may serve as examples of such interactions.

According to investigations carried out by Pfeiffer[288] a mixture of 10 non- or weakly carcinogenic PAH produced low tumor incidences. The addition of these 10 PAH to a mixture of the 2 carcinogens BaP and DBahA, however, did not modify the dose-response relationship of this mixture of carcinogens (Table 52).

Generally speaking, the combination of higher doses of different PAH seems mostly to have an inhibitory effect, whereas barely effective doses of carcinogenic PAH have a mutually increasing effect. In several cases, the regression lines of the dose-response relationship of individual PAH are steeper than those of mixtures of PAH. However, this rule is not generally applicable. So far it has not been possible to predict the effect of combinations of PAH.

### 5.2.2.3.5 Carcinogenicity of PAH-Containing Mixtures

In this connection, atmospheric particulates are surely the most interesting mixtures of substances. Instead of "atmospheric particulates" the term "air dusts" is frequently

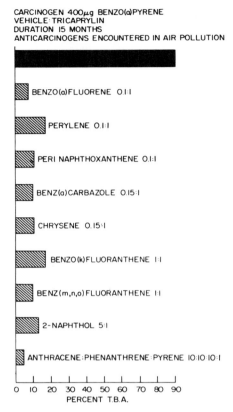

FIGURE 55. The inhibiting effect of hydrocarbons found in polluted urban air and cigarette smoke on the carcinogenicity of BaP in mice when injected in combination with BaP. (T.B.A. = tumor-bearing animals.)[261]

used. The latter expression is more precise from the linguistic point of view, but with regard to contents it is incorrect because it does not include drop-shaped substances. The simpler term "air dust" is used here in the broad sense of the term "atmospheric particulates". In most cases only the extractable components of this dust are investigated because they can be derived more easily, and, according to experience, the carcinogenic potency of airborne particulate matter is concentrated in these extracts.

Extracts of air dust were injected under the skin of laboratory animals 40 years ago to investigate the assumed tumorigenic effect of PAH contained in these extracts.[276,277,299] These investigations were carried out primarily to answer the following questions:

1. What differences exist between the carcinogenic potencies of air dusts or extracts of air dust sampled at different places and at different times?
2. What fractions or individual substances isolated from the total extract have a particularly high tumorigenic potency compared with the remainder, and to what extent do they contribute to the carcinogenicity of the total extract?
3. Are there any essential interactions of substances contained in the mixture, i.e., do inhibitory or super-additive effects occur, and how do they appear?

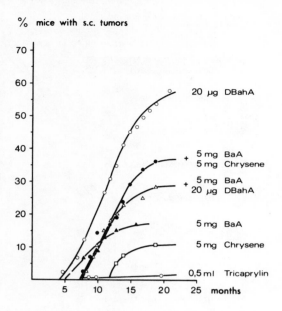

FIGURE 56. Tumor incidences in relation to duration of experiment by adding different PAH; inhibition of the carcinogenicity of dibenz[a,h]anthracene by addition of benz[a]anthracene; superadditive increase of tumor rate by combined application of chrysene and benz[a]anthracene.[300]

Due to the numerous possibilities of variation in the composition of respirable dusts, the answers to these questions cannot yet be completed. The test results presented so far contain a number of interesting data necessary for the total picture, which will be briefly outlined below.

Extracts of air dust collected in Texas City from 1965 to 1969 yielded tumor rates between 0 and 60%.[296] However, the correlation between tumor rate and BaP content of the extract, which the authors pointed out, does not seem to be conclusive.

Extracts of air dust collected in eight North American cities were investigated. Of these extracts, 3 fractions (aromatic hydrocarbons, aliphatic hydrocarbons, oxidation products) revealed that only 4 of the 32 extracts injected subcutaneously induced tumors in more than 7% of the approximately 50 animals of each group.[271] Therefore the differences in the tumor rates are significant and interpretable in only a few cases. It is remarkable, however, that all the four extracts from Los Angeles did not induce tumors; the doses were taken in accordance with the sampled amount of extract; the BaP content of the total extracts sampled in the seven other cities was up to 15 times greater (Birmingham, Cincinnati, Detroit) than in Los Angeles.

This result hints at the fundamental difficulty of determining standards for a meaningful dose comparison of PAH-containing mixtures sampled from the exhaust of an emitter or from the atmosphere. Such parameters for dosing might be in particular: air volumes, content of atmospheric particulates, percentage of soluble atmospheric particulates (amount of extract), percentage of PAH with four and more rings, and content of certain individual prominent PAH. It would be optimal to have all these data and relate them to the results of animal experiments. Results obtained in animal experiments can only be properly interpreted if the tumor incidences which have been found considerably exceed the rate of spontaneous tumors and if the tumor incidences

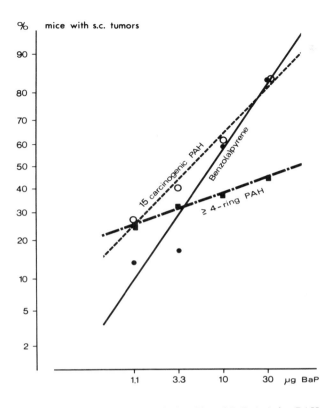

FIGURE 57. Dose-response relationships of BaP, ≥ 4-ring PAH derived from automobile exhaust condensate and a mixture of 15 carcinogenic PAH (ratio of carcinogenic PAH of the mixture as in automobile exhaust condensate).[295] (Fractionation of condensate by Grimmer.[268])

Table 52
TUMOR INCIDENCES OBSERVED AFTER SINGLE SUBCUTANEOUS INJECTION OF TWO STRONGLY CARCINOGENIC PAH (BaP AND DBahA) AND TEN NON- OR WEAKLY CARCINOGENIC PAH

| BaP | | DBahA | | BaP + DBahA | 10 PAH | 12 PAH |
|---|---|---|---|---|---|---|
| Dose (μg) | Tumors (%) | Dose (μg) | Tumors (%) | Tumors (%) | Tumors (%) | Tumors (%) |
| 3.12 | 9 | 2.35 | 37 | 48 | 6 | 41 |
| 6.25 | 35 | 4.7 | 39 | 44 | 8 | 55 |
| 12.5 | 51 | 9.3 | 44 | 61 | 6 | 61 |
| 25.0 | 57 | 18.7 | 56 | 68 | 4 | 72 |
| 50.0 | 77 | 37.5 | 65 | 69 | 13 | 68 |
| 100.0 | 83 | 75.0 | 69 | 79 | 5 | 82 |

Note: Ratio of PAH as in automobile exhaust condensate:[268] benzo[a]pyrene (1): dibenz[a,h]anthracene (0.75): benzo[e]pyrene (0.7): benz[a]anthracene (1): phenanthrene (40): anthrancene (10): pyrene (21): fluoranthene (9): chrysene (1): perylene (0.07): benzo[ghi]perylene (4.1): coronene (1).[288]

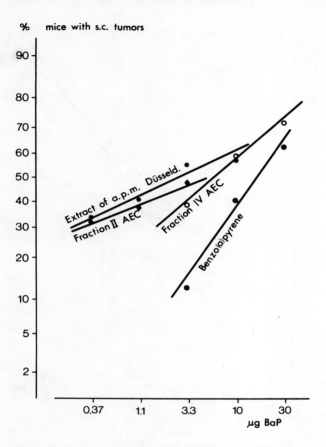

FIGURE 58. Dose-response relationships of benzo[a]pyrene, extract of airborne particulate matter (a.p.m.; Dusseldorf winter 1974/75), and fractions II and IV isolated from automobile exhaust condensate (AEC).[295] (Fractionation of condensate by Grimmer.[268])

grow with increasing doses. It would, therefore, be wrong to administer extracts from equal air volumes of an urban area and a rural area to animals: the much lower content of atmospheric particulates in rural areas would make tumor induction very improbable, so that a numerical ratio of the different carcinogenic potencies of both air qualities could not be determined.

In experiments with extracts of air dust samples in North Rhine-Westphalia, BaP served as a standard for dosing in order to determine the influence of other factors in the different extracts on the BaP-induced tumor rate. The regression lines, which characterize the course of the dose-response relationship, proved to be considerably flatter for all the extracts of air dust than for pure BaP (see example in Figure 58).

Fraction II of the automobile exhaust condensate (hydrophobic substances; for scheme of fractionation see Chapter 7) which also does not contain any insoluble particles and thus corresponds to extracts of air dust, has a similar gradient of the regression line. On the other hand, the regression line of the PAH fraction of automobile exhaust condensate is much steeper, and its angle of inclination resembles that of pure BaP. These findings correspond with the regression lines of BaP and some selected PAH determined in a separate experiment (see Figure 57). The results indicate that under the given experimental conditions the proportion in correspondence with decreasing doses, while the influence of other substances grows.

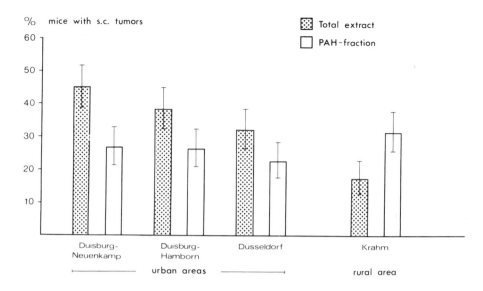

FIGURE 59. Tumor incidence at the application area after a single subcutaneous injection of extracts of airborne particulate matter (total extract and PAH fraction) collected at four sampling stations in winter 1975/76; all tumors of 4 groups of 60 animals each are taken together (4 doses on the basis of benzo[a]pyrene: 0.16 µg, 0.63 µg, 2.5 µg, 10 µg BaP); confidence interval 95%.[293]

Figure 59 shows the participation of non-PAH in the carcinogenicity of extracts of air dust sampled in three urban areas. Contrary to this result, the extract of air dust from a rural area clearly produced less tumors than the isolated PAH fraction. It is remarkable that all of the four PAH fractions which had been concentrated to an equal BaP content prior to investigation in the animal experiment induced approximately the same tumor frequency; however, the concentration of seven further PAH in these extracts being additionally analyzed also did not differ very much (see Chapter 3.5).

All the results obtained so far in experiments with extracts of air dust and automobile exhaust condensate, as well as their fractions, indicate that the carcinogenic potency of these mixtures after subcutaneous application, as well as in the skin-painting test, can be predominantly attributed to PAH. From the quantitative point of view, the PAH account for only a very small portion of the condensate or extract. Of these PAH, in turn, those with at least four rings are by far of the greatest importance, whereas PAH consisting of three rings exert only a very weak tumorigenic activity despite their high concentration in the PAH fraction.

The carcinogenic activity of fractions other than the PAH fraction or their promoting activity still has to be investigated in detail.

The earlier described inhibitory effect produced by some PAH on the carcinogenicity of individual PAH can obviously also be exerted by other substances extracted from air dust, as shown in Figure 60. It is still unclear how these inhibitory effects arise. Possibly they inhibit the enzyme activity in the tissue at the site of application, so that the PAH cannot at all, or only to a limited extent, be transformed into the carcinogenic metabolites. Furthermore, as described earlier, the possible influence of liquid carriers on the formation of PAH depots has to be taken into consideration.

Solid particles may also have a distinct dose-related inhibitory effect on BaP-induced carcinogenesis, as has been shown with aluminum hydroxide,[287] pollen,[307] air dust,[290] and, in particular, with automobile exhaust condensate[294] (Figure 61), as well as

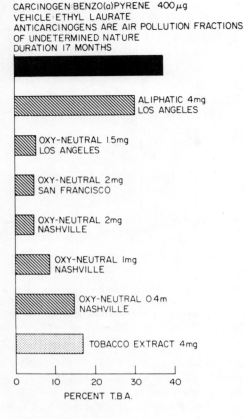

FIGURE 60. Inhibition of the carcinogenicity of BaP by fractions isolated from airborne particulate matter and cigarette smoke condensate (T.B.A. = tumor-bearing animals).[261]

lead.[289] The particle phase of automobile exhaust condensate largely obscures the carcinogenic potency of the contained PAH (Table 53).

Finally, it should be mentioned that virus treatment of mice[286] and administration of putrescine, *cis*-aconitic acid, etc. may inhibit BaP-induced carcinogenesis.[273] Further research has to determine whether, and to what extent, these inhibitory effects are due to activities that can also be observed in other test models.

Subcutaneous injection of two tobacco smoke condensates (64 injections of 50 mg each; total: 3.2 g of condensate containing 2.6 µg BaP) induced tumors at the site of application in 20% and 17.5% of the treated rats, respectively. No tumors developed in comparative animals which were injected with a 45-fold dose of BaP (116 µg).[258] Consequently, the percentage of the carcinogenic action of tobacco smoke condensate which has to be ascribed to BaP must be lower than 1%. Schmähl[297] also obtained tumor rates of between 20 and 30% when using several tobacco smoke condensates using the same experimental setup as Druckrey et al.

After fractionation of tobacco smoke condensate, Seelkopf et al.[298] detected the main carcinogenic potency in the so-called neutral fraction which contains the PAH.

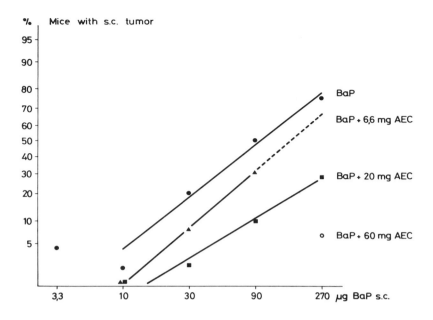

FIGURE 61. Dose-response relationships of benzo[a]pyrene (BaP) and mixtures of BaP with automobile exhaust condensate (AEC) observed in mice 30 weeks after subcutaneous injection.[294]

*5.2.2.3.6 Summary*

1. Subcutaneous administration is the most sensitive experimental model of testing PAH for tumorigenic activity.
2. Approximately 15 PAH which occur in the atmosphere or in exhaust gases of fossil fuels proved to be tumorigenic in this model. A large number of relevant environmental PAH which, according to their structure, might be carcinogenic have not yet been investigated.
3. The dose-response relationships of some thoroughly investigated PAH occasionally differ very much from each other. In those cases in which the differences in the steepness of these gradients are statistically significant, the different carcinogenic potencies of two PAH may not be expressed by a simple numerical ratio. For instance, to induce a tumor rate of 5%, 25 times more chrysene is needed than benzo[a]pyrene; for a tumor rate of 30% this ratio is about 350:1.
4. Only few results have been reported on the tumorigenic activity of mixtures of several carcinogenic PAH. According to these, the combination of PAH results in a relatively flat gradient of the dose-response relationship, i.e., in the lower dose range an additive or even superadditive tumor frequency is observed, whereas in the higher dose range the tumor frequency of a PAH may be reduced.
5. Mixtures of PAH with other substances can induce very different effects: after subcutaneous application of extracts of air dust, a very flat gradient of the dose-response relationship is normally obtained, i.e., very low doses cause a relatively high tumor rate. Extracts with a BaP content of 370 ng (annual average value at a sampling station in Duisburg in 1975: approximately 50 ng BaP per cubic meter air) induced a tumor incidence of about 30%. That part of the automobile exhaust condensate which was separated from the hydrophilic components and particles produced the same dose-response relationship as did extracts of air dust.

## Table 53
### DOSE-RESPONSE RELATIONSHIP OF AUTOMOBILE EXHAUST CONDENSATE AND ITS FRACTIONS OBSERVED IN MICE AFTER SUBCUTANEOUS ADMINISTRATION

Total condensate
D.1: 3.3% tumors
D.2: 3.4% tumors
D.3: 6.8% tumors

Fraction I (methanol phase)
D.1: 0 % tumors
D.2: 5.2% tumors
D.3: 1.7% tumors

Fraction II (cyclohexane phase 1)
D.1: 41.7% tumors
D.2: 45.6% tumors
D.3: 39.0% tumors
1/3 D.1: 42.4% tumors
D.1: 52.5% tumors

2nd batch (new condensate)
0.37 μg BaP: 31.0% tumors
1.1 μg BaP: 36.2% tumors
3.3 μg BaP: 46.6% tumors

Fraction III (cyclohexane phase 2)
D.1: 3.4% tumors
D.2: 5.1% tumors
D.3: 5.1% tumors
D.4: 16.7% tumors

Fraction IV (nitromethane phase)
D.1: 38.3% tumors
D.2: 58.6% tumors
D.3: 71.2% tumors

Fractions: IV a
D.1: 15.0% tumors
D.2: 25.4% tumors
D.3: 45.0% tumors

IVb
30.0% tumors
48.3% tumors
41.7% tumors

IVc
36.7% tumors
39.7% tumors
55.2% tumors

IVd
31.7% tumors
38.3% tumors
59.3% tumors

IVe
13.3% tumors
37.3% tumors
41.7% tumors

Combination of fractions I, III, IV a—e
D.1: 8.5% tumors
D.2: 5.2% tumors
D.3: 11.7% tumors

IVa—IVe
D.1: 41.7% tumors
D.2: 60.0% tumors
D.3: 83.1% tumors

Benzo[a]pyrene
D.1: 12.1% tumors
D.2: 40.0% tumors
D.3: 61.7% tumors

Tricaprylin (control)
0.5 mℓ: 1.7% tumors

*Note:* In general 3 doses per group of substances: the mixtures contained 3.3 μg BaP (D.1), 10 μg BaP (D.2), or 30 μg BaP (D.3) or equivalent amounts of a fraction separated from BaP; about 60 animals per group.[268,295]

[a] Applied 1 year after first experiment.

The fraction which contained the hydrophilic components and particles and accounted for about two thirds of the mass of the total condensate had a strong inhibitory effect on the considerable carcinogenic activity exerted by the isolated remaining third of the condensate. The possibility of inhibiting the carcinogenic potency of benzo[a]pyrene could be demonstrated by adding numerous organic and inorganic substances. The cause of this inhibition, however, has not yet been explained satisfactorily.

6. So far we do not have enough data to permit us to decide whether the subcutaneous test method shows the basis of the mechanism that also occurs after physiological intake of air pollutants into the lung or whether it — at least partly — reflects biochemical reactions which are essentially connected with the special conditions of this experimental model and do not have a practical value for man. The results obtained in subcutaneous tests primarily serve to detect the possible existence of carcinogenic activities, dose-response relationships, and interactions of different substances in mixtures. Use of the simple subcutaneous test model permits the determination of the possibilities of using the reaction on a relatively large scale, to provide data to thoroughly plan the considerably more complicated experiments with the respiratory tract, and to interpret the results in a more differentiating way.

### 5.2.2.4 Oral Application
#### D. Steinhoff

In animal experiments carcinogenic PAH become effective predominantly at the site of application. The animal species investigated so far show very similar reactions irrespective of their position in the phylogenetic system.[341] The local carcinogenic activity can be easily proved by topical, subcutaneous, intramuscular, intraperitoneal, and intratracheal application, as well as after operative implantation into various tissues (including the brain).[326,331]

PAH do not remain at the site of application. Kelley[333] demonstrated in mice that 85% of the strong local carcinogen dibenzo[a,i]pyrene or its metabolites were no longer detectable at the place of application 10 weeks after a single subcutaneous application of the PAH suspended in peanut oil ($^{14}$C labeled).

Irrespective of the type of application and the animal species, PAH and their metabolites are eliminated from the blood by the liver and gall bladder primarily via the gastrointestinal tract.[310,322,336,344-346,352]

The mucous layer of the gastrointestinal tract normally hampers the absorption of the lipophilic PAH. When administering 100 mg of benzo[a]pyrene in starch solution (by gavage) or 250 mg of benzo[a]pyrene in the diet to rats, only about 50% were absorbed[318] although benzo[a]pyrene penetrates quite easily the forestomach epithelium of mice, almost irrespective of the solvent used.

In the glandular stomach, on the other hand, the solvent plays a decisive role. Solvents with lipophilic properties only do not modify the penetrability of benzo[a]pyrene in the glandular stomach. Solvents with lipophilic and hydrophilic properties exert a weak enhancing effect on the absorption of benzo[a]pyrene if the lipophilic property prevails. The absorption of benzo[a]pyrene is improved considerably if the hydrophilic property is preponderant.[321,343] Substances of the first group, such as lecithin and fatty acids, frequently occur in natural foodstuff. Substances of the second group are rarely found in normal food, but they have gained an increasing importance as food additives in the last few years.[321]

The general occurrence of PAH in our environment, as well as in food (see Chapter 3.8), makes it seem possible that PAH ingested orally exert a local carcinogenic effect in the gastrointestinal tract of man. A corresponding indication may, for instance, be the almost 100% greater incidence of stomach cancer in 50- to 69-year-old men and women living in an area of high air pollution in the U.S. compared to areas with lower air pollution.[353] This information is, however, not sufficient to permit final conclusions.

Compared with the topical and subcutaneous administration to mice, which represent particularly sensitive tests for proving the local carcinogenic activity of PAH (see Chapter 5.2.2.2 and 5.2.2.3), the local effect observed after oral application is considerably weaker. The same applies to the intraperitoneal application. Thus an intraperitoneally administered dose of $4 \times 1.25$ mg BaP per rat induced local sarcomas in only 3 of 30 animals.[340] At a concentration of 4 $\mu$g/g of food and a total dose of 8 mg per mouse, benzo[a]pyrene induced tumors of the forestomach in only 8% (13 of 160 animals) of the animals over a time period of 14 months (individual or multiple papillomas which in two cases degenerated to malignancy).[319]

In other experiments concentrations of up to 11 $\mu$g BaP per g (food or drinking water) induced carcinomas of the forestomach only if benzo[a]pyrene was added to drinking water which also contained large amounts of a surfactant (3%). At a total dose of 7 to 13 mg of benzo[a]pyrene per mouse, 39 to 83% of the animals developed proliferative changes in the forestomach (mostly papillomas and carcinomas). Without simultaneous addition of a surfactant to the drinking water, even higher total doses of benzo[a]pyrene (up to 22 mg/mouse) did not induce tumors.[313,314]

Considerably higher doses of benzo[a]pyrene, however, exerted a strong tumorigenic effect even after oral intake. A 30-day feeding of benzo[a]pyrene, at a concentration of 250 $\mu$g/g of food, induced tumors of the forestomach (papillomas and carcinomas) in all the mice.[335] If the same concentration of benzo[a]pyrene was fed over a period of 140 days, lung adenomas and leukemias developed in addition to tumors of the forestomach.[338] A further number of experiments has given evidence of the systemic carcinogenic effects of orally applied PAH, for instance, in mice[309,312,319,332,350] and in golden hamsters.[327] A clear effect was already observed after two applications each of 3 mg BaP per mouse by gavage.[351] Gavage of individual doses not exceeding 200 $\mu$g BaP per mouse, however, induced only occasional, mostly benign tumors of the forestomach.[323,337,339] Application of 12 doses of 1 mg/mouse (once weekly by gavage) initiated the growth of papillomas of the forestomach and squamous cell carcinomas in 30 and 22 of the initial 30 animals, respectively, within 2 months after the last application.[325] In some strains of Sprague-Dawley rats the oral administration of benzo[a]pyrene and related compounds very easily induced carcinomas of the mammary gland ($1 \times 100$ mg in sesame oil by gavage) if female rats were treated at 50 days of age.[328]

A single gavage of 3 mg BaP per mouse and subsequent prolonged treatment of a skin area of $2 \times 2$ cm with croton oil induced skin papillomas in 14 of 18 mice (78%) in comparison with 2 of 21 mice (10%) when control animals were treated with croton oil only.[311] Subcutaneous or intraperitoneal administration of a single dose of 0.5 mg BaP per mouse in an otherwise similar experimental setup, however, did not produce a corresponding effect.[349]

Since oral application of carcinogenic PAH induced systemic carcinogenic effects in various animal species, despite their poor absorption, such an effect is also to be expected after topical, subcutaneous, intravenous, and intraperitoneal application. Depending on the experimental conditions, different tumor localizations were found in mice, rats, and golden hamsters; in mice: tumors of the ovary, the mammary gland, the lung, the kidneys, and leukemias;[312,320,334,348] in rats: leukemias and carcinomas of the mammary gland;[324,329] and in golden hamsters: malignant lymphomas.[347]

After subcutaneous application of 4 mg of benzo[a]pyrene on days 11, 13, and 15 of pregnancy, a transplacental carcinogenic effect could be observed in mice: after 28 weeks 62% of the 55 mice had lung adenomas compared with only 11% of the 89 control mice. Additional treatment of the skin with croton oil produced local papillomas in 24% of the pretreated animals and in 9% of the control mice.[315] A similar experimental setup with intraperitoneal injection of benzo[a]pyrene yielded practically the same result.[316] After oral application of benzo[a]pyrene to rats (1 × 200 mg/kg on day 21 of pregnancy) the substance penetrated the placenta in easily detectable amounts. Concentrations of 2.77 µg/g could be detected in the fetal tissue 3 hr after application.[342]

Orally ingested PAH may also effect various interactions with other carcinogens — even with compounds not chemically related — which manifest themselves in a promotion or inhibition of the carcinogenic effect.[317,330]

# REFERENCES

## Chapter 5.1

1. Druckrey, H. and Küpfmüller, K., *Dosis und Wirkung*, Editio Cantor, Aulendorf, West Germany, 1949.
2. Forth, W., Henschler, D., and Reimund, W., *Pharmakologie und Toxikologie*, B. I. Wissenschaftsverlag, Mannheim, West Germany, 1975.
3. Schmähl, D., *Entstehung, Wachstum und Chemotherapie maligner Tumoren*, Editio Cantor, Aulendorf, West Germany, 1970.

## Chapter 5.2

4. Büchner, F., *Allgemeine Pathologie*, Urban & Schwarzenberg, Berlin, 1966, 5.
5. McGinty, L., Cancer epidemic raises doubts on mineral fibres, *New Sci.*, 78, 428, 1978.

## Chapter 5.2.1

6. Ames, B. N., The detection of chemical mutagens with enteric bacteria, in *Chemical Mutagens Principles and Methods for their Detection*, Hollaender, A., Ed., Plenum Press, New York, 1971, 267.
7. Ames, B. N., Gurney, E. G., Miller, J. A., and Bartsch, H., Carcinogens as frameshift mutagens: metabolites and derivatives of 2-acetyl-aminofluorene and other aromatic amine carcinogens, *Proc. Natl. Acad. Sci. U.S.A.*, 69, 3128, 1972.
8. Ames, B. N., Durston, E. W., Yamasaki, E., and Lee, F. D., Carcinogens are mutagens: a simple test system combining liver homogenates for activation and bacteria for detection, *Proc. Natl. Acad. Sci. U.S.A.*, 70, 2281, 1973.
9. Ames, B. N., McCann, J., and Yamasaki, E., Methods for detecting carcinogens and mutagens with the *Salmonella* microsome mutagenicity test, *Mutat. Res.*, 31, 347, 1975.
10. Ames, B. N. and McCann, J., Carcinogens are mutagens: a simple test system, *IARC Sci. Publ.*, 12, 493, 1976.
11. Andrews, A. W., Thibault, L. H., and Lijinsky, W., The relationship between carcinogenicity and mutagenicity of some polynuclear hydrocarbons, *Mutat. Res.*, 51, 311, 1978.
12. Arcos, J. C. and Argus, M. F., Molecular geometry and carcinogenic activity of aromatic compounds, *Adv. Cancer Res.*, 11, 305, 1968.
13. Ashby, J. and Styles, J. A., Does carcinogenic potency correlate with mutagenic potency in the AMES assay? *Nature (London)*, 271, 452, 1978.
14. Baghirzade, M. F., Grundlagen des Talgdrüsenschwund — Phänomens, Ph.D. dissertation, University of Hamburg, West Germany, 1966.
14a. Barry, G., Cook, J. W., Hastewood, G. D. A., Hewett, C. L., Hieger, I., and Kennaway, E. L., The production of cancer by pure hydrocarbons, *Proc. R. Soc. London*, 117(3), 318, 1935.
15. Bauer, K. H., *Das Krebsproblem*, Springer Verlag, Berlin, 1963.
16. Berwald, Y. and Sachs, L., In vitro transformation of normal cells to tumor cells by carcinogenic hydrocarbons, *J. Natl. Cancer Inst.*, 35, 641, 1965.
17. Bock, F. G. and Mund, R., A survey of compounds for activity in the suppression of mouse sebaceous glands, *Cancer Res.*, 18, 887, 1958.
18. Bridges, B. A., Some general principles of mutagenicity screening and a possible framework for testing procedures, *Environ. Health Perspect.*, 6, 221, 1973.
19. Bridges, B. A., Short-term screening tests for carcinogens, review article, *Nature (London)*, 261, 195, 1976.
20. Bridges, B. A., Use of a three-tier protocol for evaluation of long-term toxic hazards particularly mutagenicity and carcinogenicity, *IARC Sci. Publ.*, No. 12, 549, 1976.
20a. Brune, H., unpublished results, 1974.
20b. Brune, H., unpublished results, 1975.
21. Brune, H. and von Hopffgarten, I., unpublished results, 1978.
22. Bryan, W. R. and Shimkin, M. B., Quantitative analysis of dose-response data obtained with three carcinogenic hydrocarbons in strain C3H male mice, *J. Natl. Cancer Inst.*, 3, 503, 1943.
23. Buecker, M., Glatt, H. R., Platt, K. L., Avnir, D., Ittah, Y., Blum, J., and Oesch, F., Mutagenicity of phenanthrene and phenanthrene K-region derivatives, *Mutat. Res.*, 66, 337, 1979.
24. Buu-Hoi, N. P., New developments in chemical carcinogenesis by polycyclic hydrocarbons and related heterocycles: review, *Cancer Res.*, 24, 1511, 1964.
25. Casto, B. C. and Dipaolo, J. A., Virus, chemicals and cancer, *Prog. Med. Virol.*, 16, 1, 1973.

26. Casto, B. C., Pieczynsky, W. J., Janosko, H., and Dipaolo, J. A., Significance of treatment interval and DNA repair in the enhancement of viral transformation by chemical carcinogens and mutagens, *Chem. Biol. Interact.*, 13, 105, 1976.
27. Casto, B. C. and DiPaolo, J. A., Biological activity of polycyclic hydrocarbons in Syrian hamster cells in vitro, in *Polycyclic Hydrocarbons and Cancer,* Vol. 1, Gelboin, H. V. and Ts'o, P. O. P., Eds., Academic Press, New York, 1978, 279.
28. Cavalieri, E., Mailander, P., and Pelfrene, A., Carcinogenic activity of anthanthrene on mouse skin, *Z. Krebsforsch.*, 89, 113, 1977.
29. Cook. J. W., Hewett, C. L., and Hieger, I., The isolation of a cancer-producing hydrocarbon from coal tar, *J. Chem. Soc.*, 395, 1933.
30. Coombs, M. M., Dixon, C., Kissonerghis, A.-M., Evaluation of the mutagenicity of compounds of known carcinogenicity, belonging to the benz[a]anthracene, chrysene and cyclopenta[a]phenanthrene series, using AMES test, *Cancer Res.*, 36, 4525, 1976.
31. Cooper, J. T. and Goldstein, S., Toxicity testing in vitro. II. Use of a microsome-cultured human fibroblast system to study the cytotoxicity of cyclophosphamide, *Can. J. Physiol. Pharmacol.*, 54, 546, 1976.
32. Dehnen, W., Pitz, N., and Tomingas, R., The mutagenicity of airborne particulate pollutants, *Cancer Lett.*, 4, 5, 1977.
33. DiPaolo, J. A., Nelson, R. L., and Donovan, P. J., In vitro transformation of Syrian hamster embryo cells by diverse chemical carcinogens, *Nature (London)*, 235, 278, 1972.
34. DiPaolo, J. A., Takano, K., and Popescu, N. C., Quantitation of chemically induced neoplastic transformation of BALB/3T3 cloned cell lines, *Cancer Res.*, 32, 2686, 1972.
35. DiPaolo, J. A., Nelson, R. L., Donovan, J., and Evans, C. H., Host-mediated in vivo in vitro assay for chemical carcinogenesis, *Arch. Pathol.*, 95, 380, 1973.
36. Dong, M., Schmeltz, I., La Voie, E., and Hoffmann, D., Aza-arene in the respiratory environment: analysis and assays for mutagenicity, in *Polynuclear Aromatic Hydrocarbons,* Jones, P. W. and Freudenthal, R. I., Eds., Raven Press, New York, 1978, 97.
37. Drake, J. W., Environmental mutagen hazards, *Science,* 187, 503, 1975.
38. van Duuren, B. L., Sivak, A., Goldsmith, B. M., Katz, C., and Melchionne, S., Initiating activity of aromatic hydrocarbons in two-stage carcinogenesis, *J. Natl. Cancer Res.*, 44, 1167, 1970.
38a. van Duuren, B. L., Sivak, A., Segal, A., Orris, L., and Langseth, L., The tumor-producing agents of tobacco leaf and tobacco smoke condensate, *J. Natl. Cancer Inst.*, 37, 519, 1966.
39. Goldsmith, B. M., Katz, C., and van Duuren, B. L., The cocarcinogenic activity of noncarcinogenic aromatic hydrocarbons, *Proc. Am. Assoc. Cancer Res.*, 14, 84, 1973.
40. Graffi, A., Fluoreszensmikroskopische Untersuchungen der Mäusehaut nach Pinselung mit Benzpyren-Benzollösungen, *Z. Krebsforsch.*, 52, 165, 1942.
41. Greim, H., Göggelmann, W., and Wolff, T., Erfahrungen mit In-vitro-Testsystemen zum Nachweis mutagener Substanzen, *Staub Reinhalt. Luft,* 38, 227, 1978.
42. Grimmer, G., Analysis of automobile exhaust condensate, *IARC Sci. Publ.*, No. 16, 29, 1977.
43. Guérin, M. and Cuzin, J., Tests cutanés chez la souris pour déterminer l'activité carcinogène des goudrons de fumée de cigarettes, *Bull. Assoc. Fr. Etude Cancer,* 48, 112, 1961.
44. Habs, M. and Schmähl, D., unpublished information, 1978.
44a. Hecht, S. S., Bondinell, W. E., and Hoffmann, D., Chrysene and methylchrysene: presence in tobacco smoke and carcinogenicity, *J. Natl. Cancer Inst.*, 53, 1121, 1974.
45. Hoffmann, D. and Wynder, E. L., Beitrag zur carcinogenen Wirkung von Dibenzopyrenen, *Z. Krebsforsch.*, 68, 137, 1966.
45a. Hoffmann, D., Bondinell, W. E., Wynder, E. L., Carcinogenicity of methylchrysenes, *Science,* 183, 215, 1974.
46. Hoffmann, D., Rathkamp, G., Nesnow, S., and Wynder, E. L., Fluoranthenes: quantitative determination in cigarette smoke formation by pyrolysis and tumor-initiating activity, *J. Natl. Cancer Inst.*, 49, 1165, 1972.
47. Horten, A. W. and Christian, G. M., Cocarcinogenic versus incomplete carcinogenic activity among aromatic hydrocarbons: contrast between chrysene and benzo[b]biphenylene, *J. Natl. Cancer Inst.*, 53, 1017, 1974.
48. Iype, P. T., Studies on chemical carcinogenesis in vitro using adult rat liver cells, *IARC Sci. Publ.*, No. 10, 119, 1974.
49. Johnson, S., Effect of thymectomy on the induction of skin tumours by dibenzanthracene, and breast tumours by dimethylanthracene, in mice of IF strain, *Br. J. Cancer,* 22, 755, 1968.
50. Kakunaga, T., A quantitative system for assay of malignant transformation by chemical carcinogens using a clone derived from BALB/3T3, *Int. J. Cancer,* 12, 463, 1973.
51. Kinoshita, N. and Gelboin, H. V., The role of aryl hydrocarbon hydroxylase (AHH) in 7,12-dimethylbenz[a]anthracene skin tumorigenesis: on the mechanism of 7,8-benzoflavone inhibition of tumorigenesis, *Cancer Res.*, 32, 1329, 1972.

52. Kracht, J., Klein, U. E., and Baghirzade, M., Erfahrungen mit dem Talgdrüsenschwundtest, Sonderdruck aus *Verh. Dtsch. Ges. Pathol.* 45, 170, 1961.
52a. Lacassagne, A., Buu-Hoi, N. P., Zajdela, I., and Vingiello, F. A., The true dibenzo[a,l]pyrene, a new potent carcinogen, *Naturwissenschaften*, 55, 43, 1968.
52b. LaVoie, E., in *3rd Batelle Symp. Polynuclear Aromatic Hydrocarbons*, Battelle Columbus Labs, Columbus, Ohio, October 1978.
53. Lavoie, E., Bedenko, V., Hirota, N., Hecht, S., and Hoffmann, D., A comparison of the mutagenicity, tumor-initiating activity and complete carcinogenicity of polynuclear aromatic hydrocarbons, in *Polynuclear Aromatic Hydrocarbons*, Jones, P. W. and Leber, P., Eds., Ann Arbor Science, Ann Arbor, Michigan, 1978, 705.
54. Lijinsky, W. and Saffiotti, U., Relation between structure and skin tumorigenic activity among hydrogenated derivatives of several PAH, *Chem. Abstr.*, 67, 1841, 1967.
55. Lijinsky, W. and Garcia, H., Skin carcinogenesis tests of hydrogenated derivatives of anthanthrene and other polynuclear hydrocarbons, *Z. Krebsforsch.*, 77, 226, 1972.
56. Marczinski-Verheugt, E. and Doerfler, W., Mutagene und teratogene Wirkungen des Zigarettenrauches *Muench. Med. Wochenschr.*, 120, 327, 1978.
57. Marquardt, H. and Heidelberger, C., Influence of "feeder cells" and inducers and inhibitors of microsomal mixed-function oxidases on hydrocarbon-induced malignant transformation of cells derived from C3H mouse prostate, *Cancer Res.*, 32, 721, 1972.
58. Marquardt, H., Sodergren, J., Grover, P. L., and Sims, P., Malignant transformation in vitro of mouse fibroblasts by 7,12-dimethylbenz[a]anthracene and 7 hydroxymethylbenz[a]anthracene and by their K-region derivatives, *Int. J. Cancer*, 13, 304, 1974.
59. Marquardt, H., Grover, P. L., and Sims, P., In vitro malignant transformation of mouse fibroblasts by non-K-region dihydrodiols derived from 7-methylbenz[a]anthracene, 7,12-dimethylbenz[a]anthracene and benzo[a]pyrene, *Cancer Res.*, 36, 2059, 1976.
60. Marquardt, H., Baker, S., Grover, P. L., and Sims, P., Malignant transformation and mutagenesis in mammalian cells induced by vicinal diol-epoxides derived from benzo[a]pyrene, *Cancer Lett.*, 3, 31, 1977.
61. Marquardt, H., Baker, S., Tierney, B., Grover, P. L., and Sims, P., The metabolic activation of 7-methylbenz[a]anthracene, the induction of malignant transformation and mutation in mammalian cells by non-K-region dihydrodiols *Int. J. Cancer*, 19, 828, 1977.
62. McCann, J., Choi, E., Yamasaki, E., and Ames, B. N., Detection of carcinogens as mutagens in the *Salmonella* microsome test: assay of 300 chemicals, *Proc. Natl. Acad. Sci. U.S.A.*, 72, 5135, 1975.
63. McCann, J., Spingarn, N. E., Kobori, J., and Ames, B. N., Detection of carcinogens and mutagens: bacterial tester strains with R factor plasmids, *Proc. Natl. Acad. Sci. U.S.A.*, 72, 979, 1975.
64. Miller, J. A. and Miller, E. C., The metabolic activation of carcinogenic aromatic amines and amides, *Prog. Exp. Tumor Res.*, 11, 273, 1969.
65. Miller, J. A., Carcinogenesis by chemicals. An overview, *Cancer Res.*, 30, 559, 1970.
66. Mohr, U., Untersuchungen über die cancerogenen Eigenschaften von 3,4-, 10,11-, 12,13-tribenzfluoranthen an Mäusen, *Arch. Hyg.*, 153, 495, 1969.
67. Montesano, R., Saint-Vincent, L., and Tomatis, L., Malignant transformation in vitro of rat liver cells by dimethylnitrosamine and N-methyl-N-nitro-N-nitrosoguanidine, *Br. J. Cancer*, 20, 215, 1973.
68. Mossanda, K., Poncelet, F., Fouassin, A., and Mercier, M., Detection of mutagenic polycyclic aromatic hydrocarbons in African smoked fish, *Food Cosmet. Toxicol.*, 17, 141, 1979.
69. Müller, E., Kanzerogene Substanzen in Wasser und Boden, *Arch. Hyg.*, 152, 23, 1968.
70. Nagao, M. and Sugimura, T., Mutagenesis: microbial systems, in *Polycyclic Hydrocarbons and Cancer*, 2, Gelboin, H. V., and Ts'o, P. O. P., Eds., Academic Press, New York, 1978, 99.
71. Norpoth, K., Qualitätssicherung bei der mikrobiologischen Fremdstoff-Testung auf Mutagenität im Routine-Laboratorium, *Staub Reinhalt. Luft*, 38, 235, 1978.
72. Norpoth, K., Djelani, G., and Gieselmann, V., Bacterial mutagenicity testing of polycyclic aromatic hydrocarbons, in *Short-Term Test Systems for Detecting Carcinogens*, Norpoth, K. and Garner, R. C., Eds., Springer-Verlag, New York, 1980, 362.
73. Oesch, F., Bentley, P., and Glatt, H. R., in *Biological Reactive Intermediates*, Jollow, D. J., Kocsis, J. J., Snyder, R., and Vainio, H., Eds., Plenum Press, New York, 1977, 181.
74. Pfeiffer, E. H., Investigations on the carcinogenic burden by air pollution in man. VII. Studies on the oncogenetic interaction of polycyclic aromatic hydrocarbons, *Zentralbl. Bakteriol. Parasitenkd. Infektionskr. Hyg. Abt. I Orig. Reihe B*, 158, 69, 1973.
75. Pienta, R. J., Poiley, J. A., and Lebherz, W. B., Morphological transformation of early passage golden Syrian hamster embryo cells derived from cryopreserved primary cultures as a reliable in vitro bioassay for identifying diverse carcinogens, *Int. J. Cancer*, 19, 642, 1977.
76. Poiley, J. A., Raineri, R., and Pienta, R. J., Two-stage malignant transformation in hamster embryo cells, *Br. J. Cancer*, 39, 8, 1979.

77. Pool, B. L., Schnelltests auf Mutagene und Carcinogene, Sonderdruck, *Nachr. Chem. Techn. Lab.*, 5, 26, 1978.
78. Pott, F., Brockhaus, A., and Huth, F., Untersuchungen zur Kanzerogenität von polycyclischen aromatischen Kohlenwasserstoffen im Tierexperiment, *Zentralbl. Bakteriol. Parasitenkd. Infektionskr. Hyg. Abt. I Orig. Reihe B.*, 157, 34, 1973.
79. Pott, F., Tomingas, R., and Huth, F., unpublished results, 1978.
80. Purchase, I. F. H., Longstaff, E., Ashby, J., Styles, J. A., Anderson, D., Lefevre, P. A., and Westwood, F. R., Evaluation of six short-term tests for detecting organic chemical carcinogens and recommendations for their use, *Nature (London)*, 264, 624, 1976.
81. Purchase, I. F. H., Longstaff, E., Ashby, J., Styles, J. A., Anderson, D., Lefevre, P. A., and Westwood, F. R., An evaluation of six short-term tests for detecting organic chemical carcinogens, *Br. J. Cancer*, 37, 873, 1978.
82. Purchase, I. F. H., Ashby, J., Styles, J. A., personal communication, 1978.
83. Richter-Reichhelm, H. B. and Emura, M., Toxizität und Carcinogenität von Benzo[a]pyren an Lungenzellkulturen und kultivierten Tracheastücken von syrischen Goldhamsterfeten. Entwicklung und erste Erfahrungen mit einem In-vitro Test-system, *Staub Reinhalt. Luft*, 38, 247, 1978.
83a. Roe, F. J. C., The development of malignant tumours of mouse skin after "initiating" and "promoting" stimuli. I. The effect of a single application of 9,10-dimethyl-1,2-benzanthracene (DMBA) with and without subsequent treatment with croton oil, *Br. J. Cancer*, 10, 61, 1956.
84. Roe, F. J. C., Effect of phenanthrene on tumor initiation by 3,4-benzopyrene, *Br. J. Cancer*, 16, 503, 1962.
85. Roe, F. J. C. and Grant, G. A., Test of pyrene and phenanthrene for incomplete carcinogenic and anticarcinogenic activity, *Br. Emp. Cancer Campaign*, 41, 59, 1964.
86. Sato, S., Seino, Y., Ohka, T., Yagahi, T., Nagao, M., Matsushima, T., and Sugimura, T., Mutagenicity of smoke condensates from cigarettes, cigars and pipe tobacco, *Cancer Lett.*, 3, 1, 1977.
87. Schramm, T., Teichmann, B., and Butschak, G., Kurzzeit-Tests als eine Stufe im Rahmen der Prüfung von Substanzen auf kanzerogene Wirkungen, *Arch. Geschwulstforsch.*, 47, 567, 1977.
88. Scribner, J. O., Tumor initiation by apparently noncarcinogenic polycyclic aromatic hydrocarbons, *J. Natl. Cancer Inst.*, 50, 1717, 1973.
89. Seemayer, N., Zelltransformation von Hamsternierenzellen in vitro unter dem Einfluss von atmosphärischen Feinstaubextrakten und dem Papovavirus SV 40, *Staub Reinhalt. Luft*, 38, 254, 1978.
90. Severson, R. F., Snook, M. E., Akin, F. J., and Chortyk, O. T., Correlation of biological activity with polynuclear aromatic hydrocarbon content of tobacco smoke fractions, in *Polynuclear Aromatic Hydrocarbons*, Jones, P. W. and Freudenthal, R. I., Eds., Raven Press, New York, 1978, 115.
91. Slaga, T., Viagi, A., Berry, D., Bracken, W., Buty, S. G., and Scribner, J. D., Skin tumor initiating ability of benzo[a]pyrene 4,5-, 7,8- and 7,8-diol-9,10-epoxides and 1,8-diol, *Cancer Lett.*, 2, 115, 1976.
92. Slaga, T. J., Bowden, G. T., Scribner, J. O., and Bouthwell, R. K., Dose-response studies on the ability of 7,12-dimethylbenz[a]anthracene and benz[a]anthracene to initiate skin tumors, *J. Natl. Cancer Inst.*, 53, 1337, 1974.
93. Styles, J. A., A method for detecting carcinogenic organic chemicals using mammalian cells in culture, *Br. J. Cancer*, 36, 558, 1977.
93a. Steiner, P. E., Carcinogenicity of multiple chemicals simultaneously administered, *Cancer Res.*, 15, 632, 1955.
93b. Steiner, P. E. and Falk, H. L., Summation and inhibition effects of weak and strong carcinogenic hydrocarbons: 1,2-benzanthracene, chrysene, 1,2,5,6-dibenzanthracene, and 20-methylcholanthrene, *Cancer Res.*, 11, 56, 1951.
94. Sugimura, T., Sato, S., Nagao, M., Yahagi, T., Matsushima, T., Seino, Y., Takeuchi, M., and Kawachi, T., Overlapping of carcinogens and mutagens, in *Fundamentals in Cancer Prevention*, Magee, P. N., Takayama, S., Sugimura, T., and Matsushima, T., Eds., University of Tokyo Press, Japan, 1976, 191.
95. Suntzeff, V., Carruthers, C., and Cowdry, E. V., The role of sebaceous glands and hair follicles in epidermal carcinogens, *Cancer Res.*, 7, 439, 1947.
96. Suntzeff, V., Croninger, A. G., Wynder, E. L., Cowdry, E. V., and Graham, E. A., Use of sebaceous-gland test of primary cigar tar fractions and of certain noncarcinogenic polycyclic hydrocarbons, *Cancer (Philadelphia)*, 10, 250, 1957.
97. Teranishi, K., Hamada, K., and Watanabe, H., Quantitative relationship between carcinogenicity and mutagenicity of polycyclic aromatic hydrocarbons in *Salmonella typhimurium* mutants, *Mutat. Res.*, 31, 97, 1975.
98. Tokiwa, H., Morita, K., Takeyoshi, H., Takahashi, K., and Ohnishi, Y., Detection of mutagenic activity in air pollutants, *Mutat. Res.*, 48, 237, 1977.
99. Wang, Y. Y., Rappaport, S. M., Sawyer, R. F., Talcott, R. E., and Wei, E. T., Direct-acting mutagens in automobile exhaust, *Cancer Lett.*, 5, 39, 1978.

100. Waravdekar, S. S. and Ranadive, K. J., Biologic testing of 3,4,9,10-dibenzpyrene, *J. Natl. Cancer Inst.*, 21, 1151, 1958.
101. Weinstein, I., Orenstein, J. M., Gebert, R., Kaighu, M. E., and Stadler, U. C., Growth and structural properties of epithelial cell cultures established from normal rat liver and chemically induced hepatomas, *Cancer Res.*, 35, 253, 1975.
102. Weinstein, D., Kath, M. L., and Kazmer, S., Chromosomal effect of carcinogens and non-carcinogens on Wi-38 after short-term exposures with and without metabolic activation, *Mutat. Res.*, 46, 297, 1977.
103. Wynder, E. L. and Hoffmann, D., The carcinogenicity of benzofluoranthenes, *Cancer (Philadelphia)*, 12, 1194, 1959.
104. Wynder, E. L. and Hoffmann, D., Objective of laboratory studies, in *Tobacco and Tobacco Smoke*, Academic Press, New York, 1968, 3; Selected laboratory methods in tobacco carcinogenesis, in *Tobacco and Tobacco Smoke*, Academic Press, New York, 1968, 133; Biological tests for tumorigenic and cilia-toxic activity, in *Tobacco and Tobacco Smoke*, Academic Press, New York, 1968, 181; Interpretation of experimental findings, in *Tobacco and Tobacco Smoke*, Academic Press, 1968, 623.

## Chapter 5.2.2.1

105. Andervont, H. B., Pulmonary tumors in mice. IV. Lung tumors induced by subcutaneous injection of 1,2,5,6-dibenzanthracene in different media and by direct contact with lung tissues, *Publ. Health Rep.*, 52, 1584, 1937.
106. Archer, V. W., Saccomanno, G., and Jones, J. H., Frequency of different histologic types of bronchogenic carcinoma as related to radiation exposure, *Cancer (Philadelphia)*, 34, 2056, 1974.
107. Blacklock, J. W. S., The production of lung tumors in rats by 3,4-benzpyrene, methylcholanthrene and the condensate from cigarette smoke, *Br. J. Cancer*, 11, 181, 1957.
108. Blacklock, J. W. S., An experimental study of the pathological effects of cigarette condensate in the lungs with special reference to carcinogenesis, *Br. J. Cancer*, 15, 745, 1961.
109. Borisjuk, J. P., On the Carcinogenesis of Smoking Products. Experimental Investigation, Dissertation, University of Kiev, U.S.S.R., 1967.
110. Brain, J., Knudson, D. E., Sorokin, S. P., and Davis, M. A., Pulmonary distribution of particles given by intratracheal instillation or by aerosol inhalation, *Environ. Res.*, 11, 13, 1976.
111. Campbell, J. A., Influences of breathing carbon monoxide and oxygen at high percentages for long periods upon development of tar cancer in mice, *J. Pathol. Bacteriol.*, 36, 243, 1933.
112. Campbell, J. A., Cancer of the skin and increase in incidence of primary tumors of the lung in mice exposed to dust obtained from tarred roads, *Br. J. Exp. Pathol.*, 15, 287, 1934.
113. Davis, B. R., Whitehead, J. K., Gill, M. E., Lee, P. N., Butterworth, A. D., and Roe, F. J. C., Response of rat lung to 3,4-benzpyrene administered by intratracheal instillation in infusine with or without carbon black, *Br. J. Cancer*, 31, 443, 1975.
114. Davis, B. R., Whitehead, J. K., Gill, M. E., Lee, P. N., Butterworth, A. D., and Roe, F. J. C., Response of rat lung to tobacco smoke condensate or fractions derived from it administered repeatedly by intratracheal instillation, *Br. J. Cancer*, 31, 453, 1975.
115. Davis, B. R., Whitehead, J. K., Gill, M. E., Lee, P. N., Butterworth, A. D., and Roe, F. J. C., Response of rat lung to inhaled tobacco smoke without prior exposure to 3,4-benzpyrene given by intratracheal instillation, *Br. J. Cancer*, 31, 469, 1975.
116. Della Porta, G., Kolb, I., and Shubik, P., Induction of tracheo-bronchial carcinomas in the Syrian Golden Hamster, *Cancer Res.*, 18, 592, 1958.
117. Dontenwill, W. and Mohr, U., Experimentelle Untersuchungen zum Problem der Carcinomentstehung im Respirationstrakt, *Z. Krebsforsch.*, 65, 56, 1962.
118. Dontenwill, W., Experimental investigations on the effect of cigarette smoke inhalation on small laboratory animals, in *Inhalation Carcinogenesis*, Hanna, M. G., Jr., Nettesheim, P., and Gilbert, J. R., Eds., U.S. Atomic Energy Commission, Oak Ridge, Tenn., 1970.
119. Drew, R. T. and Laskin, S., Environmental inhalation chambers, in *Methods of Animal Experimentation*, Vol. 4, Gay, W. J., Ed., Academic Press, New York, 1973, 1.
120. Essenberg, J. M., Cigarette smoke and the incidence of primary neoplasm of the lung in the albino mouse, *Science*, 116, 561, 1952.
121. Feron, V. J., Respiratory tract tumors in hamsters after intratracheal instillations of benzo[a]pyrene alone and with furfural, *Cancer Res.*, 32, 28, 1972.
122. Feron, V. J., De Jong, D., and Emmelot, P., Dose response correlation for the induction of respiratory tract tumors in Syrian Golden Hamster by intratracheal instillation of benzo[a]pyrene, *Eur. J. Cancer*, 9, 387, 1973.
123. Fraumeni, J. F. Jr., Respiratory carcinogenesis: an epidemiologic appraisal, *J. Natl. Cancer Inst.*, 55, 1039, 1975.

124. Gross, P., Tolker, E., Babyak, M. A., and Kashak, M., Experimental lung cancer in hamsters. Repetitive intratracheal applications of two carcinogenic hydrocarbons, *Arch. Environ. Health.*, 11, 59, 1965.
125. Guerin, M., Pulmonary tumours and buccal cancer in the rat subjected to inhalation of cigarette smoke, *Bull. Assoc. Fr. Cancer*, 46, 295, 1959.
126. Henry, M. C. and Kaufman, D. G., Clearance of benzo[a]pyrene from hamster lungs after administration on coated particles, *J. Natl. Cancer Inst.*, 51, 1961, 1973.
127. Henry, M. C., Port, C. D., Bates, R. R., and Kaufman, D. G., Respiratory tract tumors in hamsters induced by benzo[a]pyrene, *Cancer Res.*, 33, 1585, 1973.
128. Henry, M. C., Port, C. D., and Kaufman, D. G., Importance of physical properties of benzo[a]pyrene-ferric oxide mixtures in lung tumor induction, *Cancer Res.*, 35, 207, 1975.
129. Herrold, K. M. and Dunham, L. J., Induction of carcinoma and papilloma of the tracheobronchial mucosa of the Syrian hamster by intratracheal instillation of benzo[a]pyrene, *J. Natl. Cancer Inst.*, 28, 467, 1962.
130. Hilfrich, J., Bresch, H., Misfeld, J., and Mohr, U., Untersuchungen über die karzinogene Belastung des Menschen durch Luftverunreinigung. V. Tumoren des Respirationstraktes beim Syrischen Goldhamster nach intratrachealer Instillation von Benzo[a]pyren, *Zentralbl. Bakteriol. Parasitenkd. Infektionskr. Hyg. Abt. 1 Orig. Reihe B*, 158, 59, 1973.
131. Hinners, R. G., Burkart, J. K., and Contner, G. L., Animal exposure chambers in air pollution studies, *Arch. Environ. Health*, 13, 609, 1966.
132. Hinners, R. G., Burkart, J. K., and Punte, C. L., Animal inhalation chambers, *Arch. Environ. Health*, 16, 194, 1968.
133. Homburger, F., 'Smoker's larynx' and carcinoma of the larynx in Syrian hamsters exposed to cigarette smoke, *Laryngoscope*, 85, 1874, 1975.
134. Kennedy, A. R. and Little, J. B., The transport and localization of benzo[a]pyrene-hematite and hematit-$^{210}$Po in the hamster lung following intratracheal instillation, *Cancer Res.*, 34, 1344, 1974.
135. Ketkar, M., Reznik, G., Misfeld, J., and Mohr, U., Investigations on the carcinogenic burden by air pollution in man. The effect of a single dose of benzo[a]pyrene in Syrian golden hamsters, *Cancer Lett.*, 3, 231, 1977.
136. Ketkar, M., Reznik, G., Schneider, P., and Mohr, U., Investigations on the carcinogenic burden by air pollution in man. Intratracheal instillation studies with benzo[a]pyrene in bovine serum albumin in Syrian hamsters, *Cancer Lett.*, 4, 235, 1978.
137. Ketkar, M., Green, U., Schneider, P., and Mohr, U., Investigations on the carcinogenic burden by air pollution in man. Intratracheal instillation studies with benzo[a]pyrene in a mixture of Tris buffer and saline in Syrian golden hamsters, *Cancer Lett.*, 6, 279, 1979.
138. Ketkar, M. B., Haas, H., and Althoff, J., Implantation of cigarette smoke condensate in the lungs of Syrian golden hamsters, *Z. Krebsforsch.*, 94, 111, 1979.
139. Kimmerle, G., The inhalation chamber for the study of potentially carcinogenic polycyclic hydrocarbons in small laboratory animals, *IARC Sci. Publ.*, 16, 49, 1977.
140. Kobayashi, N., Production of respiratory tract tumors in hamsters by benzo[a]pyrene, *Gann*, 66, 311, 1975.
141. Kotin, R. and Wisely, D. V., Production of lung cancer in mice by inhalation exposure to influenza virus and aerosols of hydrocarbons, *Prog. Exp. Tumor Res.*, 3, 186, 1963.
142. Kuschner, M., The J. Burns Anderson Lectures. The causes of lung cancer, *Am. Rev. Respir. Dis.*, 98, 573, 1968.
143. Kuschner, M., Laskin, S., Christofano, E., and Nelson, N., Experimental carcinoma of the lung, *Proc. 3rd Natl. Cancer Conf. Philadelphia*, J. B. Lippincott, Philadelphia, 1957, 485.
144. Laskin, S., Kuschner, M., and Drew, R. T., Studies in pulmonary carcinogenesis, in *Inhalation carcinogenesis*, Hanna, M. G., Jr., Nettesheim, P., and Gilbert, J. R., Eds., U.S. Atomic Energy Commission, Oak Ridge, Tenn., 1970, 321.
145. Miller, L., Smith, W. E., and Berliner, S. W., Tests for effect of asbestos on benzo[a]pyrene carcinogenesis in the respiratory tract, *Ann. N.Y. Acad. Sci.*, 132, 489, 1965.
146. Mohr, U., Reznik-Schüller, H., Reznik, G., Grimmer, G., and Misfeld, J., Untersuchungen über die karzinogene Belastung des Menschen durch Luftverunreinigung. XIV. Wirkung von Autoabgaskondensat auf die Lungen des Syrischen Goldhamsters, *Zentralbl. Bakteriol. Parasitenkd. Infektionskr. Abt. 1 Orig. Reihe B*, 163, 425, 1976.
147. Mohr, U. and Reznik, G., Tobacco carcinogenesis, in *Pathogenesis and Therapy of Lung Cancer*, Harris, Curtis C., Ed., Marcel Dekker, New York, 10, 263, 1978.
148. Montesano, R., Saffiotti, U., and Shubik, P., The role of topical systematic factors in experimental respiratory carcinogenesis, in *Inhalation Carcinogenesis*, Hanna, M. G., Jr., Nettesheim, P., and Gilbert, J. R., Eds., U.S. Atomic Energy Commission, Oak Ridge, Tenn., 1970, 353.
149. Mori, K., Acceleration of experimental lung cancers in rats by inhalation of cigarette smoke, *Gann*, 55, 175, 1964.

150. Nagel, D. L., Stenbäck, F., Clayson, D. B., and Wallcave, L., Intratracheal instillation studies with 7H-dibenzo[c,g]carbazole in the Syrian hamster, *J. Natl. Cancer Inst.*, 57, 119, 1976.
151. Nettesheim, P., Respiratory carcinogenesis studies with the Syrian golden hamster: A review, *Prog. Exp. Tumor Res.*, 16, 185, 1972.
152. Niskanen, K. O., Observation on metaplasia of bronchial epithelium and its relation to carcinoma of the lung; pathoanatomical and experimental researches, *Acta Pathol. Microbiol. Scand., Suppl.*, 80, 1, 1949.
153. Otto, H., Experimentelle Untersuchungen an Mäusen mit passiver Zigarettenrauchbeatmung, *Frankf. Z. Pathol.*, 73, 10, 1963.
154. Pott, F., Tomingas, R., and Reiffer, F. J., Experimentelle Untersuchungen zur kanzerogenen Wirkung und Retention von Benzo[a]pyren am Applikationsort nach intratrachealer und subcutaner Injektion, *Zentralbl. Bakteriol. Parasitenkd. Infektionskr. Abt. 1 Orig. Reihe B*, 158, 97, 1973.
155. Pott, F., Mohr, U., and Brockhaus, A., Untersuchungen zur Kombinationswirkung von Benzo[a]pyren und Dibenz[a,h]anthracen mit $SO_2$ und $NO_2$. Inhalation beim Goldhamster, abstr. in *Lufthygiene und Silikoseforschung*, Vol 2, Rothe, H., Ed., Girardet, Essen, West Germany, 1978, 227.
156. Pour, P., Mohr, U., Cardesa, A., Althoff, J., and Kmoch, N., Spontaneous tumors and common diseases in two colonies of Syrian hamsters. II. Respiratory tract and digestive system, *J. Natl. Cancer Inst.*, 56, 937, 1976.
157. Roe, F. J. C., The relevance and value of studies of lung tumors in laboratory animals in research on cancer of the human lung, in *Lung Tumors in Animals*, Severi, L., Ed., Division of Cancer Research, Perugia, Italy, 1966, 101.
158. Roe, F. J. C., Inhalation tests, in *Modern Trends in Toxicology*, Vol. 1, Boyland, E. and Goulding, R., Eds., Butterworths, London, 1968, 39.
159. Saffiotti, U., Experimental respiratory tract carcinogenesis, *Progr. Exp. Tumor Res.*, 11, 302, 1969.
160. Saffiotti, U., Experimental respiratory tract carcinogenesis and its relation to inhalation exposures, in *Inhalation Carcinogenesis*, Hanna, M. G., Jr., Nettesheim, P., and Gilbert, J. R., Eds., U.S. Atomic Energy Commission, Oak Ridge, Tennessee, 1970, 27.
161. Saffiotti, U., Cefis, F., and Kolb, L. H., A method for the experimental induction of bronchogenic carcinoma, *Cancer Res.*, 28, 104, 1968.
162. Saffiotti, U., Montesano, R., Sellakumar, A. R., and Borg, S. A., Experimental cancer of the lung. Inhibition by vitamin A of the induction of tracheobronchial squamous metaplasia and squamous cell tumors, *Cancer*, 20, 857, 1967.
163. Saffiotti, U., Montesano, R., Sellakumar, A. R., and Kaufman, D. G., Respiratory tract carcinogenesis induced in hamsters by different dose levels of benzo[a]pyrene and ferric oxide, *J. Natl. Cancer Inst.*, 49, 1199, 1972.
164. Sawicki, E., Analysis of atmospheric carcinogens and their co-factors, *IARC Sci. Publ.*, 13, 297, 1976.
165. Schreiber, H., Saccomanno, G., Martin, D. H., and Brennan, L., Sequential cytological changes during development of respiratory tract tumors induced in hamsters by benzo[a]pyrene-ferric oxide, *Cancer Res.*, 34, 689, 1974.
166. Schreiber, H., Martin, D. H., and Pazmino, N., Species differences in the effect of benzo[a]pyrene-ferric oxide on the respiratory tract of rats and hamsters, *Cancer Res.*, 35, 1654, 1975.
167. Sellakumar, A. R. and Shubik, P., Carcinogenicity of 7H-dibenzo[c,g]carbazole in the respiratory tract of hamsters, *J. Natl. Cancer Inst.*, 48, 1641, 1972.
168. Sellakumar, A., Stenbäck, F., and Rowland, J., Effects of different dusts on respiratory carcinogenesis in hamster induced by benzo[a]pyrene and diethylnitrosamine, *Eur. J. Cancer*, 12, 313, 1976.
169. Shabad, L. M., Experimental cancer of the lung, *J. Natl. Cancer Inst.*, 28, 1305, 1962.
170. Shabad, L. M., Dose-response studies in experimentally induced lung tumours, *Environ. Res.*, 4, 305, 1971.
171. Shabad, L. M. and Pylev, L. N., Morphological lesions in rat lungs induced by polycyclic hydrocarbons, in *Morphology of Experimental Respiratory Carcinogenesis*, Nettesheim, P., Hanna, H. G., Jr., and Detherage, J. W., Jr., Eds., U.S. Atomic Energy Commission, Oak Ridge, Tennessee, 1970, 227.
172. Simons, P. J., Lee, P. N., and Roe, F. J. C., Squamous lesions lungs of rats exposed to tobacco-smoke-condensate fractions by repeated intratracheal instillation, *Br. J. Cancer*, 37, 965, 1978.
173. Stanton, M. F., Miller, E., Wrench, C., and Blackwell, R., Experimental induction of epidermoid carcinoma in the lungs of rats by cigarette smoke condensate, *J. Natl. Cancer Inst.*, 49, 867, 1972.
174. Stenbäck, F. and Sellakumar, A., Lung tumor induction by dibenz[a,i]pyrene in the Syrian Golden Hamster, *Z. Krebsforsch.*, 82, 175, 1974.
175. Stenbäck, F., Sellakumar, A., and Shubik, P., Magnesium oxide as a carrier dust in benzo[a]pyrene-induced lung carcinogenesis in Syrian hamsters, *J. Natl. Cancer Inst.*, 54, 861, 1975.

176. Wynder, E. L. and Hoffmann, D., *Tobacco and Tobacco Smoke,* Academic Press, New York, 1967.
177. Yamagiwa, K. and Ichikawa, K., Experimental studies on the pathogenesis of epithelial neoplasms, *Rep. Med. Fac. Univ. Tokyo,* 15, 295, 1915.
178. Yoshimoto, T., Hirao, F., Sakatani, M., Nishikawa, H., Ogura, T., and Yamamura, Y., Induction of squamous cell carcinoma in the lung of C57B1/6 mice by intratracheal instillation of benzo[a]pyrene with charcoal powder, *Gann,* 68, 343, 1977.

## Chapter 5.2.2.2

179. Arcos, J. C. and Argus, M. F., Molecular geometry and carcinogenic activity of aromatic compounds. New perspectives, *Adv. Cancer Res.,* 11, 305, 1968.
180. Arcos, J. C., Argus, M. F., and Wolf, G., *Chemical Induction of Cancer,* Vol. 1, Academic Press, New York, 1968.
181. Barry, G., Cook, J. W., Hastewood, G. D. A., Hewett, C. L., Hieger, I., and Kennaway, E. L., *Proc. R. Soc. London,* 117, 318, 1935.
182. Berenblum, J., The mechanism of carcinogenesis. A study of the significance of cocarcinogenic action and related phenomena, *Cancer Res.,* 1, 807, 1941.
183. Boutwell, R. K., Some biological aspects of skin carcinogenesis, *Prog. Exp. Tumor Res.,* 4, 207, 1964.
184. Brune, H. and Henning, S., Erzeugung von Augenlidcarcinomen bei Mäusen nach epicutaner Applikation von Methylbutyl-nitrosamin, *Z. Krebsforsch.,* 69, 307, 1967.
185. Brune, H., Habs, M., and Schmähl, D., The tumor-producing effect of automobile exhaust gas condensate and fractions thereof, *J. Environ. Pathol. Toxicol.,* 1, 737, 1978.
186. Buening, M. K., Levin, W., Wood, A. W., Chang, R. L., Lehr, R. E., Taylor, C. W., Yagi, H., Jerina, D. M., and Conney, A. H., Tumorigenic activity of benzo[e]pyrene derivatives on mouse skin and newborn mice, *Cancer Res.,* 40, 203, 1980.
187. Cavalieri, E., Mailander, P., and Pelfrene, A., Carcinogenic activity of anthracene on mouse skin, *Z. Krebsforsch.,* 89, 113, 1977.
188. Cottini, G. B. and Mazzone, G. B., The effects of 3,4-benzpyrene on human skin, *Am. J. Cancer,* 37, 186, 1939.
189. Davies, R. F., Lee, P. N., and Rothwell, K., A study of the dose response of mouse skin to cigarette smoke condensate, *Br. J. Cancer,* 30, 146, 1974.
190. Dontenwill, W., Chevalier, H.-J., Harke, H. P., Klimisch, H.-J., Brune, H., Fleischmann, B., and Keller, W., Experimentelle Untersuchungen über die tumorerzeugende Wirkung von Zigarettenrauch-Kondensaten an der Mäusehaut. VI. Mitteilung: Untersuchung zur Fraktionierung von Zigarettenrauch-Kondensat, *Z. Krebsforsch.,* 85, 155, 1976.
191. Dontenwill, W., Chevalier, H.-J., Harke, H.-P., Klimisch, H.-J., Reckzeh, G., Fleischmann, B., and Keller, W., Experimentelle Untersuchungen über die tumorerzeugende Wirkung von Zigarettenrauch-Kondensaten an der Mäusehaut. VII. Mitteilung: Einzelvergleiche von Kondensaten verschiedener modifizierter Zigaretten, *Z. Krebsforsch.,* 89, 145, 1977.
192. van Duuren, B. L., Sivak, A., Segal, A., Orris, L., and Langseth, L., The tumor-producing agents of tobacco leaf and tobacco smoke condensate, *J. Natl. Cancer Inst.,* 37, 519, 1966.
193. van Duuren, B. L., Sivak, A., Goldsmith, B. M., Katz, C., and Melchionne, S., Initiating activity of aromatic hydrocarbons in two-stage carcinogenesis, *J. Natl. Cancer Inst.,* 44, 1167, 1970.
194. van Duuren, B. L., Langseth, L., Goldsmith, B. M., and Orris, L., Carcinogenicity of epoxides, lactones and peroxy compounds. VI. Structure and carcinogenic activity, *J. Natl. Cancer Inst.,* 39, 1217, 1967.
195. Goldsmith, B. M., Katz, C., and van Duuren, B. L., The cocarcinogenic activity of noncarcinogenic aromatic hydrocarbons, *Proc. Am. Assoc. Cancer Res.,* 14, 84, 1973.
196. Grasso, P. and Crampton, R. F., The value of the mouse in carcinogenicity testing, *Food Cosmet. Toxicol.,* 10, 418, 1972.
197. Habs, M., Schmähl, D., and Misfeld, J., Local carcinogenicity of some environmentally relevant polycyclic aromatic hydrocarbons after lifelong topical application to mouse skin, *Arch. Geschwulstforsch.,* 50, 266, 1980.
198. Hecht, S. S., Bondinell, W. F., and Hoffmann, D., Chrysene and methylchrysene: presence in tobacco smoke and carcinogenicity, *J. Natl. Cancer Inst.,* 53, 1121, 1974.
199. Heston, W. E., Inheritance of susceptibility to spontaneous pulmonary tumors in mice, *J. Natl. Cancer Inst.,* 3, 79, 1942.
200. Hoffmann, D. and Wynder, E. L., Beitrag zur carcinogenen Wirkung von Dibenzopyrenen, *Z. Krebsforsch.,* 68, 137, 1966.
201. Hoffmann, D., Rathkamp, G., Nesnow, S., and Wynder, E. L., Fluoranthenes: quantitative determination in cigarette smoke formation by pyrolysis and tumor-initiating activity, *J. Natl. Cancer Inst.,* 49, 1165, 1972.

202. Hoffmann, D., Theisz, E., and Wynder, E. L., Studies on the carcinogenicity of gasoline exhaust, *J. Air Pollut. Control Assoc.*, 15, 162, 1965.
202a. Hoffmann, D., Bondinell, W. E., and Wynder, E. L., Carcinogenicity of methylchrysenes, *Science*, 183, 215, 1974.
203. Holland, J. M., Rahn, R. O., Smith, L. H., and Clark, B. R., Chang, S. S., and Stephens, T. J., Skin carcinogenicity of synthetic and natural petroleums, *J. Occup. Med.*, 21, 614, 1979.
204. Horten, A. W. and Christian, G. M., Cocarcinogenic versus incomplete carcinogenic activity among aromatic hydrocarbons: contrast between chrysene and benzo[b]biphenylene, *J. Natl. Cancer Inst.*, 53, 1017, 1974.
205. Hueper, W. C., The skin as an assay system for potential carcinogens, *Natl. Cancer Inst. Monogr.*, No. 10, 577, 1963.
206. Hueper, W. C., Kotin, P., Tabor, E. C., Payne, W. W., Falk, H., and Sawicki, E., Carcinogenic bioassays on air pollutant, *Arch. Pathol.*, 74, 89, 1962.
207. Johnson, S., Effect of thymectomy on the induction of skin tumors by dibenzanthracene, and of breast tumors by dimethylanthracene, in mice of the IF strain, *Br. J. Cancer*, 22, 755, 1968.
208. Kinoshita, N. and Gelboin, H. V., The role of aryl hydrocarbon hydroxylase (AHH) in 7,12-dimethylbenz[a]anthracene skin tumorigenesis: on the mechanism of 7,8-benzoflavone inhibition of tumorigenesis, *Cancer Res.*, 32, 1329, 1972.
209. Klar, E., Über die Entstehung eines Epithelioms beim Menschen nach experimentellen Arbeiten mit Benzpyren, *Klin. Wochenschr.*, 17, 1279, 1938.
210. Kotin, P., Falk, H. L., Mader, P., and Thomas, M., Aromatic hydrocarbons. I. Presence in the Los Angeles atmosphere and the carcinogenicity of atmospheric extracts, *AMA Arch. Ind. Health*, 153, 1954.
211. Kotin, P., Falk, H. L., and Thomas, M., Aromatic hydrocarbons. II. Presence in the particulate phase of gasoline-engine exhaust and the carcinogenicity of exhaust extracts, *AMA Arch. Ind. Health*, 164, 1954.
212. Lee, P. N. and O'Neill, J. A., The effect of time and dose applied on tumor incidence rate in benzopyrene skin painting experiments, *Br. J. Cancer*, 25, 759, 1971.
213. Levin, W., Wood, A. W., Yagi, H., Jerina, D. M., and Conney, A. H., (±)-Trans-7,8-dihydroxy-7,8-dihydrobenzo[a]pyrene: a potent skin carcinogen when applied topically to mice, *Proc. Natl. Acad. Sci. U.S.A.*, 73, 3867, 1976.
214. Levin, W., Wood, A. W., Wislocki, P. G., Kapitulnik, J., Yagi, H., Jerina, D. M., and Conney, A. H., Carcinogenicity of benzo-ring derivatives of benzo[a]pyrene on mouse skin, *Cancer Res.*, 37, 3356, 1977.
215. Lijinski, W. and Garcia, H., Skin carcinogenesis tests of hydrogenated derivatives of anthanthrene and other polynuclear hydrocarbons, *Z. Krebsforsch.*, 77, 226, 1972.
216. Maisin, J., Desmidt, P., and Jacquin, L., *C. R. Soc. Biol.*, 96, 1056, 1927.
217. Misfeld, J. and Timm, J., The tumor-producing effect of automobile exhaust gas condensate and fractions thereof, *J. Environ. Pathol. Toxicol.*, 1, 747, 1978.
218. Mohr, U., Untersuchungen über die kanzerogenen Eigenschaften von 3,4- 10,11- und 12,13-Tribenzfluoranthen an Mäusen, *Arch. Hyg.*, 153, 495, 1969.
219. Müller, E., Kanzerogene Substanzen in Wasser und Boden, *Arch. Hyg.*, 152, 1968.
220. Nakano, K., Experimental production of malignant tumors by 3,4-benzpyrene and 1,2,5,6-dibenzanthracene, *Osaka Daigaku Igaku Zasshi*, 36, 483, 1937.
221. Narat, J. K., Experimental production of malignant growths by simple chemicals, *J. Cancer Res.*, 9, 135, 1925.
222. National Academy of Sciences, *Particulate Polycyclic Organic Matter*, National Academy of Science, Washington, D.C., 1972.
223. Oberling, C., Sannié, C., Guérin, M., and Guérin, P., A propos de l'action cancérogene du benzopyrène, in: *Leeuwenhoek-Verren, 5th Conference, Rapports des Travaux*, de Bussy, Amsterdam, 1937, 57.
224. Pott, P., *Chirurgical Observations Relative to the Cataract, the Polypos of the Nose, the Cancer in the Scrotum, the Different Kinds of Ruptures, and the Mortification of the Toes and Feet*, L. Hawes, W. Clarke, and R. Collins, London, 1775, 208; reprinted in *Natl. Cancer Inst. Monogr.*, 10, 7, 1963.
225. Roe, F. J. C., The development of malignant tumors of mouse skin after "initiating" and "promoting" stimuli. I. The effect of a single application of 9,10-dimethyl-1,2-benzanthracene (DMBA) with and without subsequent treatment with croton oil, *Br. J. Cancer*, 10, 61, 1956.
226. Roe, F. J. C. and Grant, G. A., Tests of pyrene and phenanthrene for incomplete carcinogenic and anticarcinogenic activity, *Br. Emp. Cancer Campaign Res.*, 41, 59, 1964.
227. Roe, F. J. C., Effect of phenanthrene on tumor initiation by 3,4-benzopyrene, *Br. J. Cancer*, 16, 503, 1962.
228. Schmähl, D., *Entstehung, Wachstum, und Chemotherapie maligner Tumoren*, Editio Cantor, Aulendorf, West Germany, 1970, 2.

229. Schmähl, D., Schmidt, K. G., and Habs, M., Syncarcinogenic action of polycyclic hydrocarbons in automobile exhaust gas condensates, *IARC Sci. Publ.*, No. 16, 1977.
230. Schmidt, K. G., Schmahl, D., and Misfeld, J., Investigations on the carcinogenic burden of air pollution in man. VI. Experimental investigations to determine a dose-response relationship and to estimate a threshold dose of benzo[a]pyrene in the skin of two different mouse strains, *Zentralbl. Bakteriol. Parasitenkd. Infektionskr. Hyg. Abt. 1 Reihe B*, 158, 62, 1973.
231. Schmidt, K. G., Schmähl, D., Misfeld, J., and Timm, J., Experimentelle Untersuchungen zur Syncarcinogenese. VII. Mitteilung: Syncarcinogene Wirkung von polycyclischen aromatischen Kohlenwasserstoffen (PAH) im Epicutantest an der Mäusehaut, *Z. Krebsforsch.*, 87, 93, 1976.
232. Scientific Committee of the Food Safety Council, *Proposed System for Food Safety Assessment: A Comprehensive Report on the Issues of Food Ingredient Testing*, Pergamon Press, Elmsford, N.Y., 1978, 97.
233. Scribner, J. O., Tumor initiation by apparently noncarcinogenic polycyclic aromatic hydrocarbons, *J. Natl. Cancer Inst.*, 50, 1717, 1973.
234. Shubik, P., Pietra, G., and Della Porta, G., Studies of skin carcinogenesis in the Syrian golden hamster, *Cancer Res.*, 20, 100, 1960.
235. Slaga, T. J., Bowden, G. T., Scribner, J. O., and Boutwell, R. K., Dose-response studies on the ability of 7,12-dimethylbenz[a]anthracene and benz[a]anthracene to initiate skin tumors, *J. Natl. Cancer Inst.*, 53, 1337, 1974.
236. Slaga, T. J., Jecker, L., Bracken, W. M., and Weeks, C. E., The effects of weak or non-carcinogenic polycyclic hydrocarbons on 7,12-dimethylbenz[a]anthracene and benzo[a]pyrene skin tumor initiation, *Cancer Lett.*, 7, 51, 1979.
237. Smith, W. E., Sunderland, D. A., and Sigiura, K., Experimental anaylsis of the carcinogenic activity of certain petroleum products, *AMA Arch. Ind. Hyg. Occup. Med.*, 4, 299, 1951.
238. Wallcave, L., Garcia, H., Feldman, R., Lijinski, W., and Shubik, P., Skin tumorigenesis in mice by petroleum asphalts and coal-tar pitches of known polynuclear aromatic hydrocarbon content, *J. Toxicol. Appl. Pharmacol.*, 18, 41, 1971.
239. Weisburger, J. H., Bioassays and test for chemical carcinogens, *ACS Monogr.*, 173, 1, 1976.
240. Whitehead, J. K. and Rothwell, K., The mouse skin carcinogenicity of cigarette smoke condensate: fractionated by solvent partition methods, *Br. J. Cancer*, 23, 840, 1969.
241. Wood, A. M., Chang, R. L., Levin, W., Ryan, D. E., Thomas, P. E., Mah, H. D., Karle, J. M., Yagi, H., and Jerina, D. M., Mutagenicity and tumorigenicity of phenanthrene and chrysene epoxides and diol epoxides, *Cancer Res.*, 39, 4069, 1979.
242. Wynder, E. L., Fritz, L., and Furth, N., Effect of concentration of benzopyrene in skin carcinogenesis, *J. Natl. Cancer Inst.*, 19, 361, 1957.
243. Wynder, E. L. and Hoffmann, D., The carcinogenicity of benzofluoranthenes, *Cancer (Philadelphia)*, 12, 1194, 1959.
244. Wynder, E. L. and Hoffmann, D., Eds., *Tobacco and Tobacco Smoke*, Academic Press, New York, 1968.
245. Wynder, E. L. and Hoffmann, D., The epidermis and the respiratory tract as bioassay systems in tobacco carcinogenesis, *Br. J. Cancer*, 24, 574, 1970.
246. Yamagiwa, K. and Ichikawa, K., Experimentelle Studie über die Pathogenese der Epithelialgeschwülste, *Mitt. Med. Fak., Univ. Tokyo*, 15, 295, 1915.
247. Yuspa, S. H., Hennings, H., and Saffiotti, U., Cutaneous chemical carcinogenesis: past, present, and future, *J. Invest. Dermatol.*, 67, 199, 1976.

## Chapter 5.2.2.3

248. Arcos, J. C., Argus, M. F., and Wolf, G., *Chemical Induction of Cancer*, Vol. 1, Academic Press, New York, 1968, 340.
249. Asahina, S., Andrea, J., Carmel, A., Arnold, E., Bishop, Y., Joshi, S., Coffin, D., and Epstein, S. S., Carcinogenicity of organic fractions of particulate pollutants collected in New York City and administered subcutaneously to infant mice, *Cancer Res.*, 32, 2263, 1972.
250. Badger, G. M., The carcinogenic activity, *Br. J. Cancer*, 2, 309, 1948.
251. Berenblum, I., Carcinogenesis and tumor pathogenesis, *Adv. Cancer Res.*, 2, 129, 1954.
252. Brand, G., Karzinogene Wirkungsweise gewebseingelagerter Partikel, *Zentralbl. Bakteriol. Parasitenkd. Infektionskr. Hyg., Abt. I Orig. Reihe B*, 166, 159, 1978.
253. Brockhaus, A. and Pott, F., Untersuchunger zur kanzerogenen Wirkung von Luftstaubextrakten im Subkutantest bei der Maus. Vortrag in Mainz; as cited in Rothe, H., *Lufthygiene und Silikoseforschung*, Girardet, Essen, 1977, 224.
254. Bryan, W. R. and Shimkin, M. B., Quantitative analysis of dose-response data obtained with three carcinogenic hydrocarbons in strain C3H male mice, *J. Natl. Cancer Inst.*, 3, 503, 1943.

255. Burrows, H., Hieger, I., and Kennaway, E. L., The experimental production of tumors of connective tissue, *Am. J. Cancer*, 16, 57, 1932.
256. Buu-Hoi, N. P., New developments in chemical carcinogenesis by polycyclic hydrocarbons and related heterocycles: a review, *Cancer Res.*, 24, 1511, 1964.
257. Druckrey, H., Preussmann, R., Ivankovic, S., So, B. T., Schmidt, C. H., and Bücheler, J., Zur Erzeugung subcutaner Sarkome an Ratten, *Z. Krebsforsch.*, 68, 87, 1966.
258. Druckrey, H., Schmähl, D., Beuthner, H., and Muth, F., Vergleichende Prufüng von Tabakrauchkondensaten, Benzpyren und Tabakextrakt auf carcinogene Wirkung an Ratten, *Naturwissenschaften*, 47, 605, 1960.
259. van Duuren, B. L., Langseth, L., Orris, L., Teebor, G., Nelson, N., and Kuschner, M., Carcinogenicity of epoxides, lactones, and peroxy compounds. IV. Tumor response in epithelial and connective tissue in mice and rats, *J. Natl. Cancer Inst.*, 37, 825, 1966.
260. Epstein, S. S., Joshi, S., Andrea, J., Mantel, N., Sawicki, E., Stanley, T., and Tabor, E. C., Carcinogenicity of organic particulate pollutants in urban air after administration of trace quantities to neonatal mice, *Nature (London)*, 212, 1305, 1966.
261. Falk, H. L., Kotin, P., and Thompson, S., Inhibition of carcinogenesis, *Arch. Environ. Health*, 9, 169, 1964.
262. Gangolli, S. D., Grasso, P., Goldberg, L., and Hooson, J., Protein binding by food colourings in relation to the production of subcutaneous sarcoma, *Food Cosmet. Toxicol.*, 10, 449, 1972.
263. Gottschalk, R. G., Quantitative studies of tumor production in mice by benzpyrene, *Proc. Soc. Exp. Biol. Med.*, 50, 369, 1942.
264. Grasso, P., Review of tests for carcinogenicity and their significance to man, *Clin. Toxicol.*, 9, 745, 1976.
265. Grasso, P. and Crampton, R. F. The value of the mouse in carcinogenicity testing, *Food Cosmet. Toxicol.*, 10, 418, 1972.
266. Grasso, P., Gangolli, S. D., Goldberg, L., and Hooson, J., Physicochemical and other factors determining local sarcoma production by food additives, *Food Cosmet. Toxicol.*, 9, 463, 1971.
267. Grasso, P. and Goldberg, L., Subcutaneous sarcoma as an index of carcinogenic potency, *Food Cosmet. Toxicol.*, 4, 297, 1966.
268. Grimmer, G., Analysis of automobile exhaust condensates, *IARC Sci. Publ.*, No. 16, 29, 1977.
269. Heidelberger, C. and Weiss, S. M., The distribution of radioactivity in mice following administration of 3,4-benzpyrene-5-$C^{14}$ and 1,2,5,6-dibenzanthracene-9,10-$C^{14}$, *Cancer Res.*, 11, 885, 1951.
270. Hooson, J., Grasso, P., and Gangolli, S. D., Injection site tumours and preceding pathological changes in rats treated subcutaneously with surfactants and carcinogens, *Br. J. Cancer*, 27, 230, 1973.
271. Hueper, W. C., Kotin, P., Tabor, E. C., Payne, W. W., Falk, H., and Sawicki, E., Carcinogenic bioassay on air pollutants, *Arch. Pathol.*, 74, 89, 1962.
272. IARC Monographs on the Evaluation of the Carcinogenic Risk of Chemicals to Man, Certain polycyclic aromatic hydrocarbons and heterocyclic compounds, *IARC Sci. Publ.*, 3, 1973.
273. Kallistratos, G. and Fasske, E., Biologische Inaktivierung kanzerogener Stoffe, *Folia Biochim. Biol. Graeca*, 13, 94, 1976.
274. Kallistratos, G. and Pfau, A., Zur Wirkung von 3,4-Benzpyren bei grossen Säugetieren, *Naturwissenschaften*, 58, 222, 1971.
275. Kelly, M. G. and O-Gara, R. W., Induction of tumors in newborne mice with dibenz[*a,h*]anthracene and 3-methylcholanthrene, *J. Natl. Cancer Inst.*, 26, 651, 1961.
276. Leiter, J., Shimkin, M. B., and Shear, M. J., Production of subcutaneous sarcomas in mice with tars extracted from atmospheric dusts, *J. Natl. Cancer Inst.*, 3, 155, 1942-43.
277. Leiter, J. and Shear, M. J., Production of tumors in mice with tars from city air dusts, *J. Natl. Cancer Inst.*, 3, 167, 1942-43.
278. Leiter, J. and Shear, M. J., Quantitative experiments on the production of subcutaneous tumors in strain A mice with marginal doses of 3,4-benzpyrene, *J. Natl. Cancer Inst.*, 3, 455, 1943.
279. Müller, E., Kanzerogene Substanzen in Wasser und Boden. XX. Untersuchungen über die kanzerogenen Eigenschaften von 1,12-Benzperylen, *Arch. Hyg.*, 152, 23, 1968.
280. Neal, J., Trieff, M. S., and Trieff, N., Isolation of an unknown carcinogenic polycyclic hydrocarbon from carbon blacks, *Health Lab. Sci.*, 9, 32, 1972.
281. Nothdurft, H., Über die Sarkomauslösung durch Fremdkörperimplantationen bei Ratten in Abhängigkeit von der Form der Implantate, *Naturwissenschaften*, 42, 106, 1955.
282. Oppenheimer, B. S., Oppenheimer, E. T., and Stout, A. P., Sarcomas induced in rats by implanting cellophane, *Proc. Soc. Exp. Biol. Med.*, 67, 33, 1948.
283. Payne, W. W. and Hueper, W. C., The carcinogenic effect of single and repeated doses of 3,4-benzpyrene, *Am. Ind. Hyg. Assoc. J.*, 21, 350, 1960.
284. Peacock, P. R. and Beck, S., Rate of absorption of carcinogens and local tissue reaction as factors influencing carcinogenesis, *Br. J. Exp. Pathol.*, 19, 315, 1938.

285. Pfeiffer, E. H., Investigations on the carcinogenic burden by air pollution in man. VII. Studies on the oncogenetic interaction of polycyclic aromatic hydrocarbons, *Zentralbl. Bakteriol. Parasitenkd. Infektionskr. Hyg. Abt. 1 Orig. Reihe B*, 158, 69, 1973.
286. Pfeiffer, E. H. and Graf, Z., Untersuchungen über den Einfluss verschiedener Virusinfektionen auf die 3,4-Benzpyren-Kanzerogenese, *Zentralbl. Bakteriol. Parasitenkd. Infektionskr. Hyg. Abt. 1 Orig. Reihe B*, 159, 1, 1974.
287. Pfeiffer, E. H., Investigations on the carcinogenic burden by air pollution in man. XI. About the effect of aluminium hydroxide upon the benzo[a]pyrene carcinogenesis, *Zentralbl. Bakteriol. Parasitenkd. Infektionskr. Hyg. Abt. 1 Orig. Reihe B*, 160, 99, 1975.
288. Pfeiffer, E. H., Oncogenetic interaction of carcinogenic and non-carcinogenic polycyclic aromatic hydrocarbons in mice, *IARC Sci. Publ.*, No. 16, 69, 1977.
289. Pfeiffer, E. H., Über den Einfluss von Blei auf die 3,4-Benzpyren-Kanzerogenese, *Arbeit. Deutsch. Ges. Hyg. Mikrobiol.*, Abstract in press.
290. Pott, F. and Brockhaus, A., unpublished data, 1969.
291. Pott, F., Brockhaus, A., and Huth, F., Untersuchungen zur Kanzerogenität von polyzyklischen aromatischen Kohlenwasserstoffen im Tierexperiment, *Zentralbl. Bakteriol. Parasitenkd. Infektionskr. Hyg. Abt. 1 Orig. Reihe B*, 157, 34, 1973.
292. Pott, F., Tomingas, R., and Reiffer, F. J., Experimentelle Untersuchungen zur kanzerogenen Wirkung und Retention von Benzo[a]pyren am Applikationsort nach intratrachealer und subkutaner Injektion, *Zentralbl. Bakteriol. Parasitenkd. Infektionskr. Hyg. Abt. 1 Orig. Reihe B*, 158, 97, 1973.
293. Pott, F., Tomingas, R., Brockhaus, A., and Huth, F., Studies on the tumorogenicity of extracts and their fractions of airborne particulate matter with the subcutaneous test in the mouse, *Zentralbl. Bakteriol. Parasitenkd. Infektionskr. Hyg. Abt. 1 Orig. Reihe B*, 170, 17, 1980.
294. Pott, F., Tomingas, R., and Misfeld, J., Tumors in mice after subcutaneous injection of automobile exhaust condensates, *IARC Sci. Publ.*, No. 16, 79, 1977.
295. Pott, F., Tomingas, R., and Huth, F., unpublished data, 1978.
296. Rigdon, R. H. and Neal, J., Tumors in mice induced by air particulate matter from a petrochemical industrial area, *Tex. Rep. Biol. Med.*, 29, 109, 1971.
297. Schmähl, D., Vergleichende Untersuchungen an Ratten über die carcinogene Wirksamkeit verschiedener Tabakextrakte und Tabakrauchkondensate, *Arzneim. Forsch.*, 18, 814, 1968.
298. Seelkopf, C., Ricken, W., and Dhom, G., Untersuchung über die krebserzeugenden Eigenschaften des Zigarettenteers, *Z. Krebsforsch.*, 65, 241, 1963.
299. Siegel, J. and Shear, M. J., Unpublished work, 1938; as cited in Leiter, J., Shimkin, M. B., and Shear, M. J., *J. Natl. Cancer Inst.*, 3, 155, 1942-43.
300. Steiner, P. E. and Falk, H. L., Summation and inhibition effects of weak and strong carcinogenic hydrocarbons: 1,2-benzanthracene, chrysene, 1,2,5,6-dibenzanthracene, and 20-methylcholanthrene, *Cancer Res.*, 11, 56, 1951.
301. Steiner, P. E., Carcinogenicity of multiple chemicals simultaneously administered, *Cancer Res.*, 15, 632, 1955.
302. Tomatis, L., Comment on methodology and interpretation of results, *J. Natl. Cancer Inst.*, 59, 1341, 1977.
303. Tomingas, R. and Pott, F., Tierexperimentelle Untersuchungen zur Retentionsrate von polyzyklischen aromatischen Kohlenwasserstoffen in der Lunge und im subkutanen Gewebe, *Zentralbl. Bakteriol. Parasitenkd. Infektionskr. Hyg. Abt. 1 Orig. Reihe B*, 162, 18, 1976.
304. Waravdekar, S. S. and Ranadive, K. J., Biologic testing of 3,4,9,10-dibenzpyrene, *J. Natl. Cancer Inst.*, 21, 1151, 1958.
305. Weil-Malherbe, H., The elimination and carcinogenic potency of 3,4-benzpyrene in mice after subcutaneous injection of non-lipoid solutions, *Br. J. Cancer*, 1, 423, 1947.
306. Weisburger, J. H., Bioassays and tests for chemical carcinogens, *ACS Monogr.*, 173, 1, 1976.
307. Wischnewski, J., Die Wirkung von Pollen auf die Entstehung von Benzpyrentumoren im Tierversuch, Dissertation, University of Timbuktu, Düsseldorf, 1970.
308. Wynder, E. L. and Hoffmann, D., The epidermis and the respiratory tract as bioassay systems in tobacco carcinogenesis, *Br. J. Cancer*, 24, 574, 1970.

## Chapter 5.2.2.4

309. Akamatsu, Y., Ikegami, R., Watanabe, K., and Kikui, M., Induction of leukemia amyloidoses in senile C57BL mice by oral feeding of 3-methylcholanthrene in olive oil solution, *Gann*, 59, 489, 1968.
310. Berenblum, I. and Schoental, R., The metabolism of 3,4-benzpyrene in mice and rats. I. The isolation of a hydroxy and a quinone derivative and a consideration of their biological significance, *Cancer Res.*, 3, 145, 1943.

311. Berenblum, I. and Haran-Ghera, N., The induction of the initiating phase of skin carcinogenesis in the mouse by oral administration of 9,10-dimethyl-1,2-benzanthracene, 20-methylcholanthrene, 3,4-benzpyrene, and 1,2,5,6-dibenzanthracene, *Br. J. Cancer*, 11, 85, 1957.
312. Biancifiori, C., Bonser, G. M., and Caschera, F., Ovarian and mammary tumours in intact C3Hb virgin mice following a limited dose of four carcinogenic chemicals, *Br. J. Cancer*, 15, 270, 1961.
313. Borneff, J., Kanzerogene Substanzen in Wasser und Boden. XIII. Mäusefütterungsversuche mit 3,4-Benzpyren, *Arch. Hyg.*, 147, 28, 1963.
314. Borneff, J., Engelhardt, K., Griem, W., Kunte, H., and Reichert, J., Kanzerogene Substanzen in Wasser und Boden. XXII. Mäusetränkversuch mit 3,4-Benzpyren und Kaliumchromat, *Arch. Hyg.*, 152, 45, 1968.
315. Bulay, O. M. and Wattenberg, L. W., Carcinogenic effect of subcutaneous administration of benzo[a]pyrene during pregnancy on the progeny, *Proc. Soc. Exp. Biol. Med.*, 135, 84, 1970.
316. Bulay, O. M. and Wattenberg, L. W., Carcinogenic effects of polycyclic hydrocarbon carcinogen administration to mice during pregnancy on the progeny, *J. Natl. Cancer Inst.*, 46, 397, 1971.
317. Cardesa, A., Pour, P., Rustia, M., Althoff, J., and Mohr, U., The syncarcinogenic effect of methylcholanthrene and dimethylnitrosamine in Swiss mice, *Z. Krebsforsch.*, 79, 98, 1973.
318. Chang, L. H., The fecal excretion of polycyclic hydrocarbons following their administration to the rat, *J. Biol. Chem.*, 151, 93, 1943.
319. Choudroulinkov, I., Gentil, A., and Guérin, M., Etude de l'activité carcinogène du 9,10-diméthylbenzanthracène et du 3,4-benzopyrène administrés par voie digestive, *Bull. Cancer*, 54, 67, 1967.
320. Dontenwill, W., Elmenhorst, H., Harke, H. P., Reckzeh, G., Weber, K. H., Misfeld, J., and Timm, J., Experimentelle Untersuchungen über die tumorerzeugende Wirkung von Zigarettenrauch-Kondensaten an der Mäusehaut, *Z. Krebsforsch.*, 73, 265, 1970.
321. Ekwall, P., Ermala, P., Setälä, K., and Sjöblom, L., Gastric absorption of 3,4-benzpyrene. II. The significance of the solvent for the penetration of 3,4-benzpyrene into the stomach wall, *Cancer Res.*, 11, 758, 1951.
322. Falk, H. L., Kotin, P., Lee, S. S., and Nathan, A., Intermediary metabolism of benzo[a]pyrene in the rat, *J. Natl. Cancer Inst.*, 28, 699, 1962.
323. Field, W. E. H. and Roe, F. J. C., Tumor promotion in the forestomach epithelium of mice by oral administration of citrus oil, *J. Natl. Cancer Inst.*, 35, 771, 1965.
324. Flesher, J. W. and Sydnor, K. L., Carcinogenicity of derivatives of 7,12-dimethylbenz[a]anthracene, *Cancer Res.*, 31, 1951, 1971.
325. Frankfurt, O. S., Mitotic cycle and cell differentiation in squamous cell carcinomas, *Int. J. Cancer*, 2, 304, 1967.
326. Graffi, A. and Bielka, H., *Probleme der experimentellen Krebsforschung*, Akademie Verlagsgesellschaft Geest und Portig KG, Leipzig, 1959.
327. Homburger, F., Hsuek, S. S., Kerr, C. S., and Russfield, A. B., Inherited susceptibility of inbred strains of Syrian hamsters to induction of subcutaneous sarcomas and mammary and gastro-intestinal carcinomas by subcutaneous and gastric administration of polynuclear hydrocarbons, *Cancer Res.*, 32, 360, 1972.
328. Huggins, C. H. and Yang, N. C., Induction and extinction of mammary cancer, *Science*, 137, 257, 1962.
329. Huggins, C. H., Grand, L., and Oka, H., Hundred day leucemia: preferential induction in rat by pulse-doses of 7,8,12-trimethylbenz[a]anthracene, *J. Exp. Med.*, 131, 321, 1970.
330. Ito, N., Hiasa, Y., Konishi, Y., Marugami, M., The development of carcinoma in liver of rats treated with m-toluylene-diamine and the synergistic and antagonistic effects with other chemicals, *Cancer Res.*, 29, 1137, 1969.
331. Jänisch, W., Die Induktion von experimentellen Hirngeschwülsten mit cancerogenen Kohlenwasserstoffen, *Z. Krebsforsch.*, 68, 224, 1966.
332. Jull, J. W. and Russell, A., Mechanism of induction of ovarian tumors in the mouse by 7,12-dimethylbenz[a]anthracene, *J. Natl. Cancer Inst.*, 44, 841, 1970.
333. Kelley, T. F., Fate of subcutaneously injected benzo[rst]pentaphene in C57BL/6 mice, *Proc. Soc. Exp. Biol. Med.*, 133, 1402, 1970.
334. Lombard, L. S. and Vesselinovitch, S. D., Renal carcinogenesis by benzo[a]pyrene, *Proc. Am. Assoc. Cancer Res.*, 13, 49, 1972.
335. Neal, J. and Rigdon, R. H., Gastric tumors in mice fed benzo[a]pyrene: a quantitative study, *Tex. Rep. Biol. Med.*, 25, 553, 1967.
336. Peacock, P. R., Evidence regarding the mechanism of elimination of 1,2-benzpyrene, 1,2,5,6-dibenzanthracene, and anthracene from the blood-stream of injected animals, *Br. J. Exp. Pathol.*, 17, 164, 1936.
337. Peirce, W. E. H., Tumor-production by lime oil in the mouse forestomach, *Nature (London)*, 189, 497, 1961.

338. Rigdon, R. H. and Neal, J., Relationship of leukemia to lung tumors in mice fed benzo[a]pyrene, *Proc. Soc. Exp. Biol. Med.*, 130, 146, 1969.
339. Roe, F. J. C., Levy, L. S., and Carter, R. L., Feeding studies on sodium cyclamate, saccharin, and sucrose of carcinogenic and tumor promoting activity, *Food Cosmet. Toxicol.*, 8, 135, 1970.
340. Scheuer, E., Huth, F., and Pott, F., Untersuchungen zum morphologischen Erscheinungsbild experimenteller Tumoren bei Ratten nach intraperitonealer Injektion von Asbeststauben, *Arch. Geschwulstforsch.*, 41, 120, 1973.
341. Schmähl, D., *Entstehung, Wachstum uns Chemotherapie maligner Tumoren*, Editio Cantor, Aulendorf, West Germany, 1970, 2.
342. Shendrikova, I. A. and Aleksandrov, V. A., Comparative penetration of polycyclic hydrocarbons through the rat placenta into the fetus, *Byull. Eks. Biol. Med.*, 77, 77, 1974.
343. Setälä, K., Experimental chemical carcinogenesis and the influence of solvents, *Nature (London)*, 174, 873, 1954.
344. Sims, P., The metabolism of benzo[a]pyrene by rat-liver homogenates, *Biochem. Pharmacol.*, 16, 613, 1967.
345. Sims, P., The metabolism of some aromatic hydrocarbons by mouse embryo cell cultures, *Biochem. Pharmacol.*, 19, 285, 1970.
346. Sims, P., Qualitative and quantitative studies on the metabolism of aromatic hydrocarbons by rat-liver preparations, *Biochem. Pharmacol.*, 19, 795, 1970.
347. Toth, B., The induction of malignant lymphomas and other tumors by 7,12-dimethylbenz[a]anthracene in the Syrian golden hamster, *Cancer Res.*, 29, 1476, 1969.
348. Uematsu, K. and Huggins, C. H., Hydrocarbon-induced leukemia in adolescent and adult mice, *Gann*, 60, 545, 1969.
349. van Duuren, B. L., Sivak, A., and Langseth, L., The tumor-promoting activity of tobacco leaf extract and whole cigarette tar, *Br. J. Cancer*, 21, 460, 1967.
350. Walburg, H. E. and Cosgrove, G. E., Methylcholanthrene-induced neoplasms in germ-free RFM mice, *Int. J. Cancer*, 8, 338, 1971.
351. Wattenberg, L. W. and Leong, J. L., Inhibition of the carcinogenic action of benzo[a]pyrene by flavones, *Cancer Res.*, 30, 1922, 1970.
352. Weigert, F. and Mottram, J. C., The biochemistry of benzpyrene. I. A survey and new methods of analysis, *Cancer Res.*, 6, 97, 1946.
353. Winkelstein, W. and Kantor, S., Stomach cancer. Positive association with suspended particulate air pollution, *Arch. Environ. Health*, 18, 544, 1969.

## Chapter 6

## EPIDEMIOLOGY

### J. Misfeld

### TABLE OF CONTENTS

| | | |
|---|---|---|
| 6.1 | Introduction | 222 |
| 6.2 | Questions and Special Problems to be Investigated in Epidemiological Studies on Lung Cancer | 222 |
| | 6.2.1 Urban/Rural Area Studies | 223 |
| | 6.2.2 Smoker Studies | 227 |
| | 6.2.3 Work Area Studies | 230 |
| | 6.2.4 Migrant Studies | 232 |
| | 6.2.5 Global Data Studies | 233 |
| 6.3 | Summary | 234 |
| Acknowledgments | | 235 |
| References | | 235 |

## 6.1 INTRODUCTION

Epidemiology is the field of science in which the occurrence of diseases in certain groups of the human population is investigated in relation to the possible causes and effects of these diseases. Whereas epidemiological research was originally focused on contagious diseases, in the course of time epidemiological methods have also been developed for nontransmissible diseases. According to the above-stated aim, medical, sociological, and mathematical statistical methods form the essential fundamentals of epidemiological research.

As a rule, many expectations placed in the predictive value of epidemiological studies cannot be met: causal connections cannot be proved by epidemiological evidence. Correlations revealed in epidemiological studies need experimental confirmation. To a very large extent, the predictive value of epidemiological studies depends on the type of data they are based on and how these data were compiled, i.e., individual, aggregate, or global data. The data, for their part, can either be retrospective or prospective. Retrospective studies compare one group of individuals who have a common symptom (e.g., death from lung cancer) with another group without this symptom and retrospectively determine the previous existence or lack of a special factor. In prospective studies for which a considerably larger number of individuals is needed, particularly in case of rare diseases, groups are investigated which have been subjected to certain different factors while all the other circumstances have been as similar as possible. Koller[25] defined the essential prerequisites for such comparative cohorts: similarity in structure, observation, and representation. Requirements for valid epidemiological conclusions were outlined in detail by Armitage,[1] Koller,[25] McMahon et al.,[36] and Pflanz.[39]

We now turn to the problem of evaluating those epidemiological studies which are concerned with the correlation between air pollution and lung cancer. Below are discussed first the questions to be investigated, as well as possible methodological techniques, and then problems that arise. In the subsequent paragraphs the results of different studies are discussed.

## 6.2 QUESTIONS AND SPECIAL PROBLEMS TO BE INVESTIGATED IN EPIDEMIOLOGICAL STUDIES ON LUNG CANCER

At the beginning of this century an increasing lung cancer rate was observed in many countries. This trend was correlated to the increasing air pollution caused by growing industrialization. Some substances which proved to be carcinogenic in animal experiments, e.g., benzo[a]pyrene, were detected as products of incomplete combustion in exhaust gases from industrial plants and vehicles, as well as in cigarette smoke. Consequently the following hypotheses needed to be investigated:

In comparison with the "normal population" the following groups show a higher rate of lung cancer:

- Residents of "polluted" (industrialized, frequently just called "urban") areas
- Individuals who inhale products of incomplete combustion at the workplace
- Immigrants from more polluted areas (countries) in comparison with the native population
- Cigarette smokers

The different types of studies correspond to these basic hypotheses: urban/rural area studies, work area studies, immigrant studies, and smoker studies. Since the suspected factors of modification do not act independently, many studies cannot be assigned to just one type, but several aspects have been taken into consideration.

Some authors, in principle, doubt the decisive value of epidemiological studies. These fundamental doubts can often be eliminated by carrying out the collection and statistical evaluation of data with great care. Further objections and difficulties encountered, particularly in connection with "air pollution and lung cancer", will be discussed later.

The substances or chemical classes of substances which are regarded as possible carcinogens never occur separately in the air, but are components of complex mixtures. The composition of these mixtures varies from area to area. Many investigations, therefore, concentrate on measuring only benzo[a]pyrene, i.e., a single carcinogen. Carnow and Meier[6] considered it the crucial substance, and Sawicki et al.[47] regarded it as a sort of index substance. These assumptions, however, have not been proved. The different assessment of benzo[a]pyrene in animal experiments has been described in Chapters 5.2.2.2 and 5.2.2.3. Furthermore, the effect of a mixture of substances need not necessarily be an additive effect of the individual components of this mixture.

Besides the basic problems of finding cohorts which are clearly representative with respect to individual substances, it is impossible to determine the dose to which the investigated individuals have been subjected. Sampling of airborne carcinogens, which have so far only been carried out occasionally, can never produce the absolute amounts of substances which man really inhales. It is almost impossible to even estimate median values because nobody is exposed to constant environmental conditions over a period of several decades which most probably corresponds to the tumor latency period. With regard to smoking habits, the investigator has to rely on the individual's estimation of the inhalation depth. These habits, of course, do not remain unchanged over long periods of time. The attempt of some authors to avoid these difficulties by forming more homogenous, and thus smaller, cohorts increases, on the other hand, the risk of chance variations.

The problems described make it more difficult to establish dose-response relationships which are needed to assess the carcinogenic risk of air pollution to man. Attempts to infer a linear relationship between the lung cancer rate, on the one hand, and fuel and cigarette consumption, on the other hand,[6] are questionable not least because in animal experiments dose-response relationships of sigmoid (and not linear) form have been found most of the time.

Many authors, therefore, confine their survey to certain objectives. Usually they investigate only the question whether different lung cancer rates can be detected in groups subjected to different levels of pollution. As a rule these investigations have been carried out according to a uniform pattern. The population to be investigated is divided into different groups according to the objectives of the survey, for instance, into residents of urban and rural areas, smokers and nonsmokers, heavy and light smokers, etc. It is of importance that, in addition to this classification, a further differentiation is made, e.g., according to sex, race, and age. In statistical tests the question to be answered is whether differences in the lung cancer frequency exist between these groups. These tests result in probabilities which determine whether a difference is significant or not.

The objections discussed earlier are not meant to limit the value of epidemiological research, but merely indicate the limitations inherent in epidemiological methodology and warn of uncritical assessment of these studies. Some studies published in the last 25 years follow which deal with the problem of "lung cancer and air pollution (particularly by polycyclic aromatic hydrocarbons)".

### 6.2.1 Urban/Rural Area Studies

Mancuso et al.[37] investigated the cancer mortality in Ohio between 1944 and 1952. The survey was carried out according to the described method except that, instead of comparing two rates, the death rate to be expected on the basis of general mortality was calculated for each group and compared with the actually observed rate. They considered the mortality in the individual counties which, according to the population figures, were classified as large cities, towns, and rural areas, i.e., the classification followed administrative boundaries. The composition of atmospheric air was not measured. The data were obtained from official statistics based on death certificates. Statistical tests were carried out.

A significant ($p < 0.01$) gradient between urban and rural areas was demonstrated for cancer of the lung, esophagus, buccal cavity, larynx, nervous system, and bladder. It was particularly marked for cancer of the lung and esophagus. A detailed investigation of eight urban counties, especially Cuyahoga, showed that the percentage of immigrants among lung cancer cases was higher than their percentage in the population ($p < 0.001$). This result induced Mancuso[38] to conduct a further survey. In Cuyahoga, lung cancer mortality also proved to be related to the socioeconomic level, as determined by the median income and educational level: in districts with a high socioeconomic level lower lung cancer rates were observed and vice versa ($p < 0.01$).

Of interest are the author's control investigations on the reliability of these data. A review of the diagnosed tumors of the brain and nervous system confirmed this diagnosis in more than 80% of all cases,. A geographic dependence of the diagnostic reliability could not be observed. Mancuso[38] also asked how long individuals who had died of cancer of the nervous system had lived at their last residence. Most of them had lived there for more than 20 years; only 8 of 448 cases had lived there for less than 5 years.

Stocks and Campbell[49] attempted to investigate the simultaneous influences of the environment and of smoking habits. The area under investigation was divided into an urban (Liverpool), a rural (without industry), and a mixed zone. Sampling with air filters over a 6-month period at ten sites in this area confirmed that the division corresponded to the expected different air pollution (in particular by polycyclic aromatic hydrocarbons). All lung cancer cases registered between 1952 and 1954 — deaths as well as surviving patients — were taken into consideration. The patients or their relatives were interviewed about the patients' smoking habits. On the basis of data from hospital patients not suffering from lung cancer the percentage of smokers and nonsmokers in each area was estimated and death rates were calculated. The results are shown in Figure 62.

There is a clear — though not in all subgroups consistent — dependence of mortality on these two factors. Pipe smokers seem to run a considerably lower risk than cigarette smokers according to these data. This result has been confirmed by other authors. Objections may, however, be raised that, due to the high selection, the hospital patients, chosen as control by Stocks and Campbell,[49] were not representative of the total population and not suited for determining the percentage of smokers. Supposing, however, that smoking is also the cause of other diseases and smokers are consequently overrepresented among non-lung cancer patients, the influence of smoking on lung cancer development would even be underestimated. Stocks and Campbell did not carry out any statistical tests.

In a later publication, Stocks[50] based his survey on measured sampling results of 15 air filters installed at sites in Liverpool and North Wales from 1954 to 1957. On the basis of cancer mortality between 1950 and 1954 in all England and Wales, expected death rates from lung and bronchus cancer and intestinal cancer were calculated for each administrative district in which a filter had been placed and related to observed rates. Stocks and Campbell[49] calculated a correlation coefficient, each between two of the three values: air pollution (content of dust and different hydrocarbons), population density, and the ratio between observed and expected death rates. In each case the

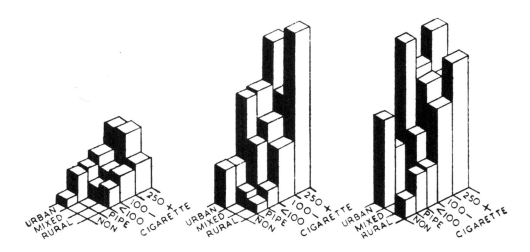

FIGURE 62. Lung cancer rate in relation to place of living and smoking habits.[49]

correlation coefficient was 0.8. For intestinal cancer, a correlation could not be established. By calculating partial correlation coefficients, Stocks[50] tried to separate the influences exerted by population density and air pollution. Since this method was based on the assumption of a linear correlation, it has to be discussed with caution.

Levin et al.[31] based their survey on a central register of cancer mortality in New York state for the period from 1949 to 1951. This register contained all observed types of cancer. The data collection was very homogenous. According to normal official classification (based on population figure and density) a division into urban and rural areas was made. The mortality from lung cancer and cancer of the buccal and pharyngeal cavity followed a gradient between urban and rural areas which was more marked for men than for women. Unfortunately the authors did not carry out statistical tests nor state absolute figures so that the significance of the results cannot be assessed. They reported that in the urban as well as the rural areas about 70% of the diagnoses were confirmed by histological examinations. They considered this an indication of homogenous diagnostic practices limiting the risk of misinterpretation. The possibility of an incorrect registration of the place of residence was also discussed. Levin et al. assumed an overestimation of the urban population by 1 to 4%, without giving, however, a satisfactory explanation of this assumption.

In two further studies, the area under investigation was divided into partial areas of high and low air pollution on the basis of sampling results. The influence of the economic level was also taken into consideration. Hagstrom et al.[17] did not find a relation between lung cancer rates in Nashville (1949 to 1960) and any of the two factors. The methods used, however, have not been documented precisely enough. It is, for instance, not evident whether sexes were investigated separately. Statistical tests were not used. Winkelstein et al.[54] observed an inverse relation between air pollution and economic level in Erie County (high economic level corresponded to low air pollution and vice versa). When dividing the population correspondingly, they obtained groups of very different size and death rates. Consequently, some rates had to be based on less than five deaths. The assessment of these standardized data is therefore difficult. Considering the causes of death from 1961 to 1963, a generally high dependence of diseases of the respiratory tract on air pollution and a low dependence on the economic level was ascertained. The correlation between lung cancer and economic level, however, is stronger. On the whole, a correlation to air pollution was seen, too, but it was not consistent within the individual classes (Table 54).

## Table 54
### AVERAGE ANNUAL DEATH RATES PER 100,000 POPULATION FROM MALIGNANT NEOPLASMS OF THE BRONCHUS, TRACHEA, AND LUNG ACCORDING TO ECONOMIC AND AIR POLLUTION LEVELS, AND AGE[54]

| Economic level | Air pollution level | | | | |
|---|---|---|---|---|---|
| | 1 (Low) | 2 | 3 | 4 (High) | Total |
| 1 (Low) | — | 189[a] | 271 | 171 | 240 |
| 2 | 191 | 206 | 144 | 241 | 190 |
| 3 | — | 182 | 160 | 154 | 173 |
| 4 | 86 | 156 | 139 | — | 126 |
| 5 (High) | 79 | 78 | 116[a] | — | 80 |
| Total | 106 | 162 | 180 | 202 | 158 |

*Note:* Survey subjects were white men, 50 to 69 years, from Buffalo, N.Y. and environs, 1959 to 1961.

[a] Rate based on less than five deaths.

## Table 55
### LUNG CANCER MORTALITY IN COUNTIES WITH AND WITHOUT PETROLEUM INDUSTRIES[3]

| | Population figure | | |
|---|---|---|---|
| Age-standardized lung cancer death rate per 100,000 in | ≤25,000 | 25,000—100,000 | >100,000 |
| Oil counties | 30.0 | 35.7 | 45.5 |
| Control counties | 27.2 | 31.1 | 39.0 |
| Ratio | 1.10 | 1.15 | 1.17 |

Two later publications should be mentioned. Blot et al.[3] investigated the relation between cancer mortality and the existence of petroleum industries, and compared 39 U.S. counties with petroleum industries to 117 counties without petroleum industries. There was, indeed a significant difference, which was particularly marked for lung cancer. The location of further chemical industry was not investigated. On the basis of his data, the author could not determine whether workers in these branches of industry were mainly affected, i.e., whether an occupational risk existed. Also of interest is the classification of the counties into three groups according to their population figure: the rate of lung cancer rose from the group with the lowest population figure to the group with the highest population figure. In the individual groups, the rates of counties with petroleum industries were higher than those of counties without petroleum industries (Table 55).

MacDonald[35] studied the air pollution in greater Houston in relation to the mortality from cancer of the respiratory organs between 1940 and 1969. Death rates were calculated for the area around each sampling station. The sites at which higher mortality occurred lay essentially along a Southeast-Northwest axis through the town which corresponds to the prevalent wind direction. The author attributed this finding to wind-distributed industrial contaminants. However, such a correlation could not be proved with data measured between 1968 and 1969.

In a retrospective study, Dean[10] investigated the influence of smoking habits and place of residence on the lung cancer mortality in Northern Ireland from 1960 to 1962. In interviews he collected all relevant data of 1040 lung cancer deaths by questioning a total of 5864 individuals, including cases of bronchitis and controls. When comparing the lung cancer rate of Belfast with that of purely rural districts, a gradient between urban and rural areas became evident for smokers as well as nonsmokers. Dean[10] also found an increase in the lung cancer rate of smokers correlated to their cigarette consumption. The differences observed after further subdivision of Northern Ireland (Belfast, environs of Belfast, towns, small towns, and purely rural districts) have to be assessed with caution because of the occasional low numbers in the individual subgroups. It was not investigated whether this division into urban and rural areas correlated with measured data of air pollution, so that this survey may at best hint at the possible influence of air pollution.

Viewed overall, the investigations of urban and rural areas present a heterogeneous picture. Some signs of a gradient between urban and rural areas certainly exist, but some authors are not convinced. In 1965, Lawther[29] stated that lung cancer rates increased while at the same time air pollution decreased. He tended to ascribe these results to misrepresentations during data collection. His opinion is, however, not supported by the surveys of Mancuso et al.[37,38] and Levin et al.[31]

Hammond and Garfinkel[22] concluded that general air pollution has very little effect, if any, on the lung cancer death rate, but that it may be true that the effects of general air pollution may combine with occupational air pollution effects to produce an excess risk. Buell et al.[4] found in a prospective study that the risk of pulmonary cancer in Los Angeles, where photochemical air pollution levels are highest, was not greater than in other major metropolitan areas of California. According to Grimmer[16] the influence of air pollution on human health and the development of certain diseases has to be considered proved.

Although the level of contamination has not yet been verified satisfactorily by exact data, there is hardly any doubt that the differentiation between cohorts in urban and rural areas really corresponds to the different content of contaminants in the atmosphere. Future epidemiological investigations on the correlation between air pollution and lung cancer should provide sufficient quantitative data on the concentration of PAH and other potential carcinogens in the atmosphere to which the individual comparative groups were exposed. These data would then permit the establishment of dose-response relationships.

The major problem in the assessment of lung cancer is the long latency period. In the cited publications sampling and measuring of particulate matter were carried out almost always at the same time (or even later) as the registration of cancer deaths. MacDonald[35] therefore attributes the obtained results to this fact, and Sawicki et al.[47] are of the same opinion. It would surely be interesting to compare the analytical results from former studies with today's lung cancer mortality.

It is difficult to interpret the dependence of lung cancer on the economic level. A correlation between economic level and air pollution might be assumed in the case of the survey of Mancuso et al.,[37] analogous to the results presented by Winkelstein et al.[54] The latter investigated these two factors separately and found a dependence on the economic level. All explanations of this relation have so far been hypothetical (correlation of the economic level to other factors: type of medical care, smoking habits, workplace).

### 6.2.2 Smoker Studies

The results published by Stocks and Campbell[49] have already been discussed and

## Table 56
## MORTALITY RATIO FOR LUNG CANCER IN SMOKERS (MALES)[19]

| Cigarette consumption per day | 1—9 | 10—19 | 20—39 | ≥40 |
|---|---|---|---|---|
| Mortality ratio | 4.6 | 7.48 | 13.14 | 16.61 |
| | | | | |
| Inhalation | Light | Moderate | Deep | |
| Mortality ratio | 8.42 | 11.45 | 14.31 | |
| | | | | |
| Age at beginning of smoking | ≥25 | 20—24 | 15—19 | <15 |
| Mortality ratio | 3.21 | 9.72 | 12.81 | 15.1 |

migrant studies in which the factor of smoking is taken into account are dealt with later. Here a study is to be discussed in which the influence of smoking on human health was investigated and the setup and scope of which are exemplary. In 1959, Hammond[18,19] started a prospective survey in the U.S., which included 1,078,894 males and females. These entered data on their personal, family, and occupational situation, case history, habits, and many other factors into questionnaires. At the death of an individual included in this study, the certificate of death was evaluated, and, in case of necessity, additional questions about the cause of death were put to the attending physician.

In the earlier publication[18] the total data potential comprised 19,208 cases of death among 422,094 males after an observation period of 34.3 months. According to the later study[19] 26,448 deaths (including 1159 lung cancer deaths) and 16,773 deaths (including 210 lung cancer deaths) occurred among 440,558 males and 562,671 females, respectively, after 44 months.

This large number of data permitted a very differentiated investigation. Hammond[18,19] distinguished according to the type of smoking (nonsmokers, pipe, cigar, and cigarette smokers), and in the case of cigarette smokers according to the depth of inhalation and the age at which smoking began. He found that these factors were not statistically independent. Smokers with a high cigarette consumption usually started to smoke at an early age and inhaled deeply.

Hammond[18,19] calculated the death rates of each risk group and formed the ratio between this rate and the rate of the corresponding nonsmoker group. This ratio normally exceeded 1. Cigarette smoking proved to be a high risk to human health, not only with respect to cancer and other diseases of the respiratory tract, but also to cardiovascular and other diseases. The ratio for lung cancer, however, stands out clearly against all the others. The risk level could be correlated to the daily cigarette consumption and the depth of inhalation, as well as the age at which smoking began (Table 56).

The data of Table 56 show a clear dose-response relationship. Referring to the American cigarettes of that time, it can be estimated that for males the consumption of 1 to 5 cigarettes per day represented the harmful threshold. This amount corresponds to about 40 to 200 ng of benzo[a]pyrene per day in the smoke condensate of these cigarettes.

Pipe and cigar smokers did not have higher death rates than nonsmokers; a result which corresponds to Stock's findings.[49] Among exsmokers of cigarettes the risk decreased in the interval of time passed since cessation of smoking. Smokers who had previously smoked 1 to 19 cigarettes per day reached the rate of nonsmokers after 10 years. This finding seems to be important for the investigation of causal factors of lung carcinogenesis. Assuming that the variation of only one influence factor (cessa-

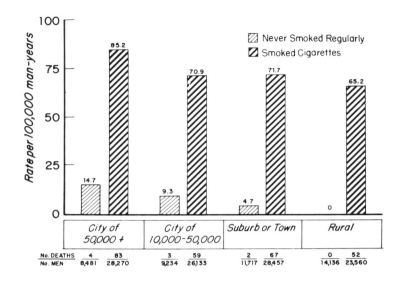

FIGURE 63. Age-standardized lung cancer rate in men in relation to smoking habits and place of residence.[24]

tion of smoking) sufficed to decrease the lung cancer rate, this result clearly hints at a causal relation.

The increase in mortality induced by smoking was observed in males as well as in females. The rate of increase was lower though in females. Hammond thought that the generally lower mortality of women can be attributed to their different smoking pattern. However, among nonsmokers women also have a significantly lower rate of lung cancer mortality than males ($p < 0.01$).

Hammond[19] formed still further groups according to other variables listed in the questionnaires. In each group the higher mortality of cigarette smokers was demonstrated. No other factor was able to upset this consistent pattern.

Although Hammond did not use any statistical tests in his survey, apart from some isolated cases, the significance of his results cannot be doubted in view of the large number of cases evaluated. Whereas some authors worked with rather small total numbers, Hammond used these only for special, occasional investigations. On the basis of his study the correlation between mortality, in particular of lung cancer, and cigarette smoking can be regarded as proved.

In a later survey, Hoffmann and Wynder[24] considered the "urban factor" in addition to smoking. Figure 63 demonstrates the importance to be attributed to smoking.

The dose-related influence of smoking on lung cancer mortality demonstrated in the references cited earlier was supported by later investigations in which the different condensate contents of cigarettes, as well as the difference between filter and nonfilter cigarettes, was taken into consideration. In continuation of their cited prospective study, Hammond et al.[21] found not only a positive correlation between the tar and nicotine content of cigarettes and mortality in general, but also in particular, mortality due to lung cancer and cardiovascular diseases. A significantly increased mortality, in particular of lung cancer, was also observed among the group of smokers of "light" cigarettes in comparison with nonsmokers.

In a recent survey, Auerbach et al.[2] investigated whether a correlation could be detected between the condensate content of cigarettes, which has been decreased in the last few years, and the frequency of histological changes of the bronchial epithelium.

For this purpose men who died without lung cancer between 1955 and 1960 (group A) were compared with men who died without lung cancer between 1970 and 1977 (group B). The authors found that the frequency of changes correlated positively to the extent of cigarette consumption, and changes occurred more frequently in group A than in group B. These findings, however, cannot be accepted uncritically because they refer only to patients who did not die of lung cancer. It has not yet been proved that the observed histological changes were the preliminary stages of malignant changes, and other modifying factors such as work area, place of residence, exposure to other contaminants, etc. were not taken into account.

### 6.2.3 Work Area Studies

The question as to whether members of certain professions tend to contract certain diseases was investigated as early as the 18th century. The first observation was reported by Percivall Pott.[40] Pott discovered a high rate of cancer of the scrotum among British chimney sweeps and related it to their occupation. Butlin[5] was the first to describe skin cancer among workers of the tar- and pitch-producing industry, and Volkmann[53] reported on skin cancer among workers in paraffin processing. As to further details of the historical development, reference is made to Falk et al.[14]

The investigation of a possible correlation between lung cancer and air pollution is promising in those fields of work in which workers are exposed to products of incomplete combustion. This applies particularly to the processing of coal and tar products.

Doll[11] investigated the death cases among retired gasworkers in London between 1939 and 1948. Based on the mortality rates of all England and Wales, and of Greater London alone, the expected rates of mortality from certain causes of death were calculated and compared with the observed rates. Some data for Greater London had to be estimated by interpolation. The differences were tested statistically for significance. A higher risk of cancer in general ($p < 0.01$) and of lung cancer in particular ($p < 0.001$) was determined in comparison with England and Wales. Compared with London, only the lung cancer risk was higher ($p < 0.01$). A total of 25 cases of lung cancer contrasted with an expected rate of 10.4 (England and Wales) and 13.8 (London). The other causes of death ranked within the scope of expectation with diseases of the nervous system and senile debility being lower than in the remaining population. The majority of the 25 lung cancer cases was not concentrated in the group which had had direct contact to coal and tar products during work.

The same author published results of a prospective study on the same subject 25 years later. He investigated the cases of death among 40- to 65-year-old workers from eight British gasworks who had been working there for more than 5 years. The period of investigation extended from 1953 to 1965 in the case of four companies and started between 1957 and 1959 for the other four companies. The workers were assigned to different classes according to the degree of contact with coal products they had during work. Based on statistical data for England and Wales, the expected death rates were calculated and related to the observed ones. Individuals working in coal processing showed a significantly higher lung cancer mortality (in the four companies observed for 12 years $p < 0.001$). Workers in other fields of work did not show this trend; in this group mortality was even lower than in the average population.

In a series of publications, Lerer et al.,[30] Lloyd et al.,[32-34] Redmond et al.,[41-44] and Robinson[46] reported on methods and results of a large-scale study on the mortality of steelworkers in Allegheny County, Pennsylvania. The investigation included 58,828 individuals who in 1953 worked in one of the seven steel mills of this county. Cases of death were registered during the years from 1953 to 1961 (in the last publication until 1966) and the cause of death was determined. The total mortality of steelworkers ranked a little lower than that of the general population. After differentiation according to

Table 57
MORTALITY OF CANCER OF THE
RESPIRATORY ORGANS (OBSERVED/
EXPECTED)[44]

| Coal processing, total | Oven workers | Workers on top of the oven (= high exposure, for more than 5 years) | Ref. |
|---|---|---|---|
| 1.75 | 2.67 | 10.00 | 34 |
| 2.05 | 3.53 | 10.83 | 44 |

occupational group and cause of death, the expected death rates were calculated on the basis of mortality among all steelworkers. The ratio of observed versus expected rates was tested for statistically significant deviations from 1. An increased cancer mortality (especially cancer of the respiratory organs) was demonstrated among coal processing workers.[33] In a later publication Lloyd[34] set out to assess this result. The increased cancer mortality proved to be restricted to the small group of workers at the coke ovens. A clear dependence on the degree of exposure to emitted substances was proved: individuals working on top of the oven (high exposure) suffered a higher mortality than those working at the oven sides. The rates were even higher if only individuals working at least 5 years in this work area were considered. These results were confirmed in a report[44] extending over 5 additional years (Table 57).

The Allegheny study is to be continued. Its intention is to publish more data in the near future including a discussion of observations made from 1953 to 1970.

The unmistakable results from the Allegheny study induced Redmond et al.[42] to start a comparison among 10 further steelworks in the U.S. and Canada in which oven workers and nonoven workers of coal processing plants were compared. Here, too, the same pattern evolved although the conclusions were based on a smaller number of data.

In another prospective study Hammond et al.[21] investigated the cancer mortality among roofers and waterproofers working with tar, pitch, and asphalt. For this purpose they observed 5939 men who in 1959 had been employed by a certain company for at least 9 years (active and retired workers), over a period of 12 years until 1971. During this time 1728 men died. The workers were assigned to groups according to the length of their employment. Expected death rates were calculated on the basis of statistical data on the general U.S. population, as was the ratio of observed versus expected rates. Statistical tests were not carried out.

In the group of workers employed for more than 20 years, an increased cancer mortality was observed, which was statistically most evident in the case of lung cancer — 99 observed cases compared with 62.32 expected — and cancer of the mouth and pharyngeal cavity — 31 observed compared with 15.92 expected. A classification according to the duration of employment suggests that the increase in lung cancer mortality exceeded the expected value depending on the length of the period of employment (Table 58). However, the cohorts were only small. For all the other causes of death, the mortality of the investigated workers lay below that of the general population.

In the American Cancer Society's prospective study, Hammond and Garfinkel[22] found that men who said they were occupationally exposed had mortality rates of lung cancer 14% greater than the nonexposed.

The discussed work-area studies gave an altogether affirmative answer to the question whether there is a relationship between lung cancer and exposure to the pyrolysis products emitted during coal and tar processing. Thus, a more consistent picture is

Table 58
OBSERVED AND EXPECTED RATES
OF DEATH FROM LUNG CANCER
(1960-1971) IN RELATION TO
DURATION OF EMPLOYMENT[21]

| Duration of employment (years) | 9—19 | 20—29 | 30—39 | ≥40 |
|---|---|---|---|---|
| Observed | 22 | 66 | 21 | 12 |
| Expected | 22.93 | 43.45 | 14.02 | 4.85 |
| Ratio | 0.92 | 1.52 | 1.50 | 2.47 |

Table 59
LUNG CANCER DEATH RATE (ANNUAL, PER 100,000 RESIDENTS)[38]

| | Nonwhite | Native white | Foreign-born white |
|---|---|---|---|
| Urban | 31 | 25 | 37 |
| Rural | 11[a] | 11 | 32 |

[a] Based on five death cases only.

derived than in studies comparing urban and rural areas. It is certainly due to the differences in design of these previously discussed studies.

These surveys also hint at an increased occurrence of cancer in organs other than the lung and respiratory tract. In particular, cancer of the bladder, kidneys, testicles, and skin, seems to be environment-related. The small numbers of individuals so far investigated do not permit the formation of final conclusions.

**6.2.4 Migrant Studies**

Since Mancuso et al.[37] detected a higher rate of lung cancer deaths among immigrants in Ohio, this phenomenon has undergone further investigation. In 1958 Mancuso[38] published a report in which he calculated and compared the death rates of native white, foreign-born white, and nonwhite (including Japanese and Chinese) residents of all Ohio and the urban and rural areas defined in his earlier paper. In all three groups a gradient between urban and rural areas was observed which was least marked among immigrants (Table 59). A classification of immigrants according to their native country did not present a consistent pattern. Mancuso[38] then compared the death rates of English and Italian immigrants with those of native white residents and with residents of England and Italy. The rates of English immigrants came close to the American rates, whereas those of Italian immigrants were near to those of their native country. However, these results seem to depend only on variations by chance.

Eastcott[13] investigated the cases of death due to cancer registered in New Zealand between 1949 and 1953 (without the Maori population). Three comparisons were made: between residents of urban and rural areas, native and British-born residents, and immigrants who had immigrated before or after their 30th year of age. In each case, expected death rates were calculated and compared with the observed rates. The groups were not divided according to sex. A gradient between urban and rural areas appeared for lung cancer only. This type of cancer was diagnosed in excess among immigrants, and of those, again most frequently in individuals who had immigrated at an age of more than 30 years (in both case $p < 0.001$). Eastcott referred to the considerably higher lung cancer rates in Great Britain in comparison to New Zealand. He admitted, however, that the different factors cannot easily be separated because most immigrants stayed in the towns.

Dean[7-9] investigated the lung cancer mortality among the white population of South Africa. Compared with other countries South Africa proved to have an extremely low lung cancer rate. Dean indicated that the South Africans are heavy smokers, and he disproved the assumption that South Africa had a lower rate of air pollution than Europe and the U.S. by measuring the pollution in South African towns. The measured data were collected during the same time period as were the investigated cancer death rates, i.e., from 1947 to 1956. During this period the cancer rate doubled for males and increased 1.5-fold for females. The calculated death rates were higher in urban areas than in rural districts, but they were based on very small numbers (occasionally less than 10 death cases). A comparison of native whites with British immigrants revealed a higher mortality among immigrants (262 observed versus 181.8 expected, significance $p < 10^{-6}$) in the age group of 45- to 64-year-old males. Whenever possible, Dean tried to discover the age of immigration in the register of inhabitants and discovered that most of the deceased had left Great Britain after 1920, whereas in the group of males above 65 years of age who did not show an increased mortality, most immigrants had arrived in South Africa before 1910, i.e., before the lung cancer rates went up in England. This result hinted at the action of a factor modifying the lung cancer rate of immigrants prior to their emigration from Great Britain. The collection of data on the time of emigration was, however, incomplete.

The author also asked 55% of the widows of deceased lung cancer patients about the smoking habits of their husbands and compared the results with the findings of an unspecified control group. Among the cancer death cases there was a higher percentage of heavy smokers.

The lung cancer death rate of British immigrants at an age of 45 to 64 years ranked between the higher rate in England and Wales and the lower rate of African-born whites. Dean found the same results for British immigrants to Australia (males above 40 years, time period: 1950 to 1958).

By questionnaire, Reid et al.[45] investigated diseases of the respiratory tract, angina, and susceptibility to infarction among natives, and British and Norwegian immigrants to the U.S. All these diseases could be correlated to the age and smoking habits of the individuals. The data, standardized according to age and smoking habits, revealed a lower risk among the two groups of immigrants. This is surprising in view of the considerably higher rates of diseases of the respiratory tract in Great Britain compared with the U.S. As a possible explanation, they proposed that the mode of data collection and evaluation might not be comparable in the two countries. Furthermore, figures standardized according to two variables are always subject to considerable chance variations.

Reid et al. did not investigate lung cancer cases. They mentioned, however, the official death rates registered between 1959 and 1961 in the U.S. and compared them with those of Norway and Great Britain (1960 to 1962). For males a consistent picture presented itself: Great Britain had the highest rate and Norway the lowest. The rates for immigrants to the U.S. ranked between those of their native country and those of the U.S.

### 6.2.5 Global Data Studies

Stocks[51,52] based his investigations on measurements of air pollution in different European countries and the corresponding official statistics on lung cancer mortality in 1960/61. The average cigarette consumption and the percentage of heavy smokers in the population was determined by random samples of the population. He calculated correlations between cigarette consumption and air pollution and stated that in areas of high air pollution the highest cigarette consumption was registered. Consequently a separation of these two factors is very difficult. Further investigations concerned a world-wide comparison of the fuel consumption in different countries in the 1950s, the cigarette consumption per capita, and the lung cancer rate, taking a possible delayed effect into consideration. Here, too, the author found a relation between the lung cancer frequency and these two factors. From his data he concluded that a period of 10 to 20 years normally elapsed from the onset of the damaging activity of these two factors to the development of cancer.

Stocks' reports[51,52] have to be interpreted with caution because his choice of mathematical methods (calculation of single and partial correlation coefficients, comparison of logs of death rates) is questionable.

Higgins[23] gave a survey of trends in the development of smoking habits and the lung cancer rate between 1940 and 1970 in the U.S. and Great Britain. Whereas in the initial years in both countries an increase in the lung cancer rate was reported, particularly in the older age groups, this trend was reversed (particularly in the younger age group) from the middle of the fifties. More detailed studies will have to check whether this result was due to the British Clean Air Act which resulted in a reduction of air pollution (in London, for instance, to 20%). In general the lung cancer rate was two to three times higher in Great Britain than in the U.S. Whereas cigarette consumption increased among women, it decreased among men; the percentage of smokers in the U.S. male population being lower than in Great Britain. Consistent correlation between the development of lung cancer rates and smoking habits could not be established. Most probably, cancer rates of the 1970s and 1980s will be more informative.

## 6.3 SUMMARY

The analysis of the cited epidemiological studies permits the formation of the following conclusions:

1. Lung cancer mortality correlates to smoking habits. Recent results show a decrease of the lung cancer rate due to the tendency to smoking "lighter" cigarettes.
2. A large number of studies suggest a correlation between the lung cancer rate and the air pollution level. There is no reliable study in which such a correlation is denied or refuted. It is not possible to assess with certainty to what extent possible additional factors might have been effective.
3. On the basis of epidemiological studies dose-response relationships can be estimated for smokers only. Thus the daily inhalation of only a few cigarettes over several decades does not seem to be without a risk of inflicting lung cancer.
4. The discussed epidemiological studies refer only to mixtures of different substances, but not to individual substances which might be responsible for inducing cancer. Based on the comparison of results of animal experiments carried out with different fractions of tar, airborne particulate matter, automobile exhaust condensate, and cigarette smoke condensate, including results of experiments with pure PAH, it can be stated that PAH — or to be more precise, some PAH — proved to be the most active group of carcinogens in these mixtures. The final evaluation of epidemiological and experimental results permits the statement that PAH can induce lung cancer in man. This statement is, however, not quantifiable.

5. Future epidemiological studies of the correlation between air pollution and lung cancer should be based on sufficient data of concentrations of PAH and other potential carcinogens in the atmosphere to which the individual groups were exposed, so that a basis for the calculation of dose-response relationships is given.

## ACKNOWLEDGMENT

We wish to thank Professor Dr. M. Pflanz for his useful advice and H. Knauf for assistance in preparing this chapter. H. Schulte and H. Wittmann are thanked for their technical assistance.

## REFERENCES

1. Armitage, P., *Statistical Methods in Medical Research,* Blackwell, Oxford, 1971.
2. Auerbach, O., Hammond, E. C., and Garfinkel, L., Changes in bronchial epithelium in relation to cigarette smoking, 1955-1960 vs. 1970-1977, *N. Engl. J. Med.,* 300, 381, 1979.
3. Blot, W. J., Brinton, L. A., Fraumeni, J. F., and Stone, B. J., Cancer mortality in U.S. counties with petroleum industries, *Science,* 198, 51, 1977.
4. Buell, P., Dunn, J. E., and Breslow, L., Cancer of the lung and Los-Angeles-type air pollution, *Cancer (Philadelphia),* 20, 2139, 1967.
5. Butlin, H. T., Cancer of the scrotum in chimney sweeps and others, Lecture II/III, *Br. Med. J.,* 2(1), 66, 1892.
6. Carnow, B. W. and Meier, P., Air pollution and pulmonary cancer, *Arch. Environ. Health,* 27, 207, 1973.
7. Dean, G., Lung cancer among white South Africans, *Br. Med. J.,* 2, 852, 1959.
8. Dean, G., The comparative mortality between the South African born white population and white immigrants lung cancer, multiple sclerosis and other diseases *Bull. Inst. Int. Stat.,* 131, 1, 1961.
9. Dean, G., Lung cancer in South Africans and British immigrants, *Proc. R. Soc. Med.,* 57, 984, 1964.
10. Dean, G., Lung cancer and bronchitis in Northern Ireland, 1960-2, *Br. Med. J.,* 1, 1506, 1966.
11. Doll, R., The causes of death among gas-workers with special reference to cancer of the lung, *Br. J. Ind. Med.,* 9, 180, 1952.
12. Doll, R., Vessey, M. P., Breasley, R. W. R., Bruckley, A. R., Fear, E. C., Fisher, R. E. W., Gammon, E. J., Gunn, W., Hughes, G. O., Lee, K., and Norman-Smith, B., Mortality of gas workers — final report of a prospective study, *Br. J. Ind. Med.,* 29, 394, 1972.
13. Eastcott, D. F., The epidemiology of lung cancer in New Zealand, *Lancet,* 1, 37, 1956.
14. Falk, H. L., Kotin, P., and Mehler, A., Polycyclic hydrocarbons as carcinogens for man, *Arch. Environ. Health,* 8, 721, 1964.
15. Gandjean, E., Epidemiologie der Luftverunreinigungen, *Schweiz. Med. Wochenschr.,* 102, 1889, 1972.
16. Grimmer, G., Untersuchungen über die carcinogene Belastung des Menschen durch Luftverunreinigung (Vortragsauszug). Wissenschaftliche Fachtagung der Universität Kaiserslautern, West Germany, June 2—4, 1976.
17. Hagstrom, R. M., Sprague, H. A., and Landau, E., The Nashville Air Pollution Study. VII. Mortality from cancer in relation to air pollution, *Arch. Environ. Health,* 15, 237, 1967.
18. Hammond, E. C., Smoking in relation to mortality and morbidity: findings in first thirty-four months of follow-up in a prospective study started in 1959, *J. Natl. Cancer Inst.,* 32, 1161, 1964.
19. Hammond, E. C., Smoking in relation to the death rates of one million men and women, *Natl. Cancer Inst. Monogr.,* No. 19, 127, 1966.
20. Hammond, E. C., Selikoff, I. J., Jawther, P. L., and Seidman, H., Inhalation of benzpyrene and cancer in man, *Ann. N.Y. Acad. Sci.,* 271, 116, 1976.
21. Hammond, E. C., Garfinkel, L., Seidman, H., and Lew, E. A., "Tar" and nicotine content of cigarette smoke in relation to death rates, *Environ. Res.,* 12, 263, 1976.
22. Hammond, E. C. and Garfinkel, L., General air pollution and cancer in the United States, *Prev. Med.,* 9, 206, 1980.
23. Higgins, I. T. T., Trends in respiratory cancer mortality in the United States and in England and Wales, *Arch. Environ. Health,* 28, 121, 1974.

24. Hoffmann, D. and Wynder, E. L., Environmental respiratory carcinogenesis, *ACS Monogr.*, 173, 324, 1976.
25. Koller, S., Einführung in die Methoden der ätiologischen Forschung — Statistik und Dokumentation, *Method. Inform. Med.*, 2, 1, 1963.
26. Kotin, P. and Falk, H. L., Air pollution and lung cancer, *Proc. Natl. Conf. Air Pollut.*, 140, 1963.
27. Kotin, P. and Falk, H. L., Polluted urban air and related environmental factors in the pathogenesis of pulmonary cancer, *Dis. Chest*, 45, 236, 1964.
28. Kotin, P., Environmental cancer, *Am. Ind. Hyg. Assoc. J.*, 27, 115, 1966.
29. Lawther, P. J., Bronchial carcinoma air pollution, smoking and lung cancer, *Trans. Med. Soc. London*, 81, 158, 1965.
30. Lerer, T. J., Redmond, C. K., Breslin, P. P., Salvin, L., and Rush, H. W., Long-term mortality study of steelworkers. VII. Mortality pattern among crane operators, *J. Occup. Med.*, 16, 608, 1974.
31. Levin, M. L., Haenszell, W., Carroll, B. E., Gerhardt, P. R., Handy, V. H., and Ingraham, S. C., Cancer incidence in urban and rural areas of New York state, *J. Natl. Cancer Inst.*, 24, 1243, 1960.
32. Lloyd, J. W. and Ciocco, A., Long-term mortality study of steelworkers. I. Methodology, *J. Occup. Med.*, 11, 299, 1969.
33. Lloyd, J. W., Lundin, F. E., Redmond, C. K., and Geiser, P. B., Long-term mortality study of steelworkers. IV. Mortality by work area, *J. Occup. Med.*, 12, 151, 1970.
34. Lloyd, J. W., Long-term mortality study of steelworkers. V. Respiratory cancer in coke plant workers, *J. Occup. Med.*, 13, 53, 1971.
35. MacDonald, E. J., Air pollution, demography, cancer: Houston, Texas. Patterns of mortality in 15 regions in Houston suggest a relationship to industrial pollution, *J. Am. Med. Women's Assoc.*, 31, 379, 1976.
36. MacMahon, B., Pugh, T. F., and Ipsen, J., *Epidemiologic Methods*, Little, Brown, Boston, 1960.
37. Mancuso, T. F., MacFarlane, E. M., and Porterfield, J. D., The distribution of cancer mortality in Ohio, *Am. J. Publ. Health*, 45, 58, 1955.
38. Mancuso, T. F., Cancer mortality among native white, foreign-born white, and non-white male residents of Ohio: cancer of the lung, larynx, bladder, and central nervous system, *J. Natl. Cancer Inst.*, 20, 79, 1958.
39. Pflanz, M., *Allgemeine Epidemiologie*, Thieme Verlag, Stuttgart, 1973.
40. Pott, P., *Chirurgical Observations Relative to the Cataract, the Polypos of the Nose, the Cancer of the Scrotum, the Different Kinds of Ruptures and the Mortification of the Toes and Feet*, Hawes, Clarke & Collings, London, 1775.
41. Redmond, C. K., Smith, E. M., Lloyd, J. W., and Rush, H. W., Long-term mortality study of steelworkers. III. Follow-up, *J. Occup. Med.*, 11, 513, 1969.
42. Redmond, C. K., Ciocco, A., Lloyd, J. W., and Rush, H. W., Long-term mortality study of steelworkers. VI. Mortality from malignant neoplasms among coke oven workers, *J. Occup. Med.*, 14, 621, 1972.
43. Redmond, C. K., Gustin, J., and Kamon, E., Long-term mortality experience of steelworkers. VIII. Mortality patterns of open hearth steelworkers, (A preliminary report), *J. Occup. Med.*, 17, 40, 1975.
44. Redmond, C. K., Reiber-Strobino, B., and Cypress, R. H., Cancer experience among coke by-product workers, *Ann. N.Y. Acad. Sci.*, 271, 102, 1976.
45. Reid, D. C., Cornfield, J., Markush, R. E., Seigel, D., Pedersen, E., and Haenszel, W., Studies of disease among migrants and native populations in Great Britain, Norway, and the United States. III. Prevalence of cardiorespiratory symptoms among migrants and native-born in the United States, *Natl. Cancer Inst. Monogr.*, No. 19, 321, 1966.
46. Robinson, H., Long-term mortality study of steelworkers. II. Mortality by level of income in whites and non-whites, *J. Occup. Med.*, 11, 411, 1969.
47. Sawicki, E., Elbert, W. C., Hauser, T. R., Fox, F. T., and Stanley, T. W., Benzo[a]pyrene content of the air of American communities, *Ind. Hyg. J.*, 21, 443, 1960.
48. Sawicki, E., Airborne carcinogens and allied compounds, *Arch. Environ. Health*, 14, 46, 1967.
49. Stocks, P. and Campbell, J. M., Lung cancer death rates among non-smokers and pipe and cigarette smokers: an evaluation in relation to air pollution by benzpyrene and other substances, *Br. Med. J.*, 2, 923, 1955.
50. Stocks, P., Air pollution and cancer mortality in Liverpool hospital region and North Wales, *Int. J. Air Pollut.*, 1, 1, 1958.
51. Stocks, P., Recent epidemiological studies of lung cancer mortality, cigarette smoking and air pollution, with discussion of a new hypothesis of causation, *Br. J. Cancer*, 20, 595, 1966.
52. Stocks, P., Lung cancer and bronchitis in relation to cigarette smoking and fuel consumption in twenty countries, *Br. J. Prev. Soc. Med.*, 21, 181, 1967.
53. Volkmann, R., Über Teer-, Paraffin- und Russkrebs, *Beiträge zur Chirurgie*, Breitkopf & Härtel, Leipzig, 1875, 370.
54. Winkelstein, W., Kantor, S., Davies, E. W., Maneri, C. S., and Mosher, W. E., The relationship of air pollution and economic status to total mortality and selected respiratory system mortality in men. I. Suspended particulates, *Arch. Environ. Health*, 14, 162, 1967.

Chapter 7

# EXTRAPOLATION OF EXPERIMENTAL RESULTS TO MAN*

## D. Schmähl and M. Habs

Since the beginning of scientific medical research, animal experiments have played a decisive role in detecting and treating diseases. In the past there were numerous spectacular successes in medical research which became possible only by using an animal experiment. The infection of guinea pigs with tubercle bacilli and subsequent treatment of the disease with "tuberculin" formed the basis for successful detection and treatment of tuberculosis. The development of antidiphtheritic serum by Emil von Behring was possible only by the immunization of wethers. Mention should also be made of Landsteiner's detection of the blood groups and the importance of rhesus monkeys in the blood typing of subgroups. In recent times, mention must be made of experiments in rats. These were designed for the induction of a disease pattern similar to Morbus Parkinson, by means of reserpine, and for the extrapolation of the experimental results to man, since these experiments initiated the detection and investigation of dopamine. Goats played a decisive role in the immune diagnostics of cardiac infarcts. We have mentioned these five examples, which may be supplemented at will, because they show that special problems demand special animal species for their solution.

Since experimental cancer research began to establish itself world-wide about 100 years ago, it has been a matter of course that here, too, use has been made of animal experiments (for review literature see D. Schmähl,[25]). Meanwhile, there is hardly any class of vertebrates that has not been used to solve special problems in cancer research. Thus, there exist investigations not only on cold-blooded animals (fish, amphibians, or frogs), on birds, on mammals, including monkeys, but also on insects and worms.

When summarizing the results obtained with various carcinogens in different animal species, it may be said, with certain reservations, that in the majority of cases the carcinogenic activity of certain substances could be proved in many animal species. Quantitative differences and differences relative to the organotropism of the carcinogenic activity have of course been encountered.[29] In some rare cases certain animal species proved to be almost resistant to the carcinogenic activity of a substance, whereas others proved to be highly sensitive.[36] With one exception (arsenic trioxide) all substances, which have definitely been shown to be carcinogenic in man, have also proved to be carcinogenic in various animal species. Therefore, we have to start from the realistic assumption that — until the contrary is proven — those substances, which so far have been identified as carcinogens in animal experiments, also have to be treated as potential carcinogens in man. In this connection the only compounds that should be considered are those on which thorough studies exist. The studies should possibly include several animal species and permit the establishment of dose-response relationships; also, the doses administered should not excessively exceed the dose needed to induce cancer in the human situation.[27]

Quite often it was the results of the animal experiments that permitted the prediction of the carcinogenic activity of a certain compound, which later proved to be carcinogenic in man (e.g. vinyl chloride, haloether, cyclophosphamide, etc.). Naturally the converse also occurred, and it was the initial observations in man that raised the suspicion of carcinogenic activity, which was later verified by animal experiments. Thus, experimental and clinical medicine have complemented each other excellently in de-

---

* See also Boyland[5] and Zapp.[39]

tecting the carcinogenic activities of chemical substances. From the historical point of view this is one of the earliest examples of multidisciplinary collaboration in cancer research.

On the basis of the facts just described[4,25] it is hard to understand why today the animal experiment is still criticized in public by certain groups who doubt its relevance as a predictive model for the detection of carcinogenic activities; it is all the more difficult to understand since there is no alternative to animal experiments. Efforts to find predictive models by means of in vitro systems and so-called short-term tests, e.g., with bacteria[1-3,20,34] have only just been started and are still in need of further thorough development in order to become reliable and conclusive methods for prescreening carcinogenic compounds.

When discussing the effects of polycyclic aromatic hydrocarbons and their significance as carcinogens in man, we ought to consider the situation in oncology that is historically unique. It is these effects that in fact initiated the development of occupational medicine, preventive medicine, and the chemical analysis of these very hydrocarbons. In 1775 the English physician Percivall Pott[22] described for the first time the carcinogenic effect of soot and coal tar in chimney sweeps. This job was generally performed by small underweight boys who had to force themselves into the narrow chimneys for sweeping. Thus their clothes and skin came constantly into contact with coal tar. As a consequence of this chronic impact of coal tar, these chimney sweeps developed papillomas on the basis of dermatitis, corium fibrosis, and hyperkeratosis. In some cases the papillomas deteriorated into malignant tumors. Preferred localizations were the flexure of the groin and the scrotum. These first observations on the carcinogenic effect of coal tar by Pott were confirmed and extended at the end of last century by the German surgeon von Volkmann,[35] who noticed the frequent occurrence of skin papillomas and carcinomas in coal tar workers in the lignite mining area of Halle, East Germany. Between 1922 and 1933, English groups working with Ernest Kennaway[17] identified polycyclic aromatic hydrocarbons in coal tar and discovered in animal experiments, predominantly with mice, that these PAH carry the carcinogenic activity of coal tar. It could be shown that these hydrocarbons account for up to 3% of the tar depending on its type and composition.

A further example of the occupational occurrence of cancer is bronchial carcinoma. It is induced by carcinogenic polycyclic aromatic hydrocarbons, thus explaining the increased occurrence of this disease in gasworkers and coking workers [10] (cf. Chapter 6). Prior to these analytical and experimental investigations concerning pure hydrocarbons obtained from tar, the Japanese cancer research workers Yamagiwa and Ichikawa[38] had proved the carcinogenic activity of coal tar by painting rabbit ears. Thus the picture arises that in the case of coal tar there were at first observations in man before analytical chemists could finally isolate hydrocarbons from coal tar and prove that some of them cause carcinogenic activity. This is a further historical example of multidisciplinary cancer research.

Beside these observations in occupational medicine there were also "experiments with human beings" serving to prove the carcinogenic activity of benzo[a]pyrene in human skin. Cottini and Mazzone[7] painted healthy and affected human skin with a 1% benzene solution of BaP daily for 4 months. The treated skin areas developed erythemas, pigmentations, desquamations of the skin, verrucae, and finally infiltrations of these pathological changes into the adjacent tissue. After discontinuation of the BaP treatment the pathological changes disappeared within less than 1 month and thus proved to be reversible. It is, however, not known what finally became of the treated individuals and whether they possibly later developed tumors at the site of application. Nevertheless, the authors referred to the observed alterations as prototypes of a precancerous process. This "experiment" seems to prove that the alterations observed in animal experiments follow similar phases as in man and are principally applicable to man, too.

Apart from the experiment just described in man there is an autoobservation by the physician Klar.[18] This investigator worked with a 0.5% solution of benzo[a]pyrene in benzene in order to induce skin tumors in mice. Although Klar, according to his own statement, took sufficient precautionary measures for self-protection, a pea-sized node developed on the dorsal side of his left forearm 3 months after termination of the experiment. This node was localized intracutaneously and could easily be moved on the surface. The histological examination proved the node to be a benign calcifying epithelioma. Although in this case the connection between exposure to benzo[a]pyrene and development of the epithelioma does not seem to be proved, the possibility still has to be considered.

According to known epidemiological, medical, and experimental data obtained in man and animals, there remains hardly any justified doubt that polycyclic aromatic hydrocarbons may be papilloma-inducing and even carcinogenic in human skin.

The only polycyclic aromatic hydrocarbon which, to our knowledge, was applied to human skin as a pure substance, i.e., benzo[a]pyrene, and all PAH-containing condensates exhibit carcinogenic activity at the site of application in man as well as in experimental animals. It is not possible to assess the doses needed to induce skin cancer in man, but doubtless the measures of industrial hygiene taken in former times[6] soon succeeded in minimizing the hazards arising from PAH-containing tars for human skin.

For ethical reasons, the necessary data on humans have to be missing. Therefore it is not possible to state whether the results obtained in skin painting tests with mice can be extrapolated to human skin as far as the order of carcinogenic potency of PAH is concerned (see also Chapters 5.2.2.2 and 5.2.2.3 concerning the problems involved in establishing such a classification). Here it is explained that the relative potency of individual PAH compared with one another often varies depending on the dose applied.

With regard to their type of action, PAH belong to the locally active carcinogens, i.e., they induce cancer predominantly at the site of application. The question arises whether PAH also act as carcinogens when they are not applied to the skin but are inhaled or taken in orally. In this case not only a local carcinogenic effect in the respiratory tract or the intestinal tract might be expected, but also a systemic resorptive effect in distant organs after distribution via blood circulation, e.g., in the kidney, brain, or liver.

It had already been observed many years ago that in countries in which people ate a lot of smoked food (e.g., Iceland, Finland), an increased frequency of stomach cancer was registered.[15] It was suggested that it was possible to correlate this phenomenon with the considerable amounts of PAH occurring in smoked food. In man, carcinoma of the colon is also supposed to be induced by PAH.[24] However, experimental investigations (e.g., with rats that were intensively fed a smoked diet for their lifetime) yielded negative results.[28] Furthermore, in other studies (cf. Chapter 5.2.2.4), clear carcinogenic activity of orally applied PAH could be observed only if high doses were administered. The relevance of the solvents used and the protective effect of the gastric mucosa are dealt with in detail in Chapter 5.2.2.4. We can at present only speculate on the role of PAH as pure substances and as PAH-containing products (mixtures) in the induction of tumors in the stomach and intestinal tract of man. There are no epidemiological studies which convincingly support the suspicion derived from individual animal experiments as to a causal relationship between the activity of PAH contained in food and tumor incidence in the stomach and intestinal tract.

There are, however, some indications of a systemic effect by PAH which were obtained in a specially designed experiment. If these substances are applied at certain times during the animal's lifetime (e.g., during day 45 and 55 in Sprague-Dawley rats), carcinomas of the mammary gland and leukemias are increasingly observed.[13,14]

As stated in Chapter 5.2.2.4, systemic carcinogenic effects, including tumors of the mammary gland, ovary, lung, kidney, and liver,[37] have repeatedly been described after various modes of application (oral, topical, subcutaneous, intravenous, and intraperitoneal) of PAH as pure substances in various animal species (mouse, rat, golden hamster). At present we still do not know to what extent the systemic action of PAH play a role in the induction of tumors in man. The necessary epidemiological data on this subject are missing, and it seems unlikely that in the next few years data will be published which will exhaustively deal with this complex of problems and permit a risk assessment for humans (cf. Chapter 6).

Another essential question is whether polycyclic aromatic hydrocarbons can be inhalation carcinogens in man. This question is of the utmost importance not only for an analysis of the carcinogenic activity of tobacco smoke, but also for the complex problem of air pollution and its possible cancer-inducing activity in the respiratory tract. Pertaining to tobacco smoke carcinogenesis, there is no doubt that, as far as carcinomas of the buccal cavity and squamous cell carcinomas of the bronchi are concerned, a local carcinogenic activity of tobacco smoke does exist. In particular, the latter type of tumor develops preferentially in individuals who inhale tobacco smoke, i.e., a direct contact of the bronchial mucosa with the inhaled smoke is necessary. Smokers who normally just "puff", but do not inhale the smoke, have to be rated similar to nonsmokers in respect to their risk of developing a bronchial carcinoma.[26]

As explained in detail in Chapter 6 there is a dose dependency between the amount of tobacco smoke inhaled and the risk of developing lung cancer in man. A reduction of the noxious dose decreases the cancer hazard. According to epidemiological investigations other PAH-containing pollutants have so far not shown a similar tendency (cf. Chapter 6).

Tobacco smoke contains carcinogenic polycyclic aromatic hydrocarbons. For instance, the content of benzo[a]pyrene in the smoke of one cigarette lies between 5 and 30 ng. If tobacco smoke condensate is applied to the skin of mice or injected into the subcutis of rats, these mice and rats develop skin carcinomas and fibrosarcomas, respectively, at the site of application. Consequently the substances contained in tobacco smoke or tobacco smoke condensate, which are responsible for this local effect, have to be substances of a locally active nature. Primary consideration has to be attributed to polycyclic aromatic hydrocarbons, of which up to now more than 15 have been shown to exhibit carcinogenic activity in animal experiments (Chapter 5.2). An analysis of the carcinogenic activity of tobacco smoke with respect to its individual components is, of course, not feasible in man; here we have to rely on animal experiments. The required animal studies have shown that the local carcinogenic activity of tobacco smoke condensates and also of automobile exhaust condensates or "smoke soot" can essentially be attributed to the polycyclic aromatic hydrocarbons contained in these materials[9,32] (cf. Chapter 3).

On the basis of existing clinical, epidemiological, and experimental observations, we may reason that, by analogy, the carcinogenic activity of tobacco smoke is essentially effected by the PAH contained therein. Thus, today, attempts to "deactivate" tobacco smoke in respect of its carcinogenic activity are aimed at a reduction of the hydrocarbon-containing "tobacco tar" of tobacco smoke.

What has been said about tobacco smoke is similarly true of carcinogenic substances in air. When considering the hydrocarbons contained in air, we have to accept that they are emitted into the air by heating installations, industrial plants, motor traffic, etc.[23] (see Chapter 3 for details). Accordingly, their concentration is higher in urban than in rural regions. When comparing bronchial cancer risks in urban and rural areas — with the proviso that the smoking habits of the population are similar — we see that the risk is higher in urban than in rural regions[8] (cf. Chapter 6). In summarizing the data of studies in which urban and rural areas are compared, a difference in the lung cancer risk of urban and rural areas becomes evident. At present the correlation with data obtained by measuring the pollutants is not satisfactory because only few data exist and most measurements were carried out at the same time as the epidemiological investigations and, therefore, do not reflect the degree of air pollution during the relevant period of exposure.

The "urbanization" factor has often been a subject of discussion because until today it has not been possible to give a clear definition of its pathognomonic relevance. Thus in towns the medical service may be better, people may go to physicians more often than in rural areas, or the tolerance or intolerance of the population to disease symptoms may vary between urban and rural population. Furthermore, it has to be accepted that, without a doubt, the air pollution in congested urban areas is more intense than in rural regions. Consequently urban residents are subjected to cancer-inducing PAH to a larger extent than people living in rural areas. Beside the inhalation of PAH contained in tobacco smoke, we here have the burden of additional inhalation of hydrocarbons from the air. This additional inhalation of hydrocarbons starts at birth and may well be considered a syncarcinogenic interaction.[4,25]

Working place studies were carried out, especially in the field of tar and coke production, in which the lung cancer risk of occupational groups subjected to larger amounts of products of incomplete combustion was investigated. These studies show a clear-cut correlation between the exposure at the working place and the risk of developing a tumor of the respiratory tract. There exist also some data on occupationally induced systemic tumors of the urinary bladder, kidney, and testis, as well as local tumors of the skin. Again the group of PAH could be detected as the most likely, relevant pollutants (cf. Chapter 6).

The decreasing frequency of bronchial cancer observed nowadays in England[12] is attributed to the fact that after the enactment of the Clear Air Act in 1956 the air in urban areas has become cleaner and thus contributed to the decrease in the frequency of bronchial cancer. These encouraging observations should induce us to ensure that in future great importance will be attached to maintaining a minimal PAH content of our air.

In a strictly scientific sense there is so far no conclusive evidence of the carcinogenic activity of PAH in the respiratory tract. Such an evidence could be obtained only if humans were used as "guinea pigs". The interlinking of results described in the previous paragraphs, however, permits the formation of the well-founded conclusion that PAH play a decisive role in the carcinogenesis in the respiratory tract of man. To date, there is no better explanation for the development of most tumors in the respiratory tract of man than to postulate that carcinogenic PAH cause this type of tumor. We have to rely on this supposition until a better possible explanation can be found.

Returning to the initial problem, i.e., the possibility of extrapolating experimental findings to the human situation, we can make the following statements:

1. The extrapolation of results obtained in animal experiments to human circumstances is possible and permissible in a qualitative respect. However, gaps still exist as to the dose-response relationship and the relative potency of individual

carcinogens in comparison with each other, i.e., in a quantitative respect.[27] The better the mode of exposure is adapted to the human situation (inhalation studies), the more valid the results will be. At the moment, it is not possible to decide which experimental animal model is best suited for predicting the effects of PAH, PAH mixtures, and PAH-containing condensates of exhaust gases or smoke gases. The advantages and disadvantages of the different models have been discussed in Chapter 5.2. On the other hand, financial reasons only permit the carrying out of intensive studies, e.g., analyses of the balance of activities (Chapter 7), in rather simple models (e.g., skin-painting test). As previously explained in Chapter 5.2.2.2, the share of individual mixtures in the total effect can be shown by establishing comparative dose-response relationships, e.g., of cigarette smoke condensate or automobile exhaust condensate and their fractions. Thus the relative potency of the carcinogenic effect can be compared. At present, the lack of experimental data does not permit confirmation of the supposition that condensates of so-called "light" cigarettes have a lower carcinogenic activity than so-called "strong" cigarettes. It is generally assumed that the relative potency of PAH, PAH-containing mixtures and condensates in comparison with each other — as shown in model animal experiments — may be extrapolated to the human situation. Since, however, a classification of PAH according to their potency leads to differences even within one animal model, depending on the dose applied, and, additionally, the potency varies from animal model to animal model, risk assessments still involve a great number of uncertainties. PAH with evident carcinogenic activity, e.g., benzo[a]pyrene, yield positive results in all models; obviously, noncarcinogenic compounds, such as e.g., pyrene, do not show any tumorigenic effects in the animal models used. In the years to come, it seems to be most important to establish clear dose-response relationships in the individual models and to investigate to what extent quantitative extrapolations of experimental results are possible. The quantitative risk assessment for man on the basis of animal experiments cannot be attempted successfully unless considerably more quantitative environmental data and epidemiological studies are compiled. As explained in Chapter 6, dose-response assessments of the bronchial cancer risk exist for humans inhaling cigarette smoke. In animal experiments clear dose-time-response relationships could be established for the carcinogenic activity of tobacco smoke condensates. In future, risk assessments should be made on the basis of quantitative environmental data and experimental findings, and the preventive elimination of noxious substances from the environment should be initiated before epidemiological data prove that the substances have already caused serious damage to health.

The relevance of so-called short-term tests for risk assessments in man is very doubtful. Limited qualitative conclusions predicting a carcinogenic effect in animal experiments can be drawn, but false-positive and, in particular, false-negative results warn against an overestimation of these tests as screening models (cf. Chapter 5.2.1). At present they do not permit the prediction of the relative potency.[21]

Investigations on the significance of metabolism on the carcinogenic action of PAH are still in their beginning. Not only are differences between the tissues of different animal species and between the tissues of animals and human beings known, but also cell-specific differences within one animal species.[19] Here, too, in most cases, the difference is probably more quantitative than basic. At present we can only speculate on the relevance of this field of investigation, e.g., for the selection of risk groups with an increased predisposition to carcinogenesis by certain noxious substances, e.g., PAH, due to altered pathways of metabolism.

2. The carcinogenesis induced by PAH on the skin of experimental animals obviously follows a similar pattern as in man. The carcinogenic effect of PAH on human skin has to be considered a proved fact.
3. So far it cannot yet be determined whether polycyclic aromatic hydrocarbons exert a carcinogenic action after oral application in man. The experimental findings have not yet produced a corresponding correlation of clinical and epidemiological investigations in man.
4. In all probability, PAH play a decisive role as inhalation carcinogens in man. The likelihood of this statement can be assumed from experiences in tobacco smoke carcinogenesis, as well as from investigations on the risk of inducing bronchial cancer by air pollution.

For a further elucidation of the possibility of extrapolating experimental findings in animals to the human situation, in particular with regard to the carcinogenic activity of hydrocarbons, it seems to be essential to develop experimental models in which the peculiarities of human behavior (e.g., smoking habits, oral or nasal respiration) can be correspondingly simulated.

It can generally be assumed that the higher the validity of experimental studies in animals is, the better the conditions of human exposure are simulated.[11,30] Consequently, inhalation studies with animals are required primarily. However, the final value of these investigations cannot yet be assessed since, in particular, the special physiological characteristics cannot be simulated. The results described in Chapter 5.2.2.1 and the difficulties of their interpretation arise partly from the fact that only few experiments using satisfactory methods have so far been carried out, and an optimization of inhalation studies seems possible.

The inhalation study carried out with benzo[a]pyrene as a pure substance (cf. Chapter 5.2.2.1) may also be interpreted as that additional noxious effects, e.g., inflammatory lesions, are necessary to induce carcinomas in the lung by PAH.

Large-scale experimental and epidemiological investigations should be carried out in the near future having special regard to emitters and emissions. Such investigations seem most promising for converting their results into practical measures of preventive medicine, such as a reduction of emission and thus of hazards arising from emissions, or the protection at the working place. Furthermore, they seem to be necessary for a determination of future scientifically founded standards since the present results and conclusions are not satisfying.[16,33]

## REFERENCES

1. Ames, B. N., Lee, F. D., and Durston, W. E., An improved bacterial test system for the detection and classification of mutagens and carcinogens, *Proc. Natl. Acad. Sci. U.S.A.*, 70, 782, 1973.
2. Ames, B. N., Durston, W. E., Yamasaki, E., and Lee, F. D., Carcinogens are mutagens: a simple test system combining liver homogenates for activation and bacteria for detection, *Proc. Natl. Acad. Sci. U.S.A.*, 70, 2281, 1973.
3. Bartsch, H., Malaveille, C., Camus, A.-M., Brun, G., and Hautefeuille, A., Validity of bacterial short-term tests for the detection of chemical carcinogens, in *Short Term Test Systems for Detecting Carcinogens*, Norpoth, K. H. and Garner, R. C., Eds., Springer Verlag, New York, 1980, 58.
4. Bauer, K.-H., *Das Krebsproblem*, Springer Verlag, Berlin, 1960.
5. Boyland, E., The correlation of experimental carcinogenesis and cancer in man, *Progr. Exp. Tumor Res.*, 11, 222, 1969.

6. Butlin, H. T., Cancer of the scrotum in chimney-sweeps and others. II. Why foreign sweeps do not suffer from scrotal cancer, *Br. Med. J.*, 2, 1, 1892.
7. Cottini, G. B. and Mazzone, G. B., The effects of 3,4-benzpyrene on human skin, *Am. J. Cancer*, 37, 186, 1939.
8. Dean, G., Lung cancer and bronchitis in Northern Ireland 1960-2, *Br. Med. J.*, 1, 1506, 1966.
9. Dontenwill, W., Chevalier, H. J., Harke, H. P., Klimisch, H. J., Brune, H., Fleischmann, B., and Keller, W., Experimentelle Untersuchungen über die tumorerzeugende Wirkung von Zigarettenrauch-Kondensaten an der Mäusehaut. VI. Mitteilung: Untersuchungen zur Fraktionierung von Zigarettenrauch-Kondensat, *Z. Krebsforsch.*, 85, 155, 1976.
10. Habs, M., Bronchialkarzinom bei Gaswerks- und Kokerei-Arbeitern, *Deutsch. Med. Wochenschr.*, 45, 1662, 1976.
11. Habs, M., Cancer prevention: responsibilities and methods of the industry, in *Advances in Medical Oncology, Research and Education*, Smith, A. and Alvarez, C. A., Eds., Pergamon Press, New York, 1979, 91.
12. Higgins, I. T. T., Trends in respiratory cancer mortality in the United Sates and in England and Wales, *Arch. Environ. Health*, 28, 121, 1974.
13. Huggins, Ch., Grand, L., and Oka, H., Hundred day leukemia: preferential induction in rat by pulse-doses of 7,12-dimethylbenzanthracene, *Exp. Med.*, 131, 321, 1970.
14. Huggins, Ch., Briziavelli, G., and Sutton, H., Rapid induction of mammary carcinoma in the rat and the influence of hormones on the tumor, *J. Exp. Med.*, 109, 25, 1959.
15. IARC-Scientific Publication, Cancer incidences in five continents, *IARC Sci. Publ.*, No. 15, 3, 1966.
16. Janyschewa, J., Zur Begründung der maximal zulässigen Konzentration von Benz[a]pyren in der atmophärischen Luft von Wohngebieten, *Gig. Sanit.*, 1, 87, 1972.
17. Kennaway, E. L., Experiments on cancer-producing substances, *Br. Med. J.*, 2, 1, 1925.
18. Klar, E., Über die Entstehung eines Epitheliome beim Menschen nach experimentellen Arbeiten mit Benzpyren, *Klin. Wochenschr.*, 17, 1279, 1938.
19. Langenbach, R., Freed, H. J., Rauch, D., and Huberman, E., Cell specificity in metabolic activation of aflatoxin B and benzo[a]pyrene to mutagens for mammalian cells, *Nature (London)*, 276, 277, 1978.
20. Marquardt, H., Malignant transformation in vitro: a model system to study mechanisms of action of chemical carcinogens and to evaluate the oncogenic potential of environmental chemicals, *IARC Sci. Publ.*, No. 12, 389, 1976.
21. Maugh, T. H., Chemical carcinogens: the scientific basis for regulation, *Science*, 201, 1200, 1978.
22. Pott, P., *Chirurgical Observations Relative to the Cataract, the Polypos of the Nose, the Cancer of the Scrotum, the Different Kinds of Ruptures and the Mortification of the Toes and Feet*, Hawes, Clarke & Collings, London, 1775.
23. Preussmann, R., Chemische Carcinogene in der menschlichen Umwelt, in *Handbuch der allgemeinen Pathologie* Vol. 2, Altmann, H.-W. et al., Eds., Springer Verlag, Berlin, 1975, 421.
24. Renwick, A. G. and Drasar, B. S., Environmental carcinogens and large bowel cancer, *Nature (London)*, 263, 234, 1976.
25. Schmähl, D., Ed., *Entstehung, Wachstum und Chemotherapie maligner Tumoren*, Editio Cantor KG, Aulendorf, West Germany, 1981.
26. Schmähl, D., Risikofaktoren beim Bronchuskarzinom, *Deutsch. Med. Wochensch.*, 17, 965, 1975.
27. Schmähl, D., Problems of dose-response studies in chemical carcinogenesis with special reference to N-nitroso compounds, in *Crit. Rev. Toxicol.*, 6(3), 257, 1979.
28. Schmähl, D. and Reiter, A., Prüfung von Räuchereiprodukten auf krebserzeugende Wirkung, *Z. Krebsforsch.*, 59, 397, 1953.
29. Schmähl, D., Thomas, C., and Scheld, G., Karzinogene Wirkung von Äthyl-butyl-nitrosamin bei Mäusen, *Naturwissenschaften*, 50, 717, 1963.
30. Schmähl, D., and Habs, M., A critical review of the present state of the testing for carcinogenic potentials, in *The Evaluation of Toxicological Data for the Protection of Public Health*, Hunter, W. J. and Smeets, J. G., Eds., Pergamon Press, Oxford, 1977, 59.
31. Schmähl, D., Habs, M., and Ivankovic, S., Carcinogenesis of N-nitrosodiethylamine (DENA) in chickens and domestic cats, *Int. J. Cancer*, 22, 552, 1978.
32. Schmidt, K. G., Schmähl, D., Misfeld, J., and Timm, J., Experimentelle Untersuchungen zur Syncarcinogenese. VII. Mitteilung: Syncarcinogene Wirkung von polycyclischen aromatischen Kohlenwasserstoffen (PAH) im Epicutantest an der Mäusehaut, *Z. Krebsforsch.*, 87, 93, 1976.
33. Shabad, L. M., On the so-called MAC (maximal allowable concentrations) for carcinogenic hydrocarbons, *Neoplasma*, 22(5), 459, 1975.
34. Stich, H. F., Lam, P., Lo, L. W., Koropatnick, D. J., and San, R. H. C., The search for relevant short term bioassays for chemical carcinogenesis: the tribulation of a modern sisyphus, *Can. J. Genet. Cytol.*, 17, 471, 1975.

35. Volkmann, R. V., *Berl. Klin. Wochenschr.*, 11, 218, 1874; cited in **Schmähl, D.**, Entstehung, Wachstum und Chemotherapie maligner Tumoren, Editio Kantor KG, Aulendorf, West Germany, 1981.
36. **Warzok, R., Schneider, J., Thust, R., Scholtze, P., and Pötzsch, H. D.**, Zur transplazentaren Tumorinduktion durch N-Äthyl-N-nitrosoharnstoff bei verschiedenen Tierarten, *Zentralbl. Allg. Pathol. Pathol. Anat.*, 121, 54, 1977.
37. **White, F. R. and Eschenbrenner, A. B.**, Note on the occurrence of hepatomas in rats following the ingestion of 1,2-benzanthracene, *J. Natl. Cancer Inst.*, 6, 19, 1945.
38. **Yamagiwa, K. and Ichikawa, K.**, Experimentelle Studie über die Pathogenese der Epithelialgeschwülste, *Mitt. Med. Fak. Univ. Tokyo,* 15, 295, 1915.
39. **Zapp, J. A.**, Extrapolation of animal studies to the human situation, *J. Toxicol. Environ. Health,* 2, 1425, 1977.

Chapter 8

# CONCLUSIONS

S. Dobbertin

The chemical class of polycyclic aromatic hydrocarbons (PAH) comprises a group of several hundred compounds including noxious substances which are strongly suspected of inducing cancer in man. PAH are formed during incomplete combustion of organic material. They are found not only in the atmosphere, but also in water, soil, plants, and foodstuff.

PAH are emitted into the air by individual heating installations, industrial furnaces, and vehicles. Small heating units, especially one-room stoves fired with coal or fuel oil, for the most part have a low rate of efficiency and a high emission of PAH. The type of fuel as well as the type and maintenance of the heater have an essential influence on the emission of PAH. There are various types of industrial firing installations. Industrial installations used for the heat and energy production seem to account only for a relatively small proportion of the PAH emission. The proportion of PAH emitted into the atmosphere by motor vehicles also is comparatively small.

The present situation regarding air quality in the Federal Republic of Germany can be summarized as follows. Of the more than 100 PAH occurring in the atmosphere, so far chiefly benzo[a]pyrene has been analyzed. This is due partly to its easy measurability and partly to its strong carcinogenic activity, detected in animal experiments about 50 years ago. The benzo[a]pyrene concentration in the atmosphere is

1. In winter about 5 to 10 times higher than in summer
2. In urban areas about 5 to 20 times higher than in rural areas
3. In West Germany about 5 to 10 times higher than in large cities of the U.S. where coal-fired one-room stoves have never been used as much

The correlation between individual heating and emissions becomes clear by taking London as an example. There the benzo[a]pyrene emission has decreased since the Clear Air Act, among other things, prohibited the use of certain kinds of coal. Obviously the changing heating habits in West Germany have contributed to a considerable improvement of the emission situation in the last few years.

The emissions of PAH by different emitters differ not only in quantity, but also in the qualitative relations in which the individual PAH are detected in them (PAH profile). These occasionally very great differences seem to become balanced in the atmosphere. However, extensive series of measurements will still be necessary to determine to which PAH the analysis can be restricted without neglect of necessary data.

The PAH are taken in by inhalation and by ingestion of food. Since the resorption of PAH via the mucosa of stomach and intestinal tract seems to be rather low, the intake by respiration plays a decisive role for man. The PAH reach the respiratory tract together with the fine particulate matter of the air and can be isolated there from the particulate matter and metabolized. Calculations show that a man living in an atmosphere which contains 10 ng of benzo[a]pyrene per $cm^3$ inhales about 3 mg benzo[a]pyrene in the course of 60 years.

In animal systems, carcinogenic PAH predominantly exert a local activity, but need metabolization to become "ultimate carcinogens". Thus topical application to mouse skin mostly yields papillomas, although squamous cell carcinomas are also formed; whereas after subcutaneous injection sarcomas develop at the site of application. The instillation of benzo[a]pyrene into the trachea of Syrian golden hamsters induces papillomas, squamous cell carcinomas, and adenocarcinomas in the respiratory tract, predominantly in the trachea. After oral administration of benzo[a]pyrene to mice, chiefly papillomas and carcinomas of the forestomach are formed. The causal correlation between application of benzo[a]pyrene or other carcinogenic PAH and tumor frequency could be proved in various experimental models. Experimental animal investigations on the combination effect of PAH did not confirm the hypothesis of an additive effect even when mixtures of pure PAH were used. General statements on the activity of PAH-containing mixtures of substances cannot yet be made. It is noteworthy that further results show that higher tumor rates are obtained in the lung and the skin if the carcinogenic PAH are applied not in one single relatively high dose, but if the total dose is divided into many smaller doses. This result is important in that it does not require the establishment of a short-term value for protection against peak concentration of air pollutants when establishing air quality standards. From the present point of view, limitation of the average annual value alone seems to be reasonable and sufficient.

Dose-response relations, or more precisely, the relationship between dose and tumor incidence could be demonstrated in various animal models. However, significant residuals of the gradient of regression lines have been found for the dose-frequency relation of some carcinogenic PAH even within one animal model. This means that the carcinogenic potency of a polycyclic aromatic hydrocarbon can be stated only with special reference to a certain experimental model and a certain dose. Nevertheless, taking the presently available experimental data into consideration, benzo[a]pyrene and dibenz[a,h]anthracene can be classified as strong carcinogenic PAH.

A risk assessment of the carcinogenic hazards evoked by air-polluting PAH in man must rely on experimental animal investigations, as well as epidemiological findings. Experimental results permit the statement that the group of PAH includes the most important carcinogenic compounds of cigarette smoke condensate, automobile exhaust condensate, and extracts of air-suspended particulate matter. Although epidemiological data cannot prove such causal correlations, they may be used in support of risk assessments, in particular since some studies show a correlation between lung cancer frequency and the level of air pollution or smoking habits.

Investigations of lung cancer frequently at certain working places, as well as in urban areas in comparison with rural areas, do not permit the formation of adequate conclusions regarding the concentrations to which the cohorts were subjected. Only for the inhalation of tobacco smoke could dose-response relations be clearly proved in man. Thus the inhalation of smoke over years from only a few cigarettes per day is not without risk. Up to today it has not been possible, however, to compare the carcinogenic potency of cigarette smoke with the carcinogenic potency of a certain pattern of urban air pollution, e.g., by relating the tumor frequency after inhalation of a certain number of cigarettes per day to the effect of certain PAH concentrations in the atmosphere.

Summarizing, we may say that although the dependence of the incidence of lung cancer in man on the PAH concentration in the atmosphere has not been proved, results obtained in animal experiments and epidemiological findings support the strong suspicion of such a correlation. The hypothesis of a correlation between the inhaled amount of PAH and lung cancer frequency permits the conclusion that one of the causes of the higher lung cancer frequency in towns is the higher concentration of PAH. Consequently limitation of the PAH air quality to the level of less polluted areas would reduce the lung cancer frequency correspondingly. In following the principle of prophylactic prevention, it seems necessary to limit the emission of carcinogenic PAH.

Up to now only the Soviet Union has taken legal steps to limit the air quality of PAH on the basis of the supposition that today a standard for benzo[a]pyrene can be scientifically determined. Thus in 1972/73 the U.S.S.R. ministry of health established a "maximum tolerable concentration" of 1 ng of benzo[a]pyrene per $cm^3$ of air. This value is justified as follows: after intratracheal application of different doses of benzo[a]pyrene to rats, the dose that did not induce cancer was determined. This "non-carcinogenic" dose was 0.1 mg for the single application and 0.02 mg for ten applications. Bringing lung surface and body weight of rats and man into relation, "the non-hazardous dose for human lungs is 3.4 mg of benzo[a]pyrene as lifelong total dose". With reference to air, the concentration was not to exceed 1 $ng/cm^3$ of air including obviously a safety margin. Such a statement is neither conclusive nor sufficiently well founded and can therefore not be considered as a scientific basis for the determination of a standard.

Thus, since there exist no scientifically proved dose-response or dose-frequency relations for PAH in man, the limitation of the air quality of carcinogenic PAH can be justified only as a preventive measure for reducing the frequency of a particularly serious disease. Due to the missing dose-response relations, the level of a certain standard can be oriented only at its technical and economic feasibility. From this point of view and with respect to the level of benzo[a]pyrene concentrations in West European and North American cities, an "orientation mark" of 10 ng BaP per $cm^3$ air has been proposed as average annual value at the hearing of experts held on the 24th of February 1978 in Berlin jointly by the German Federal Ministry of the Interior and the Environmental Protection Agency. Such a preventive value can, however, only be transitory since the relevance of benzo[a]pyrene as an index of the carcinogenic potency of all PAH contained in the atmosphere has not yet been sufficiently elucidated. In comparison with the air quality situation in other European and North American cities, the proposed value lies within a feasible range.

Since, however, the PAH content of air is caused only to a minor extent by automobile exhaust gases, it seems doubtful whether an improved additional combustion of exhaust gases of Otto engines would considerably reduce the PAH content in the atmosphere. Only the constant control of heating units and the installation of central heating plants will also improve the situation of PAH emission in congested areas.

# INDEX

## A

Acridine derivatives, 7
Acrylamide, 65
Active carbon, 44
Acute toxic effects, 158
Acute toxicity, 159
Adenocarcinomas, 135, 179, 182, 248
Adenomas, 179
Adsorption, 45, 203
Adsorption-desorption equilibrium, 42
Aerodynamic diameter of PAH particles, 130—131
Airborne PAH, subcutaneous testing with, 193
Airborne particulate matter, 2, 44—45, 50—52, 54—57, 84—101
   carcinogenic potency, 90—93, 195—199
   foodstuffs contaminated with, 109, 111—112
   PAH content of particles of different size, 85
Air dusts, 194—196, 198—199
Air filter extracts, 164
Air pollutants, 203
Air pollution, 2, 43, 84, 204, 240—241, 248
Airport soil, 102
Air-suspended particulate matter, 248
Aliphatic/aromatic separation, 46
Aliphatic hydrocarbons, 48
Alkaline saponification, 42
Alumina, 46, 48
Alveolar cell carcinoma, 135
Alevolar epithelial cells, 134
Alevolar macrophages, 136
Ames test, 11, 163—169, 171, 177
6-Aminochrysene, 143
Aminopyrine, 138
Anaplastic bronchial carcinoma, 135
Anaplastic carcinomas, 182
Angina, 233
Animal carcinogenicity, correlation with bacterial mutagenicity, 164—165, 167—169
Animal experiments, see also Animal test models; Long-term experiments; Short-term tests, 160, 237
   extrapolation of test results of, 186, 237—245
Animal fats, 43
Animal species, 237
Animal test models, 3—5, 31, 39
   carcinogenic components of automobile exhaust, 7—11
   correlation of automobile exhaust and carcinoma, 5—6
   identification of carcinogenic components, 11—12
   implantation into lung of rats, 4—5
   instillation into trachea of hamsters, 4—5, 7, 11—12, 31, 248
   oral administration, 31
   subcutaneous application in mice, 4—5, 7, 31, 247
   topical application to skin of mice, 4—5, 7, 10—12, 14—15, 31, 247
Animal tissues, 44
Anthanthrene, 36, 47—48, 142
Anthracene, 36, 47—48
Anthropogenic wastes, 105
Antibiotics, resistance to, 163
Apples, 111
Arene epoxides, reaction with glutathione, 145—146
Arene oxide hydrolase, 139
Aromatic-aliphatic substances, 7—9
Aromatic amines, 2, 39
Aryl epoxidases, 170
Aryl epoxide hydrolase, conversion of epoxides into dihydrodiols by, 139—142
Atmospheric particulates, 194—195
Automobile emissions, see Vehicle exhaust; Vehicle exhaust condensates
Automobile exhaust, see Vehicle exhaust; Vehicle exhaust condensates
Automobile exhaust condensate (AEC), see Vehicle exhaust condensate
Automobile traffic, see also Vehicle exhaust, 18—20

## B

Bacteria as indicator, 177
Bacterial mutagenicity, correlation with animal carcinogenicity, 164—165, 167—169
Balance of activity, 5—6
BaP, see Benzo[a]pyrene
BaP-induced carcinogenesis, inhibition of, 200
Barley, 112
Base components, 31
Base-pair substitutions, 163
"Bay-region" dihydrodiol epoxides, 140—142
Bay-region epoxides, 147
Beans, 44
Behavior of PAH in the the organism, 129—156
   distribution of PAH, 130—136
   intake of PAH, 130—136
   metabolism of PAH, 137—151
   methods of detection of PAH metabolites, 148—151
*Beijerinckia*, 142
Benz[e]acephenanthrylene, 39
Benz[a]anthracene, 36, 47—48, 138, 147
Benz[a]anthracene phenols, 151
Benz[a]anthracene-7,12-quinone, 143
Benzocarbazoles, 105
Benzo[b]chrysene, 47—48
5,6-Benzoflavone, 138
7,8-Benzoflavone, 139, 146
Benzo[a]fluoranthene, 38—39
Benzo[b]fluoranthene, 36, 38—39, 48
Benzo[ghi]fluoranthene, 36
Benzo[j]fluoranthene, 36, 38, 48

Benzo[k]fluoranthene, 36, 38, 48, 158
Benzo[a]fluorene, 37
Benzo[b]fluorene, 37
Benzo[d]fluorene, 37
Benzo[rst]pentaphene, 37
Benzo[ghi]perylene, 36, 47—48
Benzo[a]pyrene (BaP), 6, 8—10, 36—38, 40, 47—48, 138, 143, 146, 148, 162, 203—204, 238
  biological effect in various matrices, 15
  carcinogenic activity in relation to total carcinogenicity, 15
  concentrations in air, 93, 95—101
  concentrations in city, 16, 18
  concentrations in urban areas, 93, 95—101
  concentrations in workplace, 100
  distribution pattern, 19
  formation, 30
  fuel emittants, 15—16
  K-region of epoxide of, 147
  maximum allowable amount in smoked meat products, 111
  metabolic pathways of, 148
  reaction with DNA, 151
  temperature of pyrolysis, 29
Benzo[a]pyrene emission, 69
Benzo[a]pyrene epoxides, 139
Benzo[a]pyrene-3-yl-hydrogen sulfate, 143
Benzo[a]pyrene 4,5-oxide, 145—146
Benzo[a]pyrene-1,6-, 3,6-, and 6,12-quinone, 143
Benzo[e]pyrene, 36—38, 47—48, 142
Biological activity, see also specific topics, 157— 219
  carcinogenicity, 160—205
    long-term experiments, 178—205
    short-term tests, 161—178
  site of application, 178, 186, 190, 203
  toxicity, 158—159
Biological effect
  BaP in various matrices, 15
  carcinogenic PAH in various matrices, 15
Biosynthesis, 30
Bladder cancer, 224, 232
Blast furnace coke, 28
Blastomogenic, 160
Blowing-off effect, 84—87
Borecole, 44
Brain tumors, 224
Bronchial cancer, 241
Bronchial carcinomas, 135, 238, 240
Bronchiogenic carcinomas, 2, 179
Bronchiogenic squamous cell carcinomas, 182, 184
Brown coal, 15, 21
Brownian movement, 130
Buccal cavity cancer, 224—225
Buccal cavity carcinoma, 240

## C

$C_1$ transfer, 143
Cabbage, 44

Cancer-inducing activity, 240
Cancer-inducing effect, 28
Cancerogenic, 160
Capillary gas chromatography, 41, 50—52, 114
Carbazole, 7, 105
Carbon black, 44
Carbon black factory, 102
Carbonization, 31, 113
Carbonization process, 28—30
Carcinogenesis, 160, 243
Carcinogen-free fractions, 6—7
Carcinogenic, 160, 239
Carcinogenic activity, 40—41, 114, 142, 160, 188
  automobile exhaust condensate, 7—12
  BaP in relation to total carcinogenicity, 15
  cigarette smoke condensate, 13—14
  coal tar, 238
  diesel exhaust condensate, 77
Carcinogenic components
  automobile exhaust condensate, see Vehicle exhaust condensate
  identification of, 11—12
  isolation of, 4
Carcinogenic effect, 40, 135, 188, 238, 240, 243
  chemical analytical index, 113—122
  cigarette smoke, 83—84
  inventory of, 40
Carcinogenic fraction, 6—7
Carcinogenicity, 3, 6, 40, 140, 160—205
  correlation with mutagenicity, 163—165
  extracts of air dust, 195—196, 198—199
  long-term experiments, see also Long-term experiments, 178—205
  PAH, 30—31, 192—203
  short-term tests, see also Short-term tests, 161—178
  subcutaneous test, 190—203
  tests for, see Long-term experiments; Short-term tests
Carcinogenicity index, 123
Carcinogenic potency
  airborne PAH, 90—93
  PAH, 239
Carcinogenic properties, 39—40
Carcinogen-specific detector, 6
Carcinogenic substances, 2—4, 39, 114, 240
Carcinomas, 11, 14, 135, 187, 190, 204—205, 238
  correlation with automobile exhaust, 6
  mammary gland, 204
Cardiovascular disease, 228—229
Carotinois, 30
Catechols, 143—144
Cell transformation test, 165—171, 177
  ability to form clones, 166
  ability to form local tumors, 166
  activating components, 170
  correlation of animal carcinogenicity with, 172
  criss-cross and piling-up growth, 166
  feeder cells, 166, 170

lack of contact inhibition, 166
metabolizing cells, 166
possibility of studying effects of test substances on human cells, 170
S-9 microsomal fraction as metabolizing component, 166
Cellulose, 29, 65
Cellulose acetate, 65
Chemical analysis methods, see also specific topics, 39—42
capillary gas chromatography, 41
electron capture detector, 41
enrichment, 42, 45—49
extraction, 42—44
flame ionization detector, 41
gas chromatography, 41
height equivalents of theoretical plates, 41
high-pressure liquid chromatography, 41
isolation of PAH mixture, 42, 45—49
low-voltage mass spectrometry, 41
separation, 42, 49—50
separation methods needed, 40—41
thin-layer chromatography, 41
Chemical analytical index of carcinogenic effects of emissions and environmental samples, 113—122
Chemistry of PAH, 27—60
Chimney sweeping, 3, 238
Chlordane, 138
Choice of relevant PAH, 28—31
criterion of, 30—31
mechanisms of formation, 28—29
number of PAH detected in environment, 30
Cholanthrene-3-carboxylic acid, 145
Chromatographic methods, 46—49
Chromatographic separation of PAH mixtures, 49—50
Chronic fistulas, 161
Chronic toxic testing, 39—40
Chrysene, 36, 47—48, 142
Cigarettes, see also Cigarette smoke and smoking; Cigarette smoke condensate
main stream smoke, 77, 79, 115
side stream smoke, 77—78, 84
tobacco brand, 77
Cigarette smoke and smoking, see also Cigarettes; Lung cancer, 2—4, 30, 40—41, 64, 83, 138, 185, 228—229, 240
carcinogenic components of, 12—15
carcinogenic effect of, 83—84
Cigarette smoke condensate (CSC), 12, 14—16, 40, 47, 79, 115, 164, 188, 242, 248
BaP on carcinogenicity of, 15
Cigar smoke condensate, 164
Cigar smokers, 228
Classification of PAH, 242
Clean Air Act, 241, 247
Clones, 166
Coal, see also specific types and domestic heating, 21, 29
Coal briquets, 71—72, 116

Coal-fired heating, 114
Coal-fired power plants, 73—74
Coal heating, 3, 70—73
Coal tar, 238
Cocarcinogenic effects, 188
Cofactors, 178
Coffee, 44
Coke plant, 18—20
Coke production, 3
Coking workers, 238
Collaborative studies, 50
Collision, 130
Colloids, 44
Colon carcinoma, 239
Column bleeding, 52
Column chromatography, 50
Column-packing material, 46—47
Combustion, 29, 102
coal, 21
hard coal, 63
Complete carcinogens, 14
Concentration of product, 2
Concentration toxins, 158
Condensed systems, 35—39
Corium fibrosis, 238
Coronene, 36, 47—48
Cottonseed oil, 111
Croton oil, 204—205
Crude coconut oil, 111—112
Crude mineral oil, 47
Crude oil, 50, 52, 114, 120
Crude oil investigations, 104—105
Crude vegetable oils, 111
Cumulative effect, 158
Cumulative toxins, 158
CYC, see Cyclopenta[cd]pyrene
Cyclohexene oxide, 140
Cyclopentadieno[cd]pyrene, 36
Cyclopenta[cd]pyrene (CYC), 18, 21
distribution pattern, 20
Cytochrome P 448 with NADP, 162
Cytochrome P 450, 137—139, 148, 162

# D

Decomposition reactions, 46
Deoxyribonucleic acid, see DNA
Deposition of particles, see also Intake and distribution of PAH, 130—133, 135
Dermatitis, 238
Desquamation of skin, 238
Detection limit, 52
Detection systems for carcinogens, 3—15
analysis of percentage amounts that cause biological effects, 5—6
animal test models, see Animal test models
carcinogenic substances, 3—4
cigarette smoke, 12—15
fractionation of automobile exhaust condensate, 6—10

isolation of carcinogenic components, 4
mutagenicity tests, 11
sebaceous gland suppression, see Sebaceous gland suppression test
vehicle exhaust, see Vehicle exhaust; Vehicle exhaust condensate
Detoxification of PAH, 145
Dibenz[a,c]anthracene, 37
Dibenz[a,h]anthracene, 37, 47—48, 138, 146
Dibenz[a,h]anthracene 5,6-oxide, 140
Dibenzo[a,j]anthracene, 37
Dibenzo[b,def]chrysene, 37
Dibenzo[def,p]chrysene, 37
Dibenzo[a,e]pyrene, 37
Dibenzo[a,h]pyrene, 37
Dibenzo[a,i]pyrene, 37, 203
Dibenzo[a,l]pyrene, 37
Dibenzo[e,l]pyrene, 37
Dibenzo[fg,op]tetracene, 37
Diesel-engined automobiles and trucks, see Vehicle exhaust
Diesel engine soot, 42, 44
Diesel exhaust, see Vehicle exhaust; Vehicle exhaust condensate
Diffusion, 130
7,8-Dihydro-7,8-dihydroxybenzo[a]pyrene 9,10-oxide, 140
trans-1,2-Dihydro-1,2-dihydroxynaphthalene, 142
Dihydrodiol epoxides, 147
Dihydrodiols, conversion of epoxides by aryl epoxide hydrolase into, 139—142
cis-Dihydrodiols, 142
trans-Dihydrodiols, 138—139, 142, 151
Dihydromonols, 143—144
3,4-Dihydronaphthalene 1,2-oxide, 140
7,8-Dihydroxy-9,10-epoxy-7,8-dihydrobenzo[a]pyrene, 162
cis-1,2-Dihydroxy-3-methylcholanthrene, 145
trans-1,2-Dihydroxy-3-methylcholanthrene, 145
7β,8α-Dihydroxy-7,8,9,10-tetrahydrobenzo[a]pyrene-9α,10α-epoxide, 147
7,12-Dimethylbenz[a]anthracene, 37, 138, 144, 147
   reaction with DNA, 151
Distant heating, 70
Distribution coefficients of PAH, 46—47
Distribution of PAH, see Intake and distribution of PAH
Distribution patterns
   benzo[a]pyrene, 19
   cyclopenta[cd]pyrene, 20
DNA, 146—148, 163
DNA compounds, 151
DNA repair system, damage to, 163
Domestic heating, 69—73
   coal, 15—16, 19—20, 70—73
   oil-fired, 73
Dose dependency, 240
Dose-effect relationship, 191
Dose-frequency relationship, 248—249
Dose-response relationship, 22, 179—180, 187—188, 191—194, 197—198, 201—203, 228, 237, 241—242, 248—249

Drilling cores, 102—104
Drinking water, 43, 56, 105, 204

# E

Effect index, 123
Electricity for heating, 70
Electron capture detector (ECD), 41, 52
Embryo cells of Syrian hamster, 166, 170
Emittants, 15
Endoplasmic reticulum (ER), 137, 147—148
Engine emissions, see Vehicle exhaust
Enrichment, see also Isolation of PAH mixture, 42, 45—49
Environment, 204
Environmental carcinogens, identification of, 1—25
Environmental chemicals
   predictions on carcinogenic properties of, 166
   prescreening, 161, 177
   short-term tests for detection of carcinogenic activities of, 161—178
Environmental contamination by PAH, 84—113
   air, 84—101
   crude oil, 104—105
   foodstuffs, 109—113
   sediments, 102—104
   sewage sludge, 105—109
   soil, 101—102
   water, 105
Enzymatic activities, 162
Epicutaneous tests, 187—189
Epidemiological investigations, 2
Epidemiological studies, 186
Epidemiology, 221—236
   lung cancer, see also Lung cancer, 222—235
   prospective studies, 222
   retrospective studies, 222
Epidermal carcinomas in trachea, 12
Epoxidation, 138
Epoxide hydratases, 162—163
Epoxide hydrolase, 139—142, 147
   inducibility of, 140
   inhibition of, 140
   several with high substrate specificity, 139
   stereoselectivity, 141
   substrate specificity, 139
Epoxides, 162
   catalyzed by monooxygenase system, 137—139
   conversion into dihydrodiols by aryl epoxide hydrolase, 139—142
   conversion into phenols, 139
   spontaneous isomerization of, 142
Erythemas, 238
Esophagus cancer, 224
Europa test, 74—77
Exhaust of vehicles, see Vehicle exhaust; Vehicle exhaust condensate
Exogenous factors, 2
Extraction, 42—44
Extraction methods, 42—44, 49

fats, 43, 49
fruits, 43—44, 49
graphite-like materials, 43—44, 49
inorganic components, 43—44, 49
leaves, 43—44, 49
meat products, 43—44, 49
oils, 43, 49
seeds, 43—44, 49
soot-like materials, 43—44, 49
water samples, 43—44, 49
Extracts of air dust, carcinogenicity of, 195—196, 198—199
Extrapolation of animal experiment results to man, 186, 237—245

## F

Fats, 43, 49, 52
Feeder cells, 166, 170
Feeding, 204
Fetal tissue, 205
Fibroblastic cultures of mouse cell lines, 170
Fibrosarcomas, 240
Fibrosis, 185
Fish, 44
Fistular carcinomas, 161
Flame ionization detector (FID), 41, 50, 52, 56
Florisil, 46
Flour, 44, 112
Flue ash, 73, 112
Flue gas emission, 116—117
Fluoranthene, 36, 38—39, 47—48
Fluorene, 36
Fluorescent microscopy, 171
Food, 204
Foodstuffs, 109—113, 203
　airborne particulate matter, contamination with, 109, 111—112
　drying with smoke, 109
　maximum allowable amoung of BaP, 111
　smoked, 109—111
　smoke-dried, 112—113
　smoking process, 109
Forestomach of mice, 203—204
Forest soil, 102
Formation of PAH, see Generation of PAH
Fossil fuel, 28, 31, 114, 201
　consumption of, 70—71
Fractionation of a condensate, see Vehicle exhaust condensate
Frameshift mutations, 163
Frequency of occurrence of PAH, 31
Fruits, 43—44
Fuel consumption, 71
Fuel oil, 30, 117
Fuel oil EL heating, 70, 73
Fuel oil-fired installations, 74
Fuels, 43
Furnace exhaust gas, 44

## G

Gas chromatogram, 53
Gas-chromatographic profile analysis, 50
Gas chromatography, 41, 50—56, 87, 148, 151
　column bleeding, 52
　detection limit, 52
　glass capillary columns, 50—52, 55—56
　high-performance-columns, 50—54, 56
　nematic phases, 54
　packed columns, 54, 56
　separation stages, 52, 55
Gas heating, 70
Gas-liquid chromatography, 151
Gasoline, 76
Gasoline-air mixture in automobile exhaust, 75—76
Gasoline-engined automobiles, see Vehicle exhaust
Gastrointestinal tract, 135, 203—204
Gasworkers, 238
Gavage, 203—204
GC-MS combination, 52
Generation of PAH, 28—30, 62—64, 105
　carbonization process, 28—30
　incomplete combustion, 28—30
Glandular stomach, 203
Glass capillary columns, 50—52, 55—56
Global data studies of lung cancer, 233—234
Glucuronic acid, 143
Glucuronidases, 163
Glutathione, 145—146
Glutathione-S-epoxide transferase, 145—146, 163
Golden hamsters, 166, 180—185, 204—205
Granulomas, 185
Graphite-like materials, 43—44
Ground water, 43, 105
Guide substance, 123

## H

Hair growth cycle of mice, 171
Hard coal, 15, 21, 62—64
　briquets, 16
　flue gas, 16
Heating fuel oil, 48
Heating installations, 241, 247
Heating oils, 43
Height equivalents of theoretical plates (HETP), 41
Hepatomas, 187, 191
N-Heterocyclics, 174
High-performance-columns, 50—54, 56
High-performance liquid chromatography (HPLC), 55—56
High predictive value for selection of long-term carcinogenesis test, 162
High-pressure liquid chromatography, 41, 50, 54—55, 148
Histidine-deficient mutants, reverse mutation of, 163

Histidine-phototrophic bacteria, 163
Human bronchial epithelium, 143
Human fibroblasts of liver cells, 170
Human skin, 238—239
Hydration, 142
Hydrolase, see Epoxide hydrolase
Hydrophilic/hydrophobic separation, 46
Hydrophobic/hydrophilic substances, 7—8
2-, 8-, and 10-Hydroxybenzo[a]pyrene, 142
3-, 7-, and 9-Hydroxybenzo[a]pyrene, 142
4- or 5-Hydroxybenzo[a]pyrene, 143
5-Hydroxy-dibenz[a,h]anthracene, 147
6-Hydroxymethylbenzo[a]pyrene, 143
1- and 2-Hydroxy-3-methylcholanthrene, 145
3-Hydroxymethylcholanthrene, 145
7-Hydroxymethyl-12-methylbenz[a]anthracene, 144
Hygroscopic particles, deposition of, 133
Hyperkeratosis, 238

## I

Identification of carcinogenic components, 11—12
Identification of environmental carcinogens, 1—25
    detection systems, 3—15
    methods of analysis, 3—15
    methods or proof of existence, 2—3
    proportional activity, 21—23
    "representative PAH concentration" of a city, 15—21
Immigrant studies of lung cancer, 222, 232—233
Impaction, 130, 132, 135
Implantation methods, 4—5, 178—179, 185—186
Incomplete combustion, see Pyrolysis
Indeno[1,2,3-cd]fluoranthene, 36
Indeno[1,2,3-cd]pyrene, 36, 48
Index for carcinogenic potency of airborne PAH, 90—93
Index of carcinogenicity, 123
Index substance, 123
Inducers of monooxygenases, 146
Inducibility of epoxide hydrolase, 140
Induction of monooxygenases, 138
Induction of P 448 monooxygenases, 138
Industrial furnaces, 247
Industrial plants, 241
Inert gas, 28—29
Inflammation, 185
Inhalation carcinogens, 240, 243
Inhalation methods, 183—186
Inhalation studies, 243
Inhibition, 205
    hydrolases, 140
    monooxygenases, 139
Initiating carcinogens, 14
Inorganic components, 43—44, 49
Inorganic compounds, 2, 39
Inorganic substances, 28
Instillation into trachea, 4—5, 7, 11—12, 31, 179—183, 185—186, 248
Intake and distribution of PAH, 130—136

aerodynamic diameter, 130—131
deposition of particles, 130—133
diffusion, 130
hygroscopic particles, 133
impaction, 130, 132
particles in lung, 134
physical properties of the particles, 130—131
respiratory tract, 130—133
    lower, 130, 134
    middle, 130, 134
    upper, 130, 133—134
retention and clearance, 133—136
sedimentation, 130
total deposition, 133
Interactions, 205
Intercalation, 147
International Union of Pure and Applied Chemistry (IUPAC) rules, 31—39
Intestinal cancer, 224—225
Intestinal tract, 239
Intratracheal instillation, see Instillation into trachea
Intraperitoneal administration, 204
Intravenous application, 204
Inventory of carcinogenic effect, 40
Irradiation, 28
Isolated liver cells, 162
Isolation of carcinogenic components, 4
Isolation of carcinogenic substances, 6—7
Isolation of PAH mixture, 42, 45—49
    aliphatic/aromatic, 46
    alumina, 46, 48
    chromatographic methods, 46—49
    column-packing material, 46—47
    hydrophilic-hydrophobic, 46
    liquid-liquid partition, 45—46
    Sephadex LH 20, 46—48
    silica gel, 46, 48
Isolation schemes, see Extraction methods

## K

Kale, 44
Kale samples, 112
Keratinization, 135
Kidney cancer, 232
Kidneys, 205
Kidney tumors, 241

## L

Lake water, 43
Larynx cancer, 224
Latency period, 227
Late toxic effects, 158
Leaves, 43—44, 49
Lentils, 44
Lettering, 35
Lettuce, 44
Leukemias, 204—205, 240

Lignite, 44
Lignite-fired installations, 74
Linseed oil, 111
Lipopolysaccharides in bacterial cell wall, 163
Liquid-liquid partition, 45—46
Liver homogenates, 162
Liver microsomes, 162
Local carcinogen, 161, 239
Local carcinogenic activity, 203
Local PAH concentration, 16, 18—21
Local papillomas, 205
Local sarcomas, 204
Long-term experiments, 178—205
    oral application, 203—205
    respiratory tract application, 178—186
        implantation, 178—179, 185—186
        inhalation, 183—186
        instillation into trachea, 179—183, 185—186
    skin applications, 186—189
    subcutaneous tissue application, 190—203
        carcinogenicity of individual PAH, 192—193, 201—203
        carcinogenicity of mixtures of PAH, 193—194, 201—203
        carcinogenicity of PAH-containing mixtures, 194—203
Lower respiratory tract, 136
    deposition of PAH particles in, 130, 134
Lower respiratory tract compartment, 135—136
Low-voltage mass spectrometry (LVMS), 41, 50
L-region of PAH, 143
Lubricating oil, 43, 48, 76—77
    fresh or used, 15—16, 47—48
Lung adenomas, 187, 191, 204—205
Lung cancer, 240, 248
    dose-response relationship, 228
    epidemiological studies on, 222—235
    global data studies, 233—234
    immigrant studies, 222, 232—233
    latency period, 227
    mortality, 2—3
    mortality rate, 178, 224—228
    petroleum industries, 226
    prospective studies, 230—231
    retrospective study, 227
    smoker studies, 222, 227—230
    urban/rural area studies, 222—227, 241
    work area studies, 222, 230—232, 241
Lung tumors, 205
Lymphomas, 205

# M

Macrophages, 134, 136
Malignant cell transformation, 146, 148
Malignant transformation, 166
Malignant tumors, 160
Mammary gland, 204—205
    carcinoma, 240
Man, 204

Marijuana smoke, 138
Mass spectrometry, 50
Maximum tolerable concentration, 249
Meat, 48—49
Meat products, 43—44, 52
Mesotheliomas, 161
Metabolic activation of known carcinogenic PAH, 162—163
Metabolism of PAH, 137—151, 242
    benzo[a]pyrene, 148
    $C_1$ transfer, 143
    catechols, 143—144
    conversion of epoxides into dihydrodiols by aryl epoxide hydrolase, 139—142
    dihydromonols, 143—144
    elimination of metabolites, 203
    epoxides catalyzed by monooxygenase system, 137—139
    glucuronic acid, 143
    glutathione catalyzed by glutathione-S-epoxide transferase, 145—146
    methods of detection of metabolites, 148—151
    nucleic acids, 146—148
    phenols, 142—153
    proteins, 148
    quinones, 143
    side-chain oxidation, 144—145
    sulfuric acid, 143
Metabolites, see Metabolism of PAH; PAH metabolite detection
Metabolizing cells, 166
Metal-cutting oil, 43, 47
Method-related effects, 185
Methods of analysis, see also Detection systems for carcinogens, 3—15
Methods of PAH detection, see PAH detection
Methods of proof of existence, 2—3
7-Methylbenz[a]anthracene, 147
12-Methylbenz[a]anthracene 7-carboxylic acid, 144
3-Methylcholanthrene, 37, 138, 145—147, 162
Metyrapone, 140
Microorganisms, 142
Microranges with oxygen deficiency, 29
Microwax, 43
Middle respiratory tract, deposition of PAH particles in, 130, 134
Mineral oil, 29—30, 42
Mineral oil products, 43, 49, 52
Mixed function oxidases, 162
Mobile phase, 55
Molecular distillation, 45
Monooxygenase isoenzymes, 147
Monooxygenases, 140
    epoxides catalyzed by, 137—139
    inducers of, 146
    induction of, 138
    inhibition of, 139
    stereoselectivity, 141
Monosulfates, 142
Morbidity rate, 2
Mortality rate for lung cancer, 224—228

Motor traffic, see Vehicle exhaust; Vehicle exhaust condensate
Mouse skin, see also Topical skin application, 171, 177, 187
Multicomponent mixtures, 174
Multienzyme complex, 140
Multiple-impulse incidents, 160
Multiple lung adenomas, 11
Mutagenic activities, 177
Mutagenicity, correlation with carcinogenicity, 163—165
Mutagenicity tests, 11
Mutations, 163
Mycotoxins, 39

## N

α-Naphthoflavone, 139, 143
Naphtol[1,2,3,4-def]chrysene, 37
Nasopharyngeal space, 135
Nematic phases, 54
Nervous system tumors, 224
Nitrogen-containing aromatic compounds, 2, 7, 40
Nitrogen-containing polycyclic aromatic compounds, 14, 39
Nitrosamines, 2, 39
Nomenclature of PAH, 31—39
  base components, 31
  condensed systems, 35—39
  lettering, 35
  numbering, 35
  orientation, 31
Nonadditive action, 23
Noncarcinogenic and nonmutagenic activity, correlation between, 164—165
Nose filter, 132
Noxious environmental pollution, 28
Noxious matter, 28, 39, 105
Nucleic acids, 146—148
Numbering, 35
Number of PAH detected in environment, 30

## O

Occupational occurrence of cancer, 238
Occurrence of PAH, see also specific topics, 31, 61—128
  chemical analytical index of carcinogenic effects of emissions and environmental samples, 113—122
  environmental contamination, 84—113
  guide substance, 123
  index substance of sum of all PAH, 123
  sources of emission, 64—84
Oil-fired heating, 73
Oil-fired installation, 73
Oil fuel, 19—20
Oil heating, 16, 29, 44
Oil refinery, 102

Oils, see also specific types, 43, 49, 52
Oncogenic, 160
One-impulse incident, 160
Operative implantation, 203
Oral application, 31, 203—205, 243
Oral cavity, 135
Oral intake, 204
Organic substances, 28
Orientation, 31
Ovary, 205
Oxidative enzymes, 162

## P

PAH, see also specific topics, 2, 7—10, 174
  behavior in the organism, 129—156
  biological effect in various matrices, 15
  blowing-off effect from filter, 84—87
  cancer-inducing effect, 28
  carcinogenicity, 192—203
  chemistry, 27—60
  choice of relevant, 28—31
  correlation between carcinogenic and mutagenic activity of, 164
  correlation of animal carcinogenicity and bacterial mutagenicity, 164—165, 167—169
  correlation of animal carcinogenicity and cell-transforming activity, 172
  correlation of animal carcinogenicity and sebaceous gland suppression test, 175—176
  emittants producing, 15
  formation, see Generation of PAH
  four to seven rings, 7—11
  index for carcinogenic potency, 90—93
  inhibition of carcinogenicity, 199
  methods of detection, 39—57
  nomenclature, 31—39
  occurrence of, 61—128
  particle size, 84—85
  representative concentration of a city, see "Representative PAH concentration" of a city
  sampling from air, 84—86
  two and three rings, 7—9
PAH burden of inhabitants, 16
PAH detection, see also specific topics, 39—57
  chemical analysis, 39—42
  chromatographic separation, 49—50
  chronic toxic testing, 39—40
  enrichment, 45—49
  extraction, 42
  extraction methods, 42—44
  gas chromatography, 50—56
  high-performance liquid chromatography, 55—56
  thin-layer chromatography, 54—56
PAH-DNA hydrolysates, 148
PAH index, 95
PAH metabolite detection, 148—151
  analytics of DNA compounds, 151
  gas chromatography, 148, 151
  high-pressure liquid chromatography, 148

PAH particles, see Intake and distribution of PAH
PAH profile, see Profiles of PAH
PAH profile analysis, see Profile analysis
Palm-kernel oil, 111
Palm oil, 111
Papilloma-inducing, 239
Papillomas, 184, 187, 204, 238, 247—248
Paraffin, 43
Paraffin processing, 230
Particle-related carcinogens, 135
Particles deposited in lungs, see Intake and distribution of PAH
Partition chromatography, 45, 47, 49
Parvicellular carcinoma, 136
Passenger cars, 16, 75
Passive smoking, 83
PCB, see Polychlorinated biphenyls
Peanut oil, 111
Peas, 44
Peat, 44
Penetrability, 203
Percentage potency, 23
Period exposure, 2
Perylene, 36, 47—48
Petroleum industries and lung cancer, 226
Pharyngeal cavity cancer, 225
Phenanthrene, 36, 47—48
Phenanthrene 9,10-oxide, 140
Phenobarbital, 138, 148
Phenobarbitone, 162
Phenols, 142—143
  conversion of epoxides into, 139
Photochemical air pollution, 227
Physical properties of PAH particles, 130—131
Pigmentations of skin, 238
Pipe smokers, 228
Pipe tobacco smoke condensate, 164
Pit coal, 29
Pit coal tar, 30
Placenta, 205
Plants, 49
Plastic materials, 29
Polyamide, 65
Polychlorinated biphenyls (PCB), 138, 162
Polycyclic aromatic hydrocarbons, see PAH
Polyethylene, 29
Polynucleotides, 147
Polypropylene, 65
Polystyrene, 65
Polytetrafluorethane, 65
Polytetrafluoroethylene, 28
Polyvinyl chloride, 29, 65
Potatoes, 44
Poultry, 44
Power plants, 16
Predictions on carcinogenic properties of environmental substances, 166
Pregnancy, 205
Prescreening of environmental chemicals, 161, 177
Probit function, 22
Profile of activities, 6

Profile analysis, see also Profiles of PAH, 40, 50, 52, 56, 106, 148
Profiles of PAH, see also Profile analysis, 18, 29, 40, 64—68, 73—74, 77—79, 86—90, 113
  air, 86, 88—90, 92—93, 95
  exhaust, 87, 91
Promoting effects, 188
Promotion, 205
Promotors, 14
Proportional activity, 21—23
Proportional balance of carcinogenic activity, 2
Prospective studies, 222, 230—231
Proteins, 146
*Pseudomonas*, 142
Purine nucleotides, 147
Pyrene, 36, 47—48
Pyrimidine nucleotides, 147
Pyrolysis, 28—31, 62, 64, 94, 113, 247
Pyrolysis products, 231
Pyrolytic process, see also Pyrolysis, 31

# Q

Quality control, 166
Quinones, 143

# R

Rape-seed oil, 111
Rats, 204
Raw paraffin processing, 3
Reactive metabolites, 162
Reconstituted condensate, 7
Relative potency, 22—23, 241—242
"Representative PAH concentration" of a city, 15—21
  automobile exhaust emission, 21
  automobile traffic, 18—20
  coke plant, 18—20
  combustion of coal, 21
  local variations in different areas of city, 16, 18—21
  residential area, 16—20
  temporary variations of PAH concentration, 16—18
Residential area of city, 16—20
Resistance factor (R factor), 163
Respiratory tract, see also Lung cancer, 240, 247—248
  deposition of particles of PAH in, 130—133
  long-term experiments with, see also Long-term experiments, 178—186
  retention and clearance of deposited particles, 133—136
Respiratory tract malignancies, 2
Respiratory tract tumors, 241
Retention and clearance of deposited particles, 133—136
Retrospective studies, 222, 227

Reversed phase, 55
Risk assessment, 242, 248
River water, 43, 105
RNA, 147
Rye, 112

## S

S-9 fraction of mammalian livers, 162—163
S-9 microsomal fraction, 166
S-9 mixture, 163, 170
*Salmonella typhimurium*, different strains of, 163—164
Sandy soil, 102
Saponification of matrix, 42
Sarcomas, 187, 190
Screening system for PAH, 174
Sebaceous gland suppression (SGS) test, 7, 10—11, 165, 170—174, 177
  correlation of animal carcinogenicity with, 175—176
  screening system for PAH, 174
Sebaceous glands, 171, 177
Sediment, 44
Sedimentation, 130
Sediment investigations, 102—104
Sediment samples, 52
Seeds, 43—44, 49
Selective isolation of PAH, 45
Sensitive tests, 204
Separation, see also Isolation of PAH mixture, 42, 49—50
Separation methods needed, 40—41
Separation stages, 50, 52, 55
Sephadex LH 20, 46—49, 151
Settling, 130
Sewage sludge, 15, 44, 52, 122
  activated, 106
  fresh, 106
Short-term tests, 161—178, 242
  Ames test, 163—169, 171, 177
  cell transformation test, 165—171, 177
  metabolic activation of known carcinogens in PAH, 162—163
  sebaceous gland suppression test, 165, 170—174, 177
Side-chain oxidation, 144—145
Silica gel, 46, 48—49, 55
Silylation, 151
Single-cell proteins, 44
Single dose, 204
Site of application, 178, 186, 190, 203
Skin as site of testing, see Topical skin application
Skin cancer, 230, 232
Skin carcinoma, 6, 11, 240
Skin-painting test, 193, 242
Skin papillomas, 204, 238
Skin tumors, 241
Smoked fish, 109, 111
Smoked food, 239

Smoked foodstuffs, 109—111
Smoked ham, 109, 111
Smoked herring, 119
Smoked meat products, 111
Smoke-dried foodstuffs, 112—113
Smoker studies of lung cancer, 222, 227—230
Smoke soot, 240
Smoking habits, see also Cigarette topics, 2, 223—224, 233—234, 248
Soil, 44, 50
Soil investigations, 101—102
Solid fuel combustion, 72
Solvent, 203
Soot, 49, 112
Soot-like materials, 43—44
Sources of emission of PAH, see also specific topics, 64—84
  cigarettes, 77—84
  domestic heating, 69—73
  motor traffic, 74—77
  thermal power plants, 73—74
Soybean oil, 111
Species-specific influences, 178
Specific carcinogenic effects, 160—161
Spices, 44
Spinach, 44
Spontaneous isomerization of epoxides, 142
Spontaneous tumors, 160, 187
Squamous cell carcinoma, 135, 179, 184, 187, 204, 240, 247—248
Standardization, 166
Standardized gas chromatogram, 113—122
Statistical evaluation, 166
Stereoselectivity of monooxygenases and epoxide hydrolases, 141
Stomach cancer, 204, 239
Subcutaneous application, 4—5, 7, 31, 203—205, 247
Subcutaneous tissue, long-term experiments with, see also Long-term experiments, 190—203
Substance-related effects, 185
Sulfotransferases, 143
Sulfuric acid, 143
Summation toxins, 158
Sunflower oil, 111
Sunflower seed oil, 48
Surface soil, 102
Surface waters, 105
Surfactant, 136, 204
Systemic carcinogen, 161, 240
Systemic carcinogenic effects, 204
Systemic tumors, 187

## T

Tap water, 105
Tea, 44
Technical soots, 44
Temporary PAH concentration, 16—18
Temperature of pyrolysis, 29

Testicular cancer, 232
Testis tumors, 241
Thermal decomposition 28—29, 62—64
Thermal power plants, 73—74
Thin-layer chromatography, 41, 50, 54—56, 87
Three-tier protocol for biological testing of compounds, 162
Tobacco, 29
Tobacco brand, 77
Tobacco smoke, see also Cigarette topics, 240
  animal experiments with, 184—185
Tobacco smoke carcinogenesis, 240
Tobacco smoke condensate, 12, 80—83, 179, 200, 240
  animal experiments with, 184—185
Topical skin application, 4—5, 7, 10—12, 14—15, 31, 186—189, 204, 247
Total deposition of PAH particles, 133
Toxicity, 158—159
Toxicology, 158
Trachea, see also Instillation into trachea, 248
Tracheobronchial compartment, 135
Transformation of a cell, 160
Transformation of cell cultures, 165—170
Transplacental carcinogenic effect, 205
1,1,1-Trichloropropene oxide, 140
Triphenylene, 36
Truck exhaust, see Vehicle exhaust
Tumorigenesis, 7
Tumorigenic, 160
Tumorigenic effects, 195, 204, 242
Tumorigenic substances, 192
Tumor incidence, 14
Tumor-inducing activities, assessment of, 39—40
Tumor induction, 135, 180—183, 190—191
Tumor-producing properties, 39

## U

UDP glucuronyltransferases, 143
Ultimate carcinogens, 140—142, 147, 162
Unspecific carcinogenic effects, 161
Upper respiratory tract, deposition of PAH particles in, 130, 133—134
Urban areas
  benzo[a]pyrene concentrations, 93, 95—101
  PAH concentrations, 93—94
Urbanization factor, 241
Urban/rural area studies of lung cancer, 222—227, 241
Urban/rural populations, 241
Urinary bladder tumors, 241

## V

Vegetable oil, 45
Vehicle exhaust, 3, 29—30, 44, 50, 53—55, 74—77, 164, 241, 249
  BaP emission, 16, 75
  carcinogenic components of, 7—11
  correlation with incidence of carcinoma, 6
  diesel engines, 16, 43, 77, 121
  emission in city, 21
  gasoline-air mixture, 75—76
  gasoline engines, 11—12, 74—77, 118
  influence of engine parameters, fuel and lubricant, 75
  variation of, 74—77
Vehicle exhaust condensate, 6, 16, 40, 49, 52, 171, 188, 198, 200, 202, 240, 242, 248
  animal experiment models, 10—11
  animal test models, see Animal test models
  aromatic/aliphatic substances, 7—9
  carcinogen-free fractions, 6—7
  carcinogenic activity of, 7—9, 11—12
  carcinogenic components, 75
  carcinogenic fraction, 6—7
  comparison of benzo[a]pyrene and mixture of carcinogenic PAH extracted, 8—10
  diesel engines, 12, 15, 29
  fractionation, 6—10
  gasoline engines, 11—12, 15, 29
  hydrophobic-hydrophilic substances, 7—8
  isolation of carcinogenic substances, 6—7
  PAH with four to seven rings, 7—11
  PAH with two and three rings, 7—9
  reconstituted, 7
Vehicle exhaust gas, see Vehicle exhaust
Verrucae, 238
Viruses, 170
Vitamin A deficiency, 147

## W

Waste water, 43
Water samples, 43—44, 49, 55, 105
Weber-Fechner law, 22
Wheat, 112
Work area studies, 186
  lung cancer, 222, 230—232, 241
Working areas, benzo[a]pyrene concentrations in, 100

## Y

Yeast, 44, 52, 112